1806438571
WITHDRAWN FROM STOCK
PA AU
Llyfrgell
Library
Aberystwyth

AF616126

VOLUME EIGHTY SIX

CURRENT TOPICS IN DEVELOPMENTAL BIOLOGY

Evolution and Development

VOLUME EIGHTY SIX

CURRENT TOPICS IN DEVELOPMENTAL BIOLOGY

Evolution and Development

Edited by

WILLIAM R. JEFFERY
Department of Biology
University of Maryland, College Park
Maryland, USA

ELSEVIER

AMSTERDAM • BOSTON • HEIDELBERG • LONDON
NEW YORK • OXFORD • PARIS • SAN DIEGO
SAN FRANCISCO • SINGAPORE • SYDNEY • TOKYO
Academic Press is an imprint of Elsevier

Academic Press is an imprint of Elsevier
525 B Street, Suite 1900, San Diego, CA 92101-4495, USA
30 Corporate Drive, Suite 400, Burlington, MA 01803, USA
32, Jamestown Road, London NW1 7BY, UK
Linacre House, Jordan Hill, Oxford OX2 8DP, UK

First edition 2009

ISBN: 978-0-12-374455-5
ISSN: 0070-2153

For information on all Academic Press publications
visit our website at elsevierdirect.com

Printed and bound in USA
09 10 11 9 8 7 6 5 4 3 2 1

CONTENTS

Contributors

Marianne Bronner-Fraser
Division of Biology, California Institute of Technology, Pasadena, California, USA

Martin J. Cohn
Department of Zoology and Department of Anatomy and Cell Biology, University of Florida, Cancer/Genetics Research Complex, Gainesville, Florida, USA

B. Frank Eames
Institute of Neuroscience, University of Oregon, Eugene, Oregon, USA

Eric S. Haag
Department of Biology, University of Maryland, College Park, Maryland, USA

William R. Jeffery
Department of Biology, University of Maryland, College Park, Maryland, USA

Elena M. Kramer
Department of Organismic and Evolutionary Biology, Harvard University, Cambridge, Massachusetts, USA

J. David Lambert
Department of Biology, University of Rochester, Rochester, New York, USA

Armin P. Moczek
Department of Biology, Indiana University, Bloomington, Indiana, USA

Natalya Nikitina
Division of Biology, California Institute of Technology, Pasadena, California, USA

Rudolf A. Raff
Department of Biology, Indiana University, Bloomington, Indiana, USA and School of Biological Sciences, University of Sydney, Sydney, Australia

Tatjana Sauka-Spengler
Division of Biology, California Institute of Technology, Pasadena, California, USA

Margaret Snoke Smith*
Department of Biology, Indiana University, Bloomington, Indiana, USA

GuangJun Zhang[†]
Department of Zoology, University of Florida, Cancer/Genetics Research Complex, Gainesville, Florida, USA

* Current Address: Department of Entomology, University of Georgia, Athens, Georgia, USA
[†] Current Address: The David H. Koch Institute for Integrative Cancer Research, MIT, Cambridge, Massachusetts, USA

Preface

This is the 86th volume of *Current Topics in Developmental Biology* (CTDB). Considering that this series began in 1968, one could ask why it has taken so long for a thematic CTDB volume to appear on Evo Devo? An answer might be that Evo Devo is at once an old and a newly emerging discipline. Under the alias of evolutionary morphology or embryology, it was a popular scientific study in the 1800s, predating the surfacing of neo-Darwinism in the next century. As a new breed of experimental embryologists, and ultimately molecular embryologists, rushed to determine the secrets of development, the evolutionary perspective was temporarily left by the wayside. In retrospect, this was probably the right course: one should know the rules of development in some detail before attempting to find out how they are fashioned during evolution.

Beginning in the 1970s, there was a rebirth of interest in Evo Devo, sparked in large measure by the publication of two books: "Ontogeny and Phylogeny" by Stephen Jay Gould (1970) and "Embryos, Genes, and Evolution" by Rudolf Raff and Thomas Kaufman (1983). The latter volume, in particular, described evolution within the backdrop of new genetic and molecular discoveries showing that the rules and basic molecular tool kits used in development are fundamentally similar in all animals and plants. This launched the first phase of Evo Devo, which was devoted to understanding this deep conservation of developmental mechanisms. Although important, conservation is not the key issue in understanding the role of ontogeny in evolution. Instead, we must strive to understand the more complex issue of diversity, that is when, how, and how frequently different ontogenies arise during evolution. This activity defines the second phase of Evo Devo and is what this CTDB volume is about.

A large part of Evo Devo's second phase is understanding when and how major phenotypes evolved, and the emergence of novel biological entities during crucial evolutionary transitions, such as the transition from invertebrates to vertebrates. Two articles in the current volume are centered on this theme. Nikita, Sauka-Spengler, and Bronner-Fraser (Caltech) trace the fascinating evolution of the neural crest to the most basal vertebrates and perhaps even to invertebrate chordates. Zhang, Eames, and Cohn (University of Florida) take a similar approach to understanding the evolution and relatedness of cartilage, and its role in establishing a skeletal renaissance during vertebrate evolution. Another important part of contemporary Evo Devo depends on the comparative approach. Here emerging model systems consisting of two or more species are used to investigate complex problems, such as the diversity of body plans, the evolution of sexual reproduction, and the loss and gain of phenotypes in extreme environments. Raff and Smith (Indiana University) describe their pioneering studies on direct and indirect developing sea urchins in which the first molecular discoveries are presented for the rapid evolution of axial development. Likewise, Moczek (Indiana University) describes the evolution of horn diversity in horned beetles, a system that has immense potential for improving our understanding of microevolutionary mechanisms, and especially the role of developmental tradeoffs. When emerging models are coupled with pre-existing models—their "rich cousins" with respect to detailed developmental knowledge and molecular genetic tools—powerful new insights can be forthcoming. Thus, Kramer (Harvard University) describes a host of new land plant models linked in this way to *Arabidopsis*, Haag (University of Maryland) shows how divergence in evolution of sex determination can be studied by comparing *Caenorhabditis briggsae* to *C. elegans*, and Jeffery (University of Maryland) charts the importance of pleiotropy using the blind cavefish *Astyanax mexicanus* and zebrafish as companion species. Another important part of Evo Devo is obtaining a more complete understanding of the development of classic systems that are ripe for in depth evolutionary analysis. One of these systems, the polar lobe forming and spirally cleaving gastropod *Illyanassa*, is described here by Lambert (University of Rochester), who shows the importance of localized mRNAs and spatial signaling cues in determining this novel type of development.

The CTDB volume does not cover every contemporary issue in Evo Devo. Indeed, many important topics are not addressed. In this sampling, however, we merely hope to provide examples of how modern cutting-edge approaches are being used to investigate and generate new understanding of some central issues this field. By doing so, we endeavor to encourage, and perhaps even inaugurate, the next major phase in Evo Devo.

William R. Jeffery
College Park, MD

CHAPTER ONE

Gene Regulatory Networks in Neural Crest Development and Evolution

Natalya Nikitina, Tatjana Sauka-Spengler, *and* Marianne Bronner-Fraser

Contents

Abstract

The neural crest is a multipotent migratory embryonic cell population that is present in all vertebrates, but missing from basal chordates. In this chapter, we discuss recent work in amphioxus, ascidians, lamprey, and gnathostomes that reflects the current state of knowledge of the evolutionary origin of this fascinating cell population. We summarize recent evidence for the ongoing diversification of the neural crest in several vertebrate species, with particular reference to studies in nontraditional vertebrate model organisms.

1. Gene Regulatory Network Underlies Neural Crest Development

The neural crest, an embryonic population of migratory and multipotent precursor cells, is traditionally considered a vertebrate innovation. In fact, acquisition of the neural crest and neurogenic placodes is considered to be one of the key events in vertebrate evolution, leading to the appearance of the jaws, cranium, and sensory ganglia, which enabled the transition of

Division of Biology, California Institute of Technology, Pasadena, California, USA

Current Topics in Developmental Biology, Volume 86
ISSN 0070-2153, DOI: 10.1016/S0070-2153(09)01001-1

early vertebrates from filter feeding to active predation (Gans and Northcutt, 1983; Northcutt and Gans, 1983).

In all vertebrates examined to date, neural crest cells share some common features. These cells arise at the border between neural and non-neural ectoderm. They subsequently undergo an epithelial-to-mesenchymal transition (EMT) to detach from the neural folds or dorsal neural tube, a process that involves alterations in cell shape as well as acquisition of cell surface adhesion molecules and signaling receptors. The latter contribute to the neural crest cells' ability to migrate to diverse sites where they differentiate to form numerous different cell types. Neural crest derivatives include neurons and glia of the peripheral nervous system, bone and cartilage of the facial skeleton, as well as melanocytes and neuroendocrine cells. Interestingly, the neural crest is the only multipotent vertebrate cell type capable of giving rise to many cell types that populate different tissues and organs.

To study neural crest evolution, it is necessary to distinguish between a bona fide neural crest cell and other cell types that might superficially resemble it. Due to the lack of intermediate forms, it is not clear if all neural crest traits were acquired in a single step during the transition from nonvertebrate to vertebrate chordates or if there might have been stepwise acquisition of these properties (Donoghue *et al.*, 2008). For the purpose of this chapter, we define "neural crest" as having the entire repertoire of migratory and differentiative properties and refer to cells with subsets of these properties as "preprototypic crest." In this way, we distinguish between a migrating cell that gives rise to a single derivative that in vertebrates arises from the neural crest (e.g., pigment lineage), from a multipotent precursor that forms multiple neural crest derivatives and has both regulative and regenerative potential.

One convenient way to define the neural crest is via its regulatory state; that is, the network of the signaling molecules and transcription factors that are responsible for its induction, delamination from the neural tube, migration, and differentiation (Sauka-Spengler and Bronner-Fraser, 2006). Such a neural crest gene regulatory network (NC-GRN) confers onto this cell type the classical neural crest characteristics and provides a mechanistic explanation of how these characteristics arise in a developmental context. A framework of basic modules has been proposed to comprise this network (reviewed in Meulemans and Bronner-Fraser, 2004; Nikitina and Bronner-Fraser, 2008; Sauka-Spengler and Bronner-Fraser, 2006, 2008) and provides a solid foundation upon which questions pertaining to the evolution of the neural crest can be addressed.

These regulatory interactions can be divided hypothetically into phases. The first involves *inductive signals* that establish the neural plate border, by upregulation of transcription factors that specify the neural plate border region. These *neural plate border specifiers* in turn regulate *neural crest specifier genes* that activate or repress specific downstream targets that render the neural crest migratory and multipotent.

According to the NC-GRN, the formation of the neural crest is initiated by a set of diffusible signaling molecules (Bmp, Wnt, FGF, and Notch) that originate from either the ventral ectoderm or the paraxial mesoderm, and initiate the neural crest transcription program in a strip of cells between the neural plate and the non-neural ectoderm, the neural plate border. The early set of transcription factors, turned on in the prospective neural plate border by the combined activity of the above signaling pathways, are collectively called the neural plate border specifiers and include Pax3, Pax7, Msx1, Zic1, and AP-2 (Meulemans and Bronner-Fraser, 2004; Nikitina *et al.*, 2008). These transcription factors activate another set of genes that are expressed specifically in the prospective neural crest and play important roles in the establishment and maintenance of crucial defining characteristics of the neural crest. These neural crest specifiers include Sox8, Sox9, Sox 10, c-Myc, and Id (important for the survival of the neural crest precursors and maintenance of the pluripotency of the neural crest); Snail1 and Snail2 (play a crucial role in the epithelial–mesenchymal transformation, as well as cell cycle control and the migratory activity of the neural crest cells); and Twist (required for the correct localization of the migrating neural crest cells) (Batlle *et al.*, 2000; Bellmeyer *et al.*, 2003; Cano *et al.*, 2000; Honore *et al.*, 2003; Kim *et al.*, 2003; Soo *et al.*, 2002; Taneyhill *et al.*, 2007; Teng *et al.*, 2008). The neural crest specifiers activate transcription of several possibly interconnected modules that are responsible for the differentiation of the neural crest population into individual derivatives. Simultaneously, they turn on expression of receptors that direct migration of the differentiating neural crest cells to the appropriate destinations in the embryo. Genes belonging to the two latter categories (the neural crest effector genes) include signaling molecules, transcription factors (Mitf, trp2), molecules involved in the cell shape changes essential for the delamination and migration (Rho GTPases and cadherins) as well as cell-type-specific differentiation genes characteristic of neural crest derivatives (collagen) (reviewed in Meulemans and Bronner-Fraser, 2004; Sauka-Spengler and Bronner-Fraser, 2008).

The definition of the neural crest via this NC-GRN has limits, largely due to the fact that the network is not yet complete. Not every single gene involved in the neural crest development has as yet been identified, or can be placed accurately within the network (e.g., Meis, Blimp-1), and the exact architecture and interconnections therein are still in the process of being discovered. However, identification and testing of the core elements of the network allows its application to diverse vertebrates regardless of whether all of the elements and connections are established. This is particularly useful when applied to the formation of vertebrate-specific traits. For this purpose, an in-depth study of network components needs to be conducted exhaustively in a single vertebrate that allows precise spatial and temporal discrimination. The basal lamprey embryo has been extremely useful due to the large size, slow development, and ease of manipulations of the early embryo. Due to its basal position as an agnathan representative and its close morphological resemblance to 350-million-year-old fossils, the modern lamprey NC-GRN may provide a reasonable approximation of the ancestral vertebrate state.

2. The Evolutionary Origin of the Neural Crest

A hallmark of the vertebrate neural crest is its remarkable plasticity and ability to form many and diverse derivatives. Neural crest cells have stem cell properties, multipotency, and the ability to self-renew, at least for a limited time in their developmental history. The derivatives of a single cell are as diverse as neurons, cranial cartilage, pigment, and glial cells. This incredible versatility gives the neural crest its characteristic traits that classify it as a vertebrate novelty. Its multipotency and migratory ability render this cell type a crucial invention that contributed to the evolutionary success and diversification amongst vertebrates.

All vertebrate species, even the most basal jawless members of this group such as lampreys and hagfishes, have neural crest that is virtually indistinguishable from the neural crest of higher vertebrates in terms of multipotency, migratory behavior, and the gene regulatory network involved in its development. In fact, divergences from the basal NC-GRN appear to occur only at later stages and more distal levels of the network. These steps contribute to formation of derivative structures such as jaw or sympathetic ganglion chain. Although lamprey lack jaws and sympathetic ganglia, they do possess neural crest-derived cranial cartilage and have ganglia-like clusters of neurons scattered along the cardinal veins running in the abdominal cavity (Johnels, 1956), as well as autonomic control of the vasculature by catecholamine-containing nerve fibers, resembling sympathetic/adrenergic control in higher vertebrates. Whether these represent precursors of the homologous structures, or are simply functionally analogous structures has yet to be determined (Horigome *et al.*, 1999; McCauley and Bronner-Fraser, 2003, 2006; Ota *et al.*, 2007; Sauka-Spengler *et al.*, 2007). The evolutionary origins of the neural crest have therefore been sought among our closest chordate relatives, amphioxus and the ascidians.

The phylogenetic relationships of different chordate groups have undergone drastic reassessment in the past few years, largely due to the availability of sequenced genomes. For over a hundred years, amphioxus with its very vertebrate-like body organization was considered a sister group to vertebrates, while mostly sessile urochordates were thought of as a more distantly related side group (Wada, 2001). Early phylogenetic analyses of 18S ribosomal RNA sequences in a limited number of species confirmed amphioxus as the closest vertebrate relative (Turbeville *et al.*, 1994; Wada and Satoh, 1994), while analysis of the complete small and large ribosomal subunit DNA provided ambiguous conclusions (Winchell *et al.*, 2002). A different story began to emerge after a large data set of nuclear genes from a range of deuterostome species was examined, and the long-branch attraction artifact that results in the fast-evolving ascidian species being attracted toward the echinoderm/hemichordate outgroup was taken into account (Blair and Hedges, 2005; Breau *et al.*, 2008; Delsuc *et al.*, 2006). The new view of the chordate phylogeny that emerged demonstrated that ascidians and not cephalochordates are the true sister group of vertebrates. This conclusion received further independent support from the genome-wide analysis of the intron–exon structures in amphioxus and several vertebrate and ascidian species (Putnam *et al.*, 2008).

Consistent with the latest understanding of chordate phylogeny is the fact that amphioxus does not have anything resembling the neural crest (Holland and Holland, 2001), while migratory preprototypic neural crest cells have been discovered in several ascidian species (Jeffery, 2006; Jeffery *et al.*, 2004). Based on the experimental data currently available, two opinions as to the time of the neural crest origin have emerged in the recent years (Fig. 1.1). According to one hypothesis, the neural crest first appeared in the common ancestor of the ascidians and vertebrates, after the separation of the ancestral cephalochordate lineage (Donoghue *et al.*, 2008). Proponents of this view consider the migratory preprototypic neural crest-like cells (NCLCs) found in some of the modern ascidian species as true neural crest cells. Alternatively, these cells may represent an evolutionary experiment or an intermediate step,

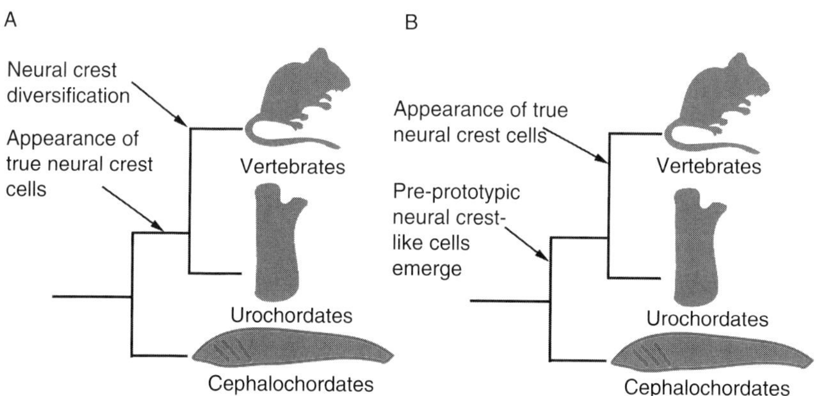

Figure 1.1 Current hypotheses of the time of neural crest origin. According to the first one (A), the neural crest first appeared in the common ancestor of the ascidians and vertebrates, and underwent diversification to form a wider range of derivatives in the vertebrate lineage. Alternatively, the true neural crest may have originated at the base of the vertebrate lineage, after the urochordate–vertebrate split (B).

whereas the origin of the true neural crest may have occurred at the base of the vertebrate lineage, after the urochordate–vertebrate split (Sauka-Spengler and Bronner-Fraser, 2006). In the absence of precise gene duplication scenarios and information on evolution of regulatory elements, it is difficult to distinguish between these possibilities. Below, we discuss the currently available data from studies on nonvertebrate chordates and the implications for the theories regarding the evolution of the neural crest.

3. Why Amphioxus Does Not Have Neural Crest

The entire subphylum Cephalochordata consist of about 30 currently living species of small burrowing filter-feeding animals that inhabit shallow tropical to cool-temperature waters (Poss and Boschung, 1996). *Branchiostoma floridae* is the best-studied species of amphioxus. Lancelets superficially closely resemble vertebrates, and yet lack many of the true vertebrate characteristics, such as the neural crest and cell types derived from the neural crest (i.e., cellular cartilage of the head, migratory pigments cells, peripheral neurons, and glia) (Holland and Holland, 2001; Morikawa *et al.*, 2001).

The process of neurulation that occurs in amphioxus is rather different from that seen in other vertebrates. Instead of the edges of the neural plate raising and fusing together to form a neural tube (as in frog, mouse, and chick), or the neural tube forming by the secondary cavitation of the neural keel (as in the lamprey and teleosts), the amphioxus ectoderm migrates medially to cover the neural plate, which then rolls up to form the neural tube (Holland *et al.*, 1996). This migratory population of the dorsal ectoderm has been considered an evolutionary precursor to the neural crest (Baker and Bronner-Fraser, 1997; Holland *et al.*, 1996); however, recent molecular data do not support this conclusion.

The phylogenetic position of the amphioxus as the basal chordate makes it a very useful model for enquiring what the ancestral preneural crest network might have looked like. To address this question, Yu *et al.* searched the amphioxus genome for the homologues of all known neural crest-inducing signals, neural plate border specifier genes, neural crest specifier genes, and some of the downstream effector genes; explored their expression in the embryonic amphioxus; and compared to that of their homologues in vertebrate model organisms (Yu *et al.*, 2008). Signaling molecules of the Bmp, FGF, Notch, and Wnt pathways that make up the top tier of the NC-GRN were expressed in the amphioxus embryo in patterns closely resembling those seen in vertebrates (Fig. 1.2). This is not surprising given that there is strong conservation of "organizer" genes and of the mechanisms responsible for establishing the dorsoventral patterning of the body axis (Yu *et al.*, 2007). The signaling pathways employed at a slightly later time

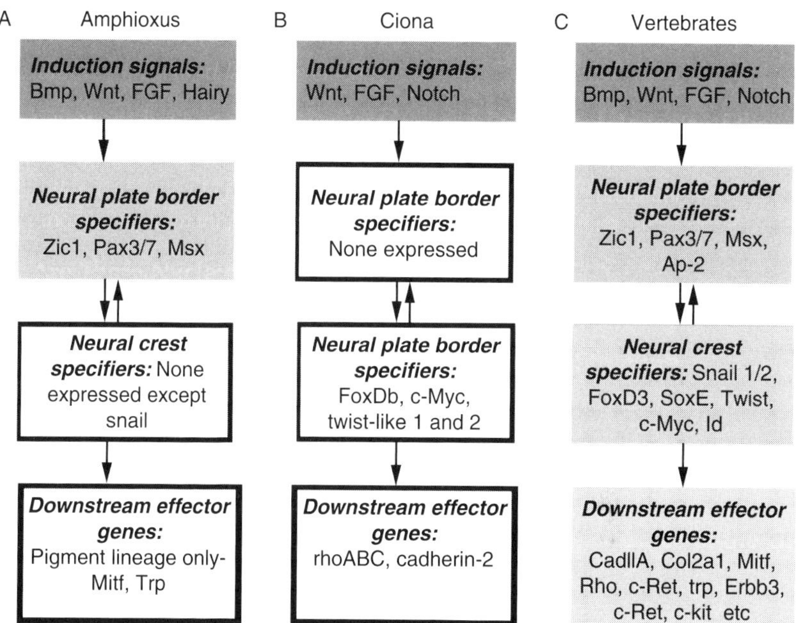

Figure 1.2 Comparison of the vertebrate neural crest gene regulatory network to the neural plate border GRN of amphioxus and trunk lateral cell network of ascidian *Ciona intestinalis*. Colored boxes indicate network modules that are conserved from amphioxus/ *Ciona* to vertebrates; black and white boxes indicate lack of evolutionary conservation of the particular module.

are instrumental in establishment of the neural plate border. These initial modules are essential for early development of the embryo and, without them, development would not proceed. These finding are consistent with the crucial function of these signaling factors in the setting up of the dorsoventral polarity and ectoderm patterning of the embryo, a function conserved across many metazoan phyla (Lowe *et al.*, 2006).

These extracellular signals in turn regulate a module of transcription factors at the neural plate border that are highly conserved across chordates. The expression patterns of the next tier of the neural crest network, the neural plate border specifiers Msx, Pax3/7, and Zic were found to be similar to the expression patterns of their vertebrate homologues. Msx transcripts were seen throughout the ectoderm and in the neural plate border at late gastrula to neurula stages, while Zic expressions marked only the neural plate border, and Pax3/7 was expressed in the NPB and throughout the neural plate (Gostling and Shimeld, 2003; Holland *et al.*, 1999; Sharman *et al.*, 1999; Yu *et al.*, 2008). Moreover, by treating amphioxus embryos with Bmp4, Yu *et al.* demonstrated that expression of these three genes is affected by the Bmp gradient (Fig. 1.2). Taken together, these data seem to suggest that the genetic mechanisms responsible for the establishment of the neural late border are conserved throughout the chordates. Interestingly, however, AP-2, which was shown to act earlier then previously though during neural plate border specification (Nikitina *et al.*, 2008), is expressed only in the amphioxus epidermal ectoderm and not in the neural plate border during late gastrula/early neurula, the time when neural plate border specification is thought to occur (Meulemans and Bronner-Fraser, 2002; Yu *et al.*, 2008). This fact indicates that neural plate border module of the NC-GRN is not fully conserved between vertebrates and the amphioxus.

In contrast to neural plate border genes, later transcription factors involved in neural crest specification in vertebrates are largely absent from the neural plate border of amphioxus or other nonvertebrate chordates. In fact, except for Snail, which makes a transient appearance at the neural plate border during neurula (and is possibly involved in the neural tube closure), none of the neural crest specifiers, are coexpressed in the amphioxus neural plate border. Thus, the transition from basal chordate to vertebrate was accompanied by the appearance a module of transcription factors at the neural plate border that help specify neural crest fate. This likely occurred either by co-option of existing cassettes to the neural plate border or by invention of novel interaction between the molecules newly expressed within the territory. Such a "neural crest kernel" may have helped confer migratory ability, multipotency, and a variety of other properties (e.g., cycle progression, delamination, motility, and fate specification). Although the mechanisms underlying this recruitment are unknown, they may have been facilitated by addition of new regulatory modules or via chromosomal rearrangements.

Similar to the neural crest specifiers, none of the downstream neural crest effector genes examined (Erbb3, Mitf) were ever expressed in this domain, and some of these genes (c-kit, P0) were not even found in the amphioxus genome (Langeland *et al.*, 1998; Meulemans and Bronner-Fraser, 2002; Meulemans *et al.*, 2003; Yu *et al.*, 2002, 2008). These data clearly demonstrated that the lack of the neural crest in the amphioxus can be explained by the fact that none of the genes that are essential for the formation of the neural crest proper are yet recruited into the neural plate border cells and placed downstream of the neural plate border specifiers. The rudimentary neural plate border network of amphioxus thus provides a snapshot of the gene regulatory foundation from which the neural crest network proper has evolved. How this occurred still remains to be elucidated, but it is possible that the *cis*-regulatory regions of the neural crest specifier genes were modified in such a way as to bring them under transcriptional control by neural plate border specifier genes.

A recent survey suggests that ~9% of genes involved in vertebrate neural crest formation evolved after the Cambrian explosion. These authors concluded that genes involved in neural crest diversification in vertebrates may have been linked to the emergence of new signaling molecules (Martinez-Morales *et al.*, 2007). These are likely to represent new downstream targets of neural crest genes that facilitate migration and differentiation processes. In contrast, other programs appear to be quite ancient. For example, the cassette used for differentiation of pigment cells in the vertebrate neural crest appears to be also be used in amphioxus. This raises the intriguing possibility that cassettes used for ancient differentiation programs may have been co-opted to early functions in the vertebrate NC-GRN. Consistent with this possibility, a population of migrating pigment cell precursors have been identified in ascidians and have been proposed to represent preprototypic neural crest. One possibility is that these precursors acquired cell cycle regulatory mechanisms that allowed them to diversify into other types of derivatives to become vertebrate neural crest. Alternatively, several independent types of cells with diverse differentiative potentials may have collectively acquired migratory and regulative ability when assembled in a presumptive population at the neural plate border.

4. Ascidians: Origins of the True Neural Crest or Parallel Evolution?

Urochordates, the true sister group of vertebrates, is a diverse group of animals comprising some 3000 species. Urochordates are traditionally divided into three classes: ascidians, larvaceans, and thaliaceans. The first two classes have a complex life cycle, which includes a swimming larval stage (tadpole) and a sessile benthic (ascidians) or swimming planktonic (larvaceans) adult stage. The swimming larvae of most species demonstrate pronounced chordate characteristics (muscular postanal tail, notochord, dorsal nerve cord), which disappear in ascidians after metamorphosis, while adult larvaceans retain much of the larval body plan. Most thaliaceans lack a tadpole stage and have an adult body plan that is somewhat similar to the adult ascidians, but modified for their holoplanktonic lifestyle (Swalla *et al.*, 2000).

All ascidians have an invariant cell lineage, which allowed mapping the fate of every embryonic cell. The neurulation occurs by a process very similar to the primary neurulation of vertebrates. The complexity of the resulting larvae, however, differs significantly among ascidian species, due to the fact that in some species differentiation of the adult structures (pigment cells, siphon primordial, branchial gill slits) is initiated during the larval development rather than after metamorphosis. Jeffery (2007) divides all ascidians into four groups based on the extent of formation of the adult structures during larval development (adultation). Perhaps not surprisingly, it is the species that form the least complex, more vertebrate-like larvae (e.g., *Ciona, Halocynthia*) that have been used extensively in embryological studies, due to their easy availability and rapid development.

The choice of model system was the prime reason that nothing resembling the neural crest was identified in ascidians, even though adult ascidians have some of the cell types that in vertebrates develop from the neural crest. For example, many adult ascidians are highly pigmented, and calcitonin-producing cells that are possibly homologous to the parafollicular C cells of the vertebrate thyroid were identified in the endostylar region of *Styela* pharynx (Thorndyke and Probert, 1979). However, these cell types appear only during or after metamorphosis in species with simple larvae. When Jeffery *et al.* chose to use *Ecteinascidia turbinata*, a species of ascidian that produces a large complex larvae with high degree of adultation, a surprising discovery was made: some of the cells from the neural tube migrated out, differentiated into pigment cells, and even expressed some neural crest markers such as HNK-1 and Zic (Jeffery *et al.*, 2004).

This newly discovered ascidian cell type, NCLCs, were then taken as a proof that neural crest originated prior to the urochordate–vertebrate split (Donoghue *et al.*, 2008; Graham, 2004; Jeffery, 2007). However, we think that while this discovery is of great importance for understanding chordate evolution, not enough evidence is currently available to distinguish whether these cells are truly neural crest or arose by convergent evolution. We do not know whether gene regulatory network responsible for the formation of these cells and their migratory ability is the same as that operating in bona fide neural crest cells of vertebrates, as only EtZic expression was examined (Jeffery *et al.*, 2004). Also, pigment cells appear to be the main, perhaps the only, derivative of these NCLCs, in striking contrast with multitude of cell types formed by vertebrate neural crest.

Since HNK-1 positive cells, assumed to be NCLCs, were found in many ascidian species, including *Ciona* (Jeffery, 2006), the question of gene regulatory network conservation was next addressed using this well-researched species. The results of this study, however, would appear not to support the homology between the ascidian NCLC and the vertebrate neural crest (Fig. 1.2). The embryonic origin of the *Ciona* HNK-1+ cells was traced to the A7.6 blastomeres, which become internalized during gastrulation and their derivatives come to lie underneath the ectoderm laterally to the neural plate (Jeffery *et al.*, 2008). In addition, neither members of the neural plate border specifier module (Ci-msxb, Ci-Pax3/7, Ci-ZicL) nor many of the neural crest specifiers (Ci-Snail, Ci-Id) appear to be expressed in A7.6 cells or their descendants, but are found instead in the cells of the neural plate border. That suggests that these cells employ a completely different developmental mechanism for mobilizing their migratory capacity and for initiating the pigment cell-like differentiation module. Given the extensive rearrangements of the upper tiers of the gene regulatory network that would have to occur to allow the evolution of the vertebrate-like neural crest from a *Ciona*-like precursor, we find the evolutionary scenario proposed by Jeffery *et al.* (2008) extremely unlikely.

It is, however, too early to conclude that ascidian NCLC is an independent evolutionary experiment where the pigment cells specific differentiation module was combined with some sort of module for migration, resulting in the formation of a migratory pigment cell that shares some of the neural crest characteristics. Extensive examination of the NC-GRN gene expression patterns in the NCLC of *E. turbinata* and other ascidian species with complex larvae may reveal a deeper conservation than what was found in *Ciona*, while more detailed lineage tracing experiments in different urochordate species may uncover that NCLC can give rise to derivatives other than pigment cells. It seems that urochordates hold the key to understanding the origins of the neural crest.

5. Neural Crest in Different Vertebrate Species: Evolution in Progress?

The differentiation ability of the neural crest is remarkable, with up to 100 different human cell types that are known to be neural crest-derived. Interestingly, basal vertebrates such as lamprey display a more limited repertoire of neural crest derivatives than more derived vertebrate species. On the other hand, unusual taxon-specific cell types (carapace in turtles, diverse pigment cells in fish and lizards) have been identified. Understanding the molecular interactions responsible for the diversification and ongoing evolution of the neural crest-derived cell types is the next frontier for the evolutionary biology of the neural crest.

6. Conclusion and Future Perspectives

We have used regulatory state to help define the vertebrate neural crest and gain insights into its evolution. By analyzing diverse chordates both embryologically and via genome sequence information, clues as to the evolutionary origin of this complex population are emerging. Importantly, comparative studies amongst chordates will help address how regulatory changes may have facilitated neural crest evolution. Future studies must be geared toward refining the sequence of deployment and interconnections within the NC-GRN as well as identifying important posttranscriptional modifications which may modulate the network and the process of derivative formation. In addition, it is important to address the nature and function of downstream genes involved in EMT, migration, and the formation of diverse neural crest derivatives.

REFERENCES

Baker, C. V., and Bronner-Fraser, M. (1997). The origins of the neural crest. Part II. An evolutionary perspective. *Mech. Dev.* **69,** 13–29.

Batlle, E., Sancho, E., Franci, C., Dominguez, D., Monfar, M., Baulida, J., and Garcia De Herreros, A. (2000). The transcription factor snail is a repressor of E-cadherin gene expression in epithelial tumour cells. *Nat. Cell Biol.* **2,** 84–89.

Bellmeyer, A., Krase, J., Lindgren, J., and LaBonne, C. (2003). The protooncogene c-myc is an essential regulator of neural crest formation in xenopus. *Dev. Cell* **4,** 827–839.

Blair, J. E., and Hedges, S. B. (2005). Molecular phylogeny and divergence times of deuterostome animals. *Mol. Biol. Evol.* **22,** 2275–2284.

Breau, M. A., Pietri, T., Stemmler, M. P., Thiery, J. P., and Weston, J. A. (2008). A nonneural epithelial domain of embryonic cranial neural folds gives rise to ectomesenchyme. *Proc. Natl. Acad. Sci. USA* **105,** 7750–7755.

Cano, A., Perez-Moreno, M. A., Rodrigo, I., Locascio, A., Blanco, M. J., del Barrio, M. G., Portillo, F., and Nieto, M. A. (2000). The transcription factor snail controls epithelial–mesenchymal transitions by repressing E-cadherin expression. *Nat. Cell Biol.* **2,** 76–83.

Delsuc, F., Brinkmann, H., Chourrout, D., and Philippe, H. (2006). Tunicates and not cephalochordates are the closest living relatives of vertebrates. *Nature* **439,** 965–968.

Donoghue, P. C., Graham, A., and Kelsh, R. N. (2008). The origin and evolution of the neural crest. *Bioessays* **30,** 530–541.

Gans, C., and Northcutt, R. G. (1983). Neural crest and the origin of vertebrates: A new head. *Science* **220,** 268–273.

Gostling, N. J., and Shimeld, S. M. (2003). Protochordate Zic genes define primitive somite compartments and highlight molecular changes underlying neural crest evolution. *Evol. Dev.* **5,** 136–144.

Graham, A. (2004). Evolution and development: Rise of the little squirts. *Curr. Biol.* **14,** R956–R958.

Holland, L. Z., and Holland, N. D. (2001). Evolution of neural crest and placodes: Amphioxus as a model for the ancestral vertebrate? *J. Anat.* **199,** 85–98.

Holland, N. D., Panganiban, G., Henyey, E. L., and Holland, L. Z. (1996). Sequence and developmental expression of AmphiDll, an amphioxus Distal-less gene transcribed in the ectoderm, epidermis and nervous system: Insights into evolution of craniate forebrain and neural crest. *Development* **122,** 2911–2920.

Holland, L. Z., Schubert, M., Kozmik, Z., and Holland, N. D. (1999). AmphiPax3/7, an amphioxus paired box gene: Insights into chordate myogenesis, neurogenesis, and the possible evolutionary precursor of definitive vertebrate neural crest. *Evol. Dev.* **1,** 153–165.

Honore, S. M., Aybar, M. J., and Mayor, R. (2003). Sox10 is required for the early development of the prospective neural crest in Xenopus embryos. *Dev. Biol.* **260,** 79–96.

Horigome, N., Myojin, M., Ueki, T., Hirano, S., Aizawa, S., and Kuratani, S. (1999). Development of cephalic neural crest cells in embryos of *Lampetra japonica*, with special reference to the evolution of the jaw. *Dev. Biol.* **207,** 287–308.

Jeffery, W. R. (2006). Ascidian neural crest-like cells: Phylogenetic distribution, relationship to larval complexity, and pigment cell fate. *J. Exp. Zool. B Mol. Dev. Evol.* **306,** 470–480.

Jeffery, W. R. (2007). Chordate ancestry of the neural crest: New insights from ascidians. *Semin. Cell Dev. Biol.* **18,** 481–491.

Jeffery, W. R., Strickler, A. G., and Yamamoto, Y. (2004). Migratory neural crest-like cells form body pigmentation in a urochordate embryo. *Nature* **431,** 696–699.

Jeffery, W. R., Chiba, T., Krajka, F. R., Deyts, C., Satoh, N., and Joly, J. S. (2008). Trunk lateral cells are neural crest-like cells in the ascidian *Ciona intestinalis*: Insights into the ancestry and evolution of the neural crest. *Dev. Biol.* **324**(1), 152–160.

Johnels, A. G. (1956). On the peripheral autonomic nervous system of the trunk region of *Lampetra planeri*. *Acta Zool. (Stockholm)* **37,** 251–286.

Kim, J., Lo, L., Dormand, E., and Anderson, D. J. (2003). SOX10 maintains multipotency and inhibits neuronal differentiation of neural crest stem cells. *Neuron* **38,** 17–31.

Langeland, J. A., Tomsa, J. M., Jackman, W. R. Jr., and Kimmel, C. B. (1998). An amphioxus snail gene: Expression in paraxial mesoderm and neural plate suggests a conserved role in patterning the chordate embryo. *Dev. Genes Evol.* **208,** 569–577.

Lowe, C. J., Terasaki, M., Wu, M., Freeman, R. M. Jr., Runft, L., Kwan, K., Haigo, S., Aronowicz, J., Lander, E., Gruber, C., Smith, M., Kirschner, M., and Gerhart, J. (2006). Dorsoventral patterning in hemichordates: Insights into early chordate evolution. *PLoS Biol.* **4,** e291.

Martinez-Morales, J. R., Henrich, T., Ramialison, M., and Wittbrodt, J. (2007). New genes in the evolution of the neural crest differentiation program. *Genome Biol.* **8,** R36.

McCauley, D. W., and Bronner-Fraser, M. (2003). Neural crest contributions to the lamprey head. *Development* **130,** 2317–2327.

McCauley, D. W., and Bronner-Fraser, M. (2006). Importance of SoxE in neural crest development and the evolution of the pharynx. *Nature* **441,** 750–752.

Meulemans, D., and Bronner-Fraser, M. (2002). Amphioxus and lamprey AP-2 genes: Implications for neural crest evolution and migration patterns. *Development* **129,** 4953–4962.

Meulemans, D., and Bronner-Fraser, M. (2004). Gene-regulatory interactions in neural crest evolution and development. *Dev. Cell* **7,** 291–299.

Meulemans, D., McCauley, D., and Bronner-Fraser, M. (2003). Id expression in amphioxus and lamprey highlights the role of gene cooption during neural crest evolution. *Dev. Biol.* **264,** 430–442.

Morikawa, K., Tsuneki, K., and Ito, K. (2001). Expression patterns of HNK-1 carbohydrate and serotonin in sea urchin, amphioxus, and lamprey, with reference to the possible evolutionary origin of the neural crest. *Zoology (Jena)* **104,** 81–90.

Nikitina, N. V., and Bronner-Fraser, M. (2008). Gene regulatory networks that control the specification of neural-crest cells in the lamprey. *Biochim. Biophys. Acta.* doi:10.1016/j.bbagrm.2008.03.006.

Nikitina, N. V., Sauka-Spengler, T., and Bronner-Fraser, M. (2008). Dissecting early regulatory relationships in the lamprey neural crest gene regulatory network. *Proc. Natl. Acad. Sci. USA* **105,** 20083–20088.

Northcutt, R. G., and Gans, C. (1983). The genesis of neural crest and epidermal placodes: A reinterpretation of vertebrate origins. *Q. Rev. Biol.* **58,** 1–28.

Ota, K. G., Kuraku, S., and Kuratani, S. (2007). Hagfish embryology with reference to the evolution of the neural crest. *Nature* **446,** 672–675.

Poss, S., and Boschung, H. T. (1996). Lancelets (Cephalochordata: Branchiostomatidae): How many species are valid? *Israel J. Zool.* **42**(Suppl.), 13–66.

Putnam, N. H., Butts, T., Ferrier, D. E., Furlong, R. F., Hellsten, U., Kawashima, T., Robinson-Rechavi, M., Shoguchi, E., Terry, A., Yu, J. K., Benito-Gutierrez, E. L., Dubchak, I., *et al.* (2008). The amphioxus genome and the evolution of the chordate karyotype. *Nature* **453,** 1064–1071.

Sauka-Spengler, T., and Bronner-Fraser, M. (2006). Development and evolution of the migratory neural crest: A gene regulatory perspective. *Curr. Opin. Genet. Dev.* **16,** 360–366.

Sauka-Spengler, T., and Bronner-Fraser, M. (2008). A gene regulatory network orchestrates neural crest formation. *Nat. Rev. Mol. Cell Biol.* **9,** 557–568.

Sauka-Spengler, T., Meulemans, D., Jones, M., and Bronner-Fraser, M. (2007). Ancient evolutionary origin of the neural crest gene regulatory network. *Dev. Cell* **13,** 405–420.

Sharman, A. C., Shimeld, S. M., and Holland, P. W. (1999). An amphioxus Msx gene expressed predominantly in the dorsal neural tube. *Dev. Genes Evol.* **209,** 260–263.

Soo, K., O'Rourke, M. P., Khoo, P. L., Steiner, K. A., Wong, N., Behringer, R. R., and Tam, P. P. (2002). Twist function is required for the morphogenesis of the cephalic neural tube and the differentiation of the cranial neural crest cells in the mouse embryo. *Dev. Biol.* **247,** 251–270.

Swalla, B. J., Cameron, C. B., Corley, L. S., and Garey, J. R. (2000). Urochordates are monophyletic within the deuterostomes. *Syst. Biol.* **49,** 52–64.

Taneyhill, L. A., Coles, E. G., and Bronner-Fraser, M. (2007). Snail2 directly represses cadherin6B during epithelial-to-mesenchymal transitions of the neural crest. *Development* **134,** 1481–1490.

Teng, L., Mundell, N. A., Frist, A. Y., Wang, Q., and Labosky, P. A. (2008). Requirement for Foxd3 in the maintenance of neural crest progenitors. *Development* **135,** 1615–1624.

Thorndyke, M. C., and Probert, L. (1979). Calcitonin-like cells in the pharynx of the ascidian Styela clava. *Cell Tissue Res.* **203,** 301–309.

Turbeville, J. M., Schulz, J. R., and Raff, R. A. (1994). Deuterostome phylogeny and the sister group of the chordates: Evidence from molecules and morphology. *Mol. Biol. Evol.* **11,** 648–655.

Wada, H. (2001). Origin and evolution of the neural crest: A hypothetical reconstruction of its evolutionary history. *Dev. Growth Differ.* **43,** 509–520.

Wada, H., and Satoh, N. (1994). Details of the evolutionary history from invertebrates to vertebrates, as deduced from the sequences of 18S rDNA. *Proc. Natl. Acad. Sci. USA* **91,** 1801–1804.

Winchell, C. J., Sullivan, J., Cameron, C. B., Swalla, B. J., and Mallatt, J. (2002). Evaluating hypotheses of deuterostome phylogeny and chordate evolution with new LSU and SSU ribosomal DNA data. *Mol. Biol. Evol.* **19,** 762–776.

Yu, J. K., Holland, N. D., and Holland, L. Z. (2002). An amphioxus winged helix/forkhead gene, AmphiFoxD: Insights into vertebrate neural crest evolution. *Dev. Dyn.* **225,** 289–297.

Yu, J. K., Satou, Y., Holland, N. D., Shin, I. T., Kohara, Y., Satoh, N., Bronner-Fraser, M., and Holland, L. Z. (2007). Axial patterning in cephalochordates and the evolution of the organizer. *Nature* **445,** 613–617.

Yu, J. K., Meulemans, D., McKeown, S. J., and Bronner-Fraser, M. (2008). Insights from the amphioxus genome on the origin of vertebrate neural crest. *Genome Res.* **18,** 1127–1132.

CHAPTER TWO

Evolution of Vertebrate Cartilage Development

GuangJun Zhang,[*,#] B. Frank Eames,[†] *and* Martin J. Cohn[*,‡]

Contents

Abstract

Major advances in the molecular genetics, paleobiology, and the evolutionary developmental biology of vertebrate skeletogenesis have improved our understanding of the early evolution and development of the vertebrate skeleton.

[*] Department of Zoology, University of Florida, Cancer/Genetics Research Complex, Gainesville, Florida, USA
[†] Institute of Neuroscience, University of Oregon, Eugene, Oregon, USA
[‡] Department of Anatomy and Cell Biology, University of Florida, Cancer/Genetics Research Complex, Gainesville, Florida, USA
[#] Current address: The David H. Koch Institute for Integrative Cancer Research, MIT, Cambridge, Massachusetts, USA

Current Topics in Developmental Biology, Volume 86
ISSN 0070-2153, DOI: 10.1016/S0070-2153(09)01002-3

These studies have involved genetic analysis of model organisms, human genetics, comparative developmental studies of basal vertebrates and nonvertebrate chordates, and both cladistic and histological analyses of fossil vertebrates. Integration of these studies has led to renaissance in the area of skeletal development and evolution. Among the major findings that have emerged is the discovery of an unexpectedly deep origin of the gene network that regulates chondrogenesis. In this chapter, we discuss recent progress in each these areas and identify a number of questions that need to be addressed in order to fill key gaps in our knowledge of early skeletal evolution.

1. Introduction

The vertebrate skeleton consists of two predominant tissue types: cartilage and bone. Although generally considered a vertebrate character, cartilage is found across a broad range of animal taxa, indicating a long and complex evolutionary history (Hall, 2005). Cartilage differs from bone in several ways; cartilage has a lower metabolic rate, is mostly avascular, and contains different cellular and extracellular components that give it unique structural properties. Classically, true cartilage was defined by three criteria (1) it contains chondrocytes suspended in rigid matrix, (2) the matrix has a high content of collagen, and (3) the matrix is rich in acidic polysaccharides (Person and Mathews, 1967). The proposal that the cartilage of some vertebrates, such as lampreys and hagfishes, is noncollagenous led to a revision of this definition to substitute "fibrous proteins" for "collagen" (Cole and Hall, 2004a); however, recent work has shown that these jawless fishes also have collagen-based cartilage (Ohtani *et al.*, 2008; Zhang and Cohn, 2006; Zhang *et al.*, 2006). Such studies of cartilage in nontetrapod lineages have revealed that a deeply conserved genetic system underlies a diverse array of cartilage types. These discoveries have enhanced our understanding of the early evolution of cartilage and raised new questions about the homologies of animal connective tissues. Here, we review these advances in the context of skeletal developmental genetics and the evolutionary history of vertebrates, and discuss how changes to developmental and genomic programs may have contributed to the origin of the vertebrate skeleton.

2. Skeletal Cell Lineage Determination and the Skeletogenic Gene Network

Vertebrate cartilage and bone are composed of three major cell lineages, chondrocytes, osteoblasts, and osteoclasts. The former two cell types are derived from common mesenchymal progenitor cells, whereas osteoclasts are of hematopoietic origin. After condensation, mesenchymal cells start to differentiate into chondrocytes. These chondrocytes may remain as cartilage throughout life, or the cartilage template may undergo hypertrophy and eventually be replaced by bone, a process termed endochondral ossification. Alternatively, the mesenchymal cells may differentiate directly into bone, through a process termed intramembranous ossification, as seen in the membrane bones of the skull, such as the calvaria. In both intramembranous and endochondral ossification, osteoblasts first aggregate as mesenchymal condensations (Karsenty and Wagner, 2002; Yang and Karsenty, 2002; Zelzer and Olsen, 2003). The cell fate decisions made by aggregating mesenchymal cells are regulated by a skeletogenic gene network (Fig. 2.1), and understanding the hierarchy, regulation, and function of these factors is critical to our discussion of the evolution of skeletogenic mechanisms. Below, we review the major components of this network and describe their functions and interactions during embryonic development of the skeleton.

2.1. Sox9

As cells in mesenchymal condensations begin to differentiate into chondrocytes, the earliest marker of chondrogenesis is *Sox9*, a member of the vertebrate SoxE family that contains a high-mobility-group (HMG)-box

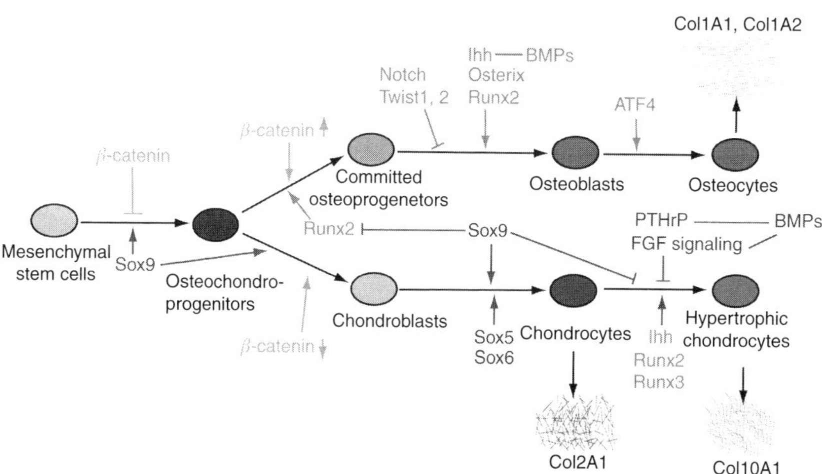

Figure 2.1 Schematic representation of gene network that directs mesenchymal cells along chondrogenic (bottom) and osteogenic (top) differentiation pathways. Arrows indicate positive regulation, lines indicate interaction, and bars indicate negative regulation. Data represented in this schematic are taken from multiple sources cited in the text. The scheme depicts hierarchical arrangement of genes in the network and does not necessarily indicate direct transcriptional regulation at each step.

DNA-binding domain (Fig. 2.1) (Healy *et al.*, 1996; Wright *et al.*, 1995). *Sox9* directly regulates expression of two genes that code for major matrix proteins, type II collagen (Col2α1) and aggrecan, and is required for expression of genes that encode minor matrix proteins, including type IX and XI collagen (Lefebvre and de Crombrugghe, 1998; Lefebvre *et al.*, 1997; Liu *et al.*, 2000; Ng *et al.*, 1997; Zhang *et al.*, 2003b; Zhou *et al.*, 1998). Haploinsufficiency of *Sox9* in humans underlies campomelic dysplasia, a congenital malformation of the skeleton characterized by shortening and bowing of the limbs, and similar anomalies occur in mice with loss-of-function mutation in *Sox9* (Foster *et al.*, 1994; Wagner *et al.*, 1994). Reciprocal experiments involving ectopic expression of *Sox9* in chick embryos can induce dermomyotomal or neural crest-derived cells to form cartilage (Healy *et al.*, 1999; Eames *et al.*, 2004). Sox9 function is enhanced by Sox5 and Sox6, which can bind to Sox9 and act as cofactors in the activation of *Col2α1* (Ikeda *et al.*, 2004; Lefebvre and de Crombrugghe, 1998; Lefebvre *et al.*, 1998, 2001; Smits and Lefebvre, 2003; Stolt *et al.*, 2006). The Sox5/6/9 trio also has been shown to bind S100A1 and S100B, two novel targets that mediate the trio's ability to inhibit chondrocyte differentiation (Saito *et al.*, 2007). Sox9 can form complexes with the CREB-binding protein CBP/P300, and the association of these proteins may be required for chondrocyte-specific expression of *Col2α1* (Tsuda *et al.*, 2003). Interestingly, the chondrogenic activity of TGFβ/Bmp signaling (described below) may be mediated, at least in part, by the ability of Smad3 to promote binding of Sox9 with the CBP/P300 coactivator (Furumatsu *et al.*, 2005). These interactions may account for the ability of Sox9 to activate *Col2α1* in some cell lineages (e.g., limb bud, sclerotome, and cranial neural crest) but not others (e.g., genital ridge).

2.2. Runx2

The vertebrate *Runx2* gene [also known as *PEBP2A (polyoma enhancer-binding protein 2A)*, *Osf2 (osteoblast-specific factor 2)*, *AML3 (acute myelogenous leukemia 3)*, and *Cbfa1 (core-binding factor alpha 1)*] is an ortholog of the fly *runt* gene and a master regulator of osteoblast differentiation (Fig. 2.1) (van Wijnen *et al.*, 2004). In addition to its role in osteoblast differentiation (Ducy *et al.*, 1997, 1999; Komori *et al.*, 1997; Otto *et al.*, 1997), *Runx2* is required for chondrocyte hypertrophy (Fig. 2.1). In *Runx2*-null mice, the entire skeleton remains cartilaginous due to the maturational arrest of osteoblasts, and there is a failure of chondrocyte hypertrophy (Inada *et al.*, 1999; Kim *et al.*, 1999; Takeda *et al.*, 2001). Reciprocally, ectopic expression of *Runx2* in chick head mesenchyme can drive excess bone formation and ectopic chondrocyte hypertrophy (Eames *et al.*, 2004). Haploinsufficiency of *Runx2* in humans causes cleidocranial dysplasia, a rare skeletal malformation characterized by short stature, distinctive facial features and narrow, sloping shoulders associated with defective or absent clavicles (Mundlos and Olsen, 1997a,b; Mundlos *et al.*, 1996). *Runx1* and *Runx3*, two genes closely related to *Runx2*, also are expressed in chondrocytes and participate in the progression of chondrocytes to the hypertrophic stage (Karsenty, 2008; Levanon *et al.*, 2001; Lian *et al.*, 2003; Smith *et al.*, 2005; Stricker *et al.*, 2002; Wang *et al.*, 2005).

2.3. Interaction of Sox9 and Runx2

Several lines of evidence have shown that in many cases, condensed mesenchymal cells have chondrogenic and osteogenic potential, since they express both *Sox9* and *Runx2* (Bi *et al.*, 1999; Ducy *et al.*, 1997; Eames and Helms, 2004; Otto *et al.*, 1997; Yamashiro *et al.*, 2004). Moreover, cultured embryonic cells may form both bone and cartilage (Fang and Hall, 1997; Toma *et al.*, 1997; Wong and Tuan, 1995). Inactivation of *Sox9* in the cranial neural crest-derived mesenchymal cells blocks cartilage differentiation, but this also leads to ectopic expression of osteoblast-specific genes such as *Runx2*, *Osterix*, and *Col1α1*(Mori-Akiyama *et al.*, 2003). Conversely, it was reported that in *Osterix* mutants, ectopic chondrocytes formed at the expense of the bone collar in long bones and in some membrane bones (Nakashima *et al.*, 2002). These data support the idea that the common skeletal mesenchymal progenitors have three possible differentiation fates in the skeleton, chondrogenesis, intramembranous ossification or endochondral ossification (it is noteworthy, however, that these mesenchymal cells also can take on other, nonskeletal cell fates, such as adipose tissue) (Karsenty, 2003; Karsenty and Wagner, 2002). In mesenchymal osteochondrogenic progenitors, removal of *Sox9* will abolish cartilage and endochondral bone formation, indicating that *Sox9* is required for skeletal differentiation (Akiyama *et al.*, 2005). Experiments in chick embryos demonstrated that higher levels of *Sox9* will commit cells to chondrogenesis, whereas higher levels of *Runx2* will push them toward osteogenesis (Fig. 2.1) (Eames *et al.*, 2004). *Sox9* has been shown to be dominant to *Runx2* (Zhou *et al.*, 2006), which suggests that if these transcription factors are expressed at similar levels, then skeletal progenitor cells may differentiate preferentially into cartilage.

2.4. Parathyroid hormone-related protein and Indian hedgehog

During long bone growth, chondrocyte proliferation and differentiation is tightly regulated by a negative feedback loop between Indian hedgehog (Ihh) and parathyroid hormone-related protein (PTHrP) (Fig. 2.1) (Karp *et al.*, 2000; Lanske *et al.*, 1996; St-Jacques *et al.*, 1999; Vortkamp *et al.*, 1996). PTHrP is a peptide hormone that is secreted by the most distal perichondrium, and its G protein-coupled receptor, *PPR*, localizes to the proliferative prehypertrophic zone. PTHrP acts to maintain proliferation and to inhibit differentiation (St-Jacques *et al.*, 1999). In humans, activating mutations of *PPR* cause Jansen's metaphyseal chondrodysplasia, which involves delayed skeletal differentiation and abnormal growth plates (Schipani *et al.*, 1995). Loss-of-function mutations in *PTHrP* in mice result in dwarfism due to accelerated hypertrophy (Karaplis *et al.*, 1994; Lanske *et al.*, 1996). *Ihh* is expressed along with *PPR* in the prehypertrophic zone and controls expression of *PTHrP* (Vortkamp *et al.*, 1996). Deletion of *Ihh* results in reduced chondrocyte proliferation and failure of perichondral osteoblast formation, ultimately leading to dwarfism. In the *Ihh*-null mutants, *PTHrP* expression is lost (Razzaque *et al.*, 2005; St-Jacques *et al.*, 1999), and *Ihh* overexpression results in upregulation of *PTHrP*, promoting proliferation and delaying hypertrophy. PTHrP feeds back to negatively regulate *Ihh* expression. This Ihh–PTHrP feedback loop maintains the balance between proliferation and differentiation (Kronenberg, 2006). Very recent work has shown that Ihh can promote chondrocyte hypertrophy independently of PTHrP (Mak *et al.*, 2008). Bapx1 (Nk3.2) is a downstream target of Ihh–PTHrP loop and, at least in part, mediates chondrocyte hypertrophy (Provot *et al.*, 2006). Interestingly, Runx2 and Runx3 can induce *Ihh* expression (St-Jacques *et al.*, 1999; Yoshida *et al.*, 2004) and Ihh can feed back to inhibit *Runx2* expression through the PKA pathway (Iwamoto *et al.*, 2003; Li *et al.*, 2004).

2.5. Wnt signaling

The canonical Wnt pathway is a key regulator for mesenchymal cell lineage determination (Fig. 2.1). *Wnt* genes are vertebrate orthologs of the *Drosophila wingless* gene, and there are 19 known *Wnt* genes in humans (Logan and Nusse, 2004; Miller, 2002). This group of secreted molecules is highly conserved in metazoan animals ranging from cnidarians to humans, and they have critical functions both in normal development and tumorigenesis (Kusserow *et al.*, 2005; Lee *et al.*, 2006; Logan and Nusse, 2004; Prud'homme *et al.*, 2002). Wnt proteins that bind to Frizzed receptors transduce the input into the cell together with the coreceptor, LDL receptor-related protein 5/6 (LRP5/6). There are at least three intracellular pathways for Wnt signaling; the canonical pathway mediated by β-catenin, the Ca–PKC pathway, and the planar cell polarity pathway (Miller, 2002). Interestingly, Sox9 also interacts with β-catenin. Sox9 can inhibit β-catenin-dependent promoter activation through the interaction between HMG-box and Armadillo repeats. Sox9 also promotes degradation of β-catenin by ubiquitation or the proteasome pathway (Akiyama *et al.*, 2004).

Canonical Wnt signaling has been implicated in skeletal development (Bodine *et al.*, 2004; Boyden *et al.*, 2002; Gong *et al.*, 2001; Hartmann and Tabin, 2001; Kato *et al.*, 2002; Little *et al.*, 2002; Rawadi *et al.*, 2003).

Several lines of evidence have revealed that the canonical Wnt pathway regulates skeletogenic cell fate determination through a cell-autonomous mechanism to induce osteoblast differentiation and to repress chondrocyte differentiation (Fig. 2.1) (Day *et al.*, 2005; Glass *et al.*, 2005; Hill *et al.*, 2005; Hu *et al.*, 2005; Rodda and McMahon, 2006). When β-catenin is conditionally removed from skeletogenic mesenchyme using the *Prx1-Cre* allele, osteoblast differentiation arrests, and neither cortical nor membrane bone forms (although this can be rescued by Ihh and Bmp2). Similar phenotypes were found when β-catenin was deleted from the skeletal primordium using *Dermo1-Cre* and *Col2a1-Cre* mouse lines, in which ectopic chondrocytes formed at the expense of osteoblasts (Day *et al.*, 2005; Hu *et al.*, 2005). Moreover, micromass cell culture experiments showed that β-catenin levels can control the expression of *Sox9* and *Runx2 in vitro* (Day *et al.*, 2005). Collectively, β-catenin controls early osteochondroprogenitor differentiation into chondrocytic or osteoblastic lineages. High levels of β-catenin lead to osteogenic differentiation and low levels lead to chondrogenic differentiation (Day *et al.*, 2005; Hill *et al.*, 2005). The process is summarized in Fig. 2.1. These studies suggest that variation in skeletal composition, both developmentally and evolutionarily, may be accomplished by tinkering with the temporal and spatial expression of canonical Wnt signals.

2.6. Fibroblast growth factor signaling

Fibroblast growth factors (Fgfs) and their receptors are also critical regulators of chondrocyte proliferation and differentiation (Fig. 2.1). In humans and mice, there are 22 Fgf genes and 4 Fgf receptors (Fgfr), many of which are involved in skeletal development, including those that signal through Fgfr1, Fgfr2, and Fgfr3 (Ornitz and Marie, 2002). Fgf9 has been shown to regulate differentiation of hypertrophic chondrocytes and to direct vascularization of the limb skeleton (Hung *et al.*, 2007). *Fgf18* is expressed in the perichondrium, and it signals to the chondrocytes through Fgfr3. *Fgfr1* is found in prehypertrophic and hypertrophic zone, and *Fgfr2* and *Fgfr3* are expressed, respectively, in perichondral cells and in the proliferating zone. Each of the three receptors has a unique function. Human genetic studies first revealed the importance of Fgf signaling in skeletal development, when *Fgfr3* mutations were shown to underlie achondroplasia, hypochondroplasia, and thanatophoric dysplasia (Olsen *et al.*, 2000). In *Fgfr3*-null mice, the proliferative rate is accelerated, which causes the chondrocyte column length to be increased (Colvin *et al.*, 1996; Deng *et al.*, 1996). Moreover, activating mutations in mouse *Fgfr3* cause reduced proliferation and increased apoptosis of chondrocytes (Sahni *et al.*, 1999). These studies suggested that Fgfr3 is a negative regulator of proliferation in the growth plate, and this process is mediated through STAT1–P21 pathway (Sahni *et al.*, 1999). As with Fgfr3, conditional removal of Fgfr1 in chondrocytes results in expansion of the hypertrophic chondrocyte zone, indicating that Fgfr1 is also a negative regulator of proliferation (Jacob *et al.*, 2006).

2.7. Bone morphogenetic protein signaling

Bone morphogenetic proteins (BMPs) and their receptors play multiple roles in chondrocyte differentiation and proliferation, and have been reviewed extensively elsewhere (Li and Cao, 2006; Pogue and Lyons, 2006). *Bmp7* is found mainly in the proliferating chondrocytes, whereas *Bmp2–Bmp5* are expressed primarily in the perichondrium (Lyons *et al.*, 1995; Minina *et al.*, 2001), although hypertrophic chondrocytes also express *Bmp2* and *Bmp6* (Solloway *et al.*, 1998). These distinctive expression patterns suggest that each of these Bmps has a unique function. The relationship of Bmp and Indian hedgehog is somewhat unclear. Although *in vitro* experiments in chick and mouse and *in vivo* studies in chick showed that Bmp receptor IA is an upstream regulator of *Ihh*, other *in vivo* and *in vitro* studies in mouse failed to detect changes in *Ihh* following activation of Bmp receptors or treatment with Bmp protein (Kobayashi *et al.*, 2005; Seki and Hata, 2004; Zhang *et al.*, 2003a). Different experimental approaches also have led to curious findings regarding the function of BmpR1A and BmpR1B. Studies in the chick limb suggested that BmpR1A and BmpR1B may have very different functions (Zou *et al.*, 1997), although more recent studies in mice found them to be interchangeable (Kobayashi *et al.*, 2005). Kobayashi *et al.* (2005) used multiple experimental strategies to overexpress *BmpR1A* in chondrocytes and found that BmpR1A has different roles at different stages of cartilage development. According to their findings, constitutive activation of *BmpR1A* stimulates chondrocyte hypertrophy and also promotes differentiation of prechondrogenic mesenchyme into chondrocytes.

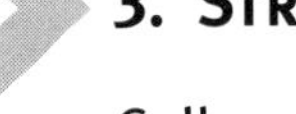

3. Structure of Vertebrate Cartilage Matrix

3.1. Collagens

Most of the connective tissues of vertebrates are formed from extracellular fibers, matrix, and ground substance. For example, up to 90% of the dry weight of cartilage is extracellular matrix (Hardingham and Fosang, 1992). In jawed vertebrates, cartilage extracellular matrix typically is composed of mucopolysaccharides (in the form of proteoglycans) deposited within a meshwork of collagen fibers (Bruckner and van der Rest, 1994). Collagens are the main components of animal extracellular matrix (Exposito *et al.*, 2002), and the expansion of this gene family within the vertebrate clade coincided with evolution of a broad range of vertebrate skeletal tissues. For example, 29 different collagen genes have been identified in humans thus far

(Soderhall *et al.*, 2007), and the resultant proteins can be divided into two major groups, fibrillar and nonfibrillar collagens. The fibrillar collagen proteins, in which multiple collagen fibrils are assembled into collagen fibers, are further divisible into three clades, designated A, B, and C (Aouacheria *et al.*, 2004). Clade A collagens are the major fibril-forming collagens, including types I, II, III, and V (Aouacheria *et al.*, 2004). Clade A fibril procollagens consist of an N-propeptide, an N-telopeptide, a triple helix, a C-telopeptide, and a C-propetide (from N- to C-terminus). The triple helix domain consists of a Gly–X–Y triplet repeat, with X and Y usually being proline and hydroxyproline. The propeptide is removed during the maturation of collagen through posttranslational processing by N- and C-proteinase (Exposito *et al.*, 2002; Kadler *et al.*, 1996). Type II collagen is encoded by *Col2α1*, and nearly 40 years ago this was shown to be the major matrix protein found in cartilage (Miller and Matukas, 1969). Each type II collagen fibril is made of three identical chains that provide tensile strength and a scaffolding network for proteoglycans (van der Rest and Garrone, 1991). Cartilage also contains minor collagens type IX and XI, which belong to the clade B fibrillar collagen family and participate in the process of fibril formation (Eyre *et al.*, 2004; Kadler *et al.*, 1996; Li *et al.*, 1995). Different types of cartilage are characterized by different combinations and quantities of collagen proteins. In addition, the profile of collagen expression can be dynamic during skeletal development. During long bone development, for example, the major matrix protein found in proliferative cartilage is type II collagen, whereas type X collagen is most abundant during the hypertrophic stage and type I collagen dominates bony matrix (Olsen *et al.*, 2000).

3.2. Proteoglycans

Proteoglycans are the second-most abundant proteins (after the fibrillar collagens) in cartilage matrix. Glycosaminoglycan side chains of proteoglycans become heavily sulfated, which increases their retention of water, giving cartilage its characteristic resistance to compression. Chondroitin sulfate was shown to be the predominant glycosaminoglycan in cartilage, and one of its substrates, aggrecan, was found to be the most abundant cartilage proteoglycan (Doege *et al.*, 1991). Deposition of aggrecan has been considered a hallmark of chondrogenesis (although it is also present in aorta, intervertebral disks, and tendons) (Schwartz *et al.*, 1999). Aggrecan not only contributes to the physical properties of cartilage, but also it protects cartilage collagen from degradation by stabilizing collagen protein (Pratta *et al.*, 2003). In addition to the large aggregating proteoglycan aggrecan, there are many small leucine-rich proteoglycans in cartilage, including biglycan, decorin, fibromodulin, lumican, and epiphycan, which have a variety of functions in cartilage development and maintenance (Iozzo, 1998; Knudson and Knudson, 2001). Chondrocytes also express cell surface proteoglycans, such as syndecans and glypican, which can bind growth factors during cell–cell and cell–matrix interactions (Iozzo, 1998; Song *et al.*, 2007).

4. Evolutionary History of the Vertebrate Skeleton

For extant deuterostomes, mapping the key characters of skeletogenesis onto a phylogeny provides a window into the distribution and pattern of skeletal evolution (Fig. 2.2), but what does the fossil record reveal about the evolution of cartilage and bone within vertebrates? Obviously, most preserved specimens will reflect the existence of mineralized tissues, since they are most easily fossilized, but some samples reveal unmineralized cartilage as well. Although studies of invertebrates indicate that cartilage had an earlier origin than did bone in metazoans, it is less clear which of these tissues appeared first in vertebrate skeletal evolution. Conodonts lacked a dermal skeleton and early descriptions of bone in conodonts have been disputed, although their dental elements were rich in dentine and enamel (Donoghue *et al.*, 2006). The 530-million-year-old fossil *Haikouella* is one of the earliest examples of unmineralized vertebrate cartilage, and comparison with modern lamprey cartilage shows striking morphological similarity (Mallatt and Chen, 2003). Jawless fishes dominate the vertebrate fossil record through the upper Paleozoic, and most possessed a heavily armored dermoskeleton, a character that has been lost in lampreys and hagfishes (Sansom *et al.*, 2005). Histological and microscopic studies of dermoskeletons have identified a variety of tissue types, including bone, dentine, and enamel, although neither cartilage nor perichondral bone have been observed (Donoghue and Sansom, 2002; Donoghue *et al.*, 2006; Patterson *et al.*, 1977). Most crown-group vertebrates show few similarities between the mineralized tissues of the teeth and those of the skeleton. Interestingly, the dermal skeletons of early vertebrates were composed of both "dental" and "skeletal" tissue types, and the presence of dentine and enamel in dermal armor has led some investigators to suggest that the evolutionary origin of teeth may be traced to the dermal skeleton (Smith and Johanson, 2003). The earliest examples of mineralized endoskeletons are found in galeaspids and pteraspidomorphs (Donoghue and Sansom, 2002; Donoghue *et al.*, 2006; Janvier, 1996; Stensio, 1927). Galeaspids had dermal armor of unmineralized cartilage and acellular bone. In heterostracans, the dermal skeleton contained dentine, acellular bone, and enameloid tissues. Cellular bone is found in the dermal skeletons of osteostracans, which was combined with dentine in their head shields. The dermoskeleton of thelodonts consisted of scales that were made up of dentine and also may have contained acellular bone (Donoghue and Smith, 2001; Donoghue

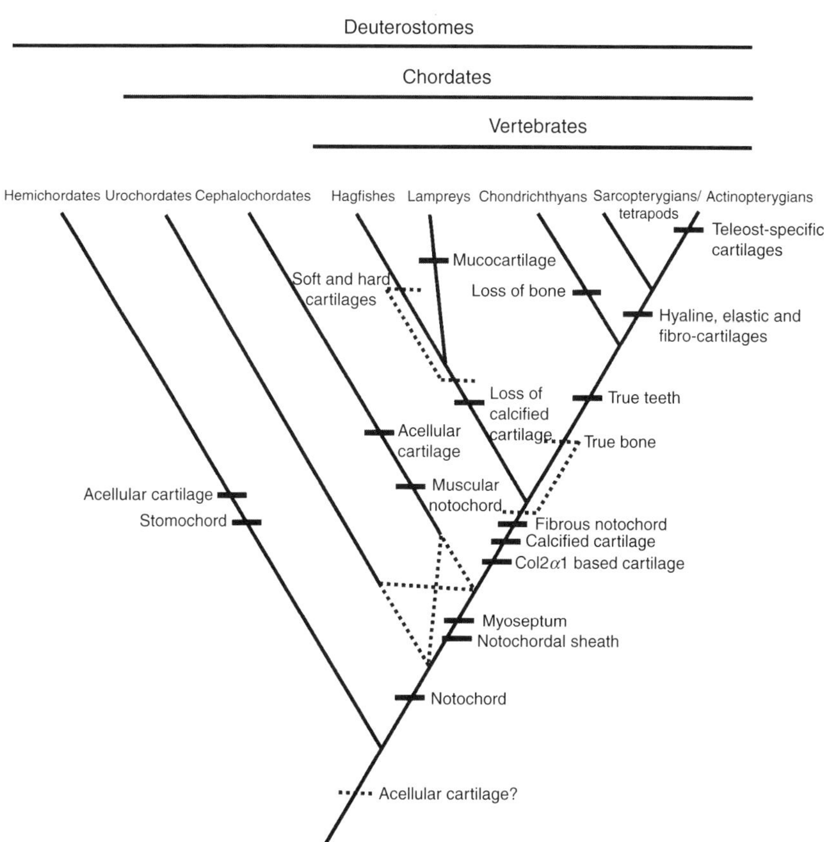

Figure 2.2 Phylogenetic distribution of key skeletogenic characters in deuterostomes. Dotted lines at the base of the cephalochordate and urochordate branches indicate ambiguous positions and these may be transposed. Dotted horizontal bar at base of tree indicates a possible early origin of acellular cartilage in stem deuterostomes (see Rychel and Swalla, 2007). Alternatively, acellular cartilage may have arisen independently in hemichordates and cephalochordates. Dotted horizontal bar in cyclostome (hagfish + lamprey) clade indicates uncertainty regarding the origin of classically defined "hard" and "soft" cartilage (see Cole, 1905; Parker, 1883; Zhang and Cohn, 2006 for further details).

et al., 2006). The almost exclusively cartilaginous skeletons of extant cyclostomes and sharks have been misinterpreted as evidence that cartilage predated bone in vertebrate evolution; however, this is a derived condition that followed an evolutionary loss of bone (Carroll, 1988; Daniel, 1934; Donoghue and Sansom, 2002; Goodrich, 1930; Hall, 1975; Janvier, 1996; Maisey, 1988; Moss, 1977; Orvig, 1951; Romer, 1985; Smith and Hall, 1990). The fossil record of sharks shows abundant evidence of exoskeletal bone (Coates and Sequeira, 2001; Hall, 1975; Maisey, 1988; Moss, 1977; Zangerl, 1966) and limited examples of endoskeletal bone (Coates *et al.*, 1998). Indeed, true bone has persisted in some extant chondrichthyans, in subchondral linings, neural arches, and dermal denticles (Bordat, 1987; Eames *et al.*, 2007; Kemp and Westrin, 1979; Moss, 1970, 1977; Peignoux-Deville *et al.*, 1982; Reif, 1980; Sire and Huysseune, 2003). Thus far, despite the rich diversity of skeletal tissues in the fossil record, the question of whether the earliest vertebrate skeletons were cartilaginous, bony, or both remains unclear.

5. Diversification of Cartilaginous Tissues

A major challenge has been the classification of different cartilage types at the molecular and biochemical levels, and understanding the interrelationships among this diverse family of tissues. Depending on relative amounts of cells and extracellular matrix, there are generally four kinds of cartilage in vertebrates and invertebrates: matrix-rich cartilage, cell-rich cartilage, vesicular cartilage, and acellular cartilage, although skeletal tissues with an intermediate or mosaic composition have been identified in some vertebrates, such as the cartilage-like chondroid tissues, which possess characters of both bone and cartilage (Cole and Hall, 2004a). Whether these four cartilage types evolved independently or diversified from a single type of ancestral connective tissue is unknown (Fig. 2.2). The similarities in matrix composition, histological properties, gene expression profiles, and cell biology of notochord cells and chondrocytes have led some to propose that vertebrate cartilage may have evolved from the notochord of early chordates (Stemple, 2004; Zhang and Cohn, 2006). Alternatively, vertebrate cartilage may have its origins in the secretion of acellular matrix by ectodermal cells. Acellular cartilage, which lacks chondrocytes, has been found in hemichordates, cephalochordates, and vertebrates (e.g., rays) (Cole and Hall, 2004b; Meulemans and Bronner-Fraser, 2007; Rychel and Swalla, 2007; Rychel *et al.*, 2006; Wright *et al.*, 2001). Rychel *et al.* proposed that ectodermally derived acellular cartilage is an ancestral mode of pharyngeal cartilage development in deuterostomes (Fig. 2.2) (Rychel and Swalla, 2007; Rychel *et al.*, 2006). The conservation of cartilage matrix genes in invertebrates could be interpreted as evidence for an unexpectedly deep origin of cartilage, or may simply reflect the limited number of tools in the genetic toolkit for making cartilaginous tissues. According to the latter idea, the molecular program for chondrogenesis has a single origin, but the tissue itself may have evolved many times. Resolving this question will require comparative studies of the molecular mechanisms of chondrogenesis across metazoa. In the next two sections, we review the diversity of cartilaginous tissues in vertebrates and invertebrates.

5.1. Cartilage variation within vertebrates

5.1.1. Tetrapods

Cartilage exists in a variety of forms in vertebrates (Fig. 2.2). In tetrapods, cartilage is broadly divisible into three major subtypes: hyaline cartilage, elastic cartilage, and fibrocartilage (Hall, 2005). *Hyaline cartilage* is the primary component of the endoskeleton and serves as the scaffold for bone that develops by endochondral ossification. Sometimes termed "true cartilage," hyaline cartilage derives its structural integrity mainly from glycosaminoglycans and type II collagen fibrils. *Elastic cartilage*, such as that found in the mammalian ear pinnae and epiglottis, is also rich in glycosaminoglycans and collagen proteins, but additionally contains thick bundles of elastic fibrils and elastin-rich extracellular matrix (Naumann *et al.*, 2002). This combination of matrix proteins gives elastic cartilage the toughness of hyaline cartilage but with increased elasticity. *Fibrocartilage* is found at the attachment points of tendons and ligaments, in intervertebral disks, and at the pubic symphysis. Fibrocartilage matrix contains large amounts of type I collagen, which makes it both tensile and tough (Benjamin and Evans, 1990; Benjamin and Ralphs, 2004; Eyre and Wu, 1983). Even in tetrapods, some cartilage can demonstrate intermediate tissue properties that do not adhere to this tidy classification scheme. For example, secondary cartilage, which forms from osteoblast precursors at stressed joint regions, is similar to hyaline cartilage, but expresses high amounts of type I collagen (Fang and Hall, 1997; Fukada *et al.*, 1999; Fukuoka *et al.*, 2007; Ishii *et al.*, 1998).

5.1.2. Teleosts

Teleost fishes exhibit an even richer diversity of cartilage types (Fig. 2.2). According to one classification scheme, there are five "cell-rich" cartilages and three "matrix-rich" cartilages (Benjamin, 1989, 1990). The "cell-rich" cartilages, which are defined by cells or lacunae making up >50% of a cartilage tissue's volume, include (1) hyaline-cell cartilage, (2) cell-rich hyaline cartilage, (3) fibrocell cartilage, (4) elastic/cell-rich cartilage, and (5) Schaffer's Zellknorpel. *Hyaline-cell cartilage*, which is found in the lips, rostral folds, and other cranial cartilages, is characterized by compact chromophobic chondrocytes and hyaline cytoplasm with little matrix (Benjamin, 1989). Hyaline-cell cartilage is divisible into three subtypes; fibro/hyaline has greater quantities of collagen than elastin, elastic/hyaline contains more elastin in the matrix, and lipo/hyaline contains adipocytes as well as chondrocytes. *Cell-rich hyaline cartilage* is more cellular than hyaline-cell cartilage, with lacunae occupying more than half of the total volume. Parts of neurocranium and Meckel's cartilage belong in this category

(Benjamin, 1990). *Fibrocell cartilage* is a highly cellular (nonhyaline) fibrocartilage that is rich in collagen, lacks a perichondrium, and is commonly found on articular surfaces. *Elastic/cell-rich cartilage*, which is usually found in the barbels and maxillary oral valves, is dense with elastin, the cells are not hyaline, and these elements are surrounded by a thick fibrous perichondrium (Benjamin, 1990). The fifth type of "cell-rich" cartilage is known as Schaffer's Zellknorpel and occurs in teleost gill filament rays and the basal plate. Zellknorpel chondrocytes are more chromophilic than those of hyaline-cell cartilage and are shrunken within large lacunae (Benjamin, 1990).

The "matrix-rich cartilages" of teleosts are defined by cells or lacunae making up <50% total volume. In teleosts, like tetrapods, the "matrix-rich cartilages" are divisible into three subtypes (1) matrix-rich hyaline cartilage, (2) fibrocartilage, and (3) elastic cartilage. Each of these cell-rich and matrix-rich cartilages can be found in the cranial and postcranial skeletons, with the exception of the cranially restricted Zellknorpel (Benjamin *et al.*, 1992). Scleral cartilage is particularly interesting, as it has been described as a composite structure, in which a central zone of cell-rich hyaline cartilage is surrounded by a cortex of matrix-rich hyaline cartilage (Benjamin and Ralphs, 1991; Franz-Odendaal *et al.*, 2007). This classification system is based on histological/structural characters, and little is known about their molecular composition or development. The observation that teleosts have a broader variety of cartilage tissue types than do tetrapods may relate to the larger number of matrix (and other skeletogenic) genes that resulted from the teleost genome duplication event. Accordingly, the increased number of gene expression combinations that are possible in teleosts may underlie the diversity of cartilage types. Alternatively, similar patterns of gene expression in chondrogenic tissues may yield different structural and histological patterns due to differences in the local environment during chondrogenesis. Molecular characterization of the different cartilaginous tissues of teleosts is needed to uncover the developmental basis of their diversity.

5.1.3. Chondrichthyans

Chondrichthyan skeletons are almost entirely cartilaginous; however, their cartilage undergoes extensive mineralization (Dean and Summers, 2006; Eames *et al.*, 2007; Hall, 2005). The majority of the shark skeleton appears to be true hyaline cartilage, staining strongly for sulfated proteoglycans and type II collagen (Eames *et al.*, 2007). The cartilaginous nature of chondrichthyan skeletons is likely a derived condition that followed an evolutionary loss of bone (Fig. 2.2). Catsharks, for example, retain true bone in their neural arches, and the fossil record of sharks shows evidence of both exoskeletal and endochondral bones (Coates *et al.*, 1998; Kemp and Westrin, 1979; Moss, 1970, 1977; Peignoux-Deville *et al.*, 1982). Biochemical studies showed that shark and skate cartilage may contain type I collagen in addition to type II collagen (Mizuta *et al.*, 2003; Moss, 1977;

Rama and Chandrakasan, 1984); however, antibodies to type I collagen did not react to shark cartilage immunohistochemically (Eames *et al.*, 2007). It has been suggested that biochemical identification of type I collagen in shark cartilage may have resulted from contamination from shark bone (Eames *et al.*, 2007). Cartilage development within chondrichthyans has not received the level of scrutiny provided to teleost skeletal tissues, and a comprehensive and comparative analysis of gene expression, regulation and function is needed.

As an aside, the widely held belief that sharks do not develop tumors is false (neoplasias in sharks have been known for over 150 years) and there is no scientific basis to support the notion that consumption of crude shark cartilage affects tumor development in cancer patients (reviewed in Ostrander *et al.*, 2004). Some general features of cartilage (not restricted to sharks) that may contribute to the rarity of tumor invasion into cartilaginous tissues are that it is hypoxic, has poor vascularity, produces collagenase inhibitors, and may contain antiangiogenic factors.

5.1.4. Cyclostomes

Cartilaginous skeletons are also present in both extant groups of jawless (agnathan) vertebrates, lampreys and hagfishes. Lamprey and hagfish have mucocartilage (Fig. 2.2) and were described as lacking collagen (Wright *et al.*, 2001). Instead, their matrix was reported to contain the elastin-like molecules *lamprin* and *myxinin*, respectively (Wright *et al.*, 2001). Recent molecular developmental studies have overturned the idea that agnathans lack collagenous cartilage by demonstrating that both lampreys and hagfishes do indeed have type II collagen-based cartilage (Ohtani *et al.*, 2008; Zhang and Cohn, 2006; Zhang *et al.*, 2006). Lamprey cartilages are found mainly in the cranial region. The postcranial skeleton is limited to paired axial cartilage nodules (termed arcualia) and caudal fin rays (Morrison *et al.*, 2000). In the head of larval lamprey, the proteoglycan-rich mucocartilage occurs as a transient, avascular cartilage that is surrounded by perichondrium (Hall, 2005). During metamorphosis, mucocartilage is transformed into the pistal and tongue cartilages (Hall, 2005). In the nineteenth century, two kinds of cartilages, "soft" and "hard," were identified in lampreys (Parker, 1883). The hard cartilage is similar structurally to mammalian hyaline cartilage. In hagfishes, cartilage is also present in the cranium and median fin rays, although they lack the paired arcualia found along the trunks of lampreys. Like lampreys, hagfish were reported to contain soft and hard cartilages (Cole, 1905). Cole (1905) described hagfish "soft" cartilage as containing large hypertrophic chondrocytes that stain with hematoxylin (blue) and are surrounded by a thin extracellular matrix, whereas "hard" cartilage contains smaller chondrocytes that are surrounded by an abundance of extracellular matrix. Biochemical analysis also supported the two types of hagfish cartilage, designated type I and type II cartilage, with only type I containing myxinin and type II being more similar to adult lamprey cartilage (Wright *et al.*, 1984). Neither lamprey nor hagfish cartilage is mineralized, but lamprey cartilage can be calcified *in vitro* (Langille and Hall, 1988). Interestingly, calcified cartilage was reported in the fossil lamprey *Euphanerops*, suggesting that mineralized cartilage in this group persisted at least to the Devonian (Janvier and Arsenault, 2002). These recent analyses of extant and extinct agnathans suggest that cartilage containing high amounts of type II collagen and sulfated proteoglycans was present in the common ancestor of jawed and jawless vertebrates (Fig. 2.2).

5.2. Invertebrate cartilage

Cartilaginous tissues are not restricted to the vertebrates; examples of cellular and/or acellular cartilage exist in such diverse taxa as cephalochordates, hemichordates, annelids, mollusks, brachiopods, arthropods, and cnidaria (see Cole and Hall, 2004a for a detailed review). Some of these tissues bear striking similarities to vertebrate cartilage at the structural, morphological, and histological levels. In general, there are three kinds of cartilages found in invertebrates: central cell-rich cartilage, vesicular cartilage with large vesicles or vacuoles, and acellular cartilage. Within deuterostomes, fibrillar collagens are expressed in the developing acellular cartilage of hemichordates and cephalochordates, in the cellular cartilage and the notochord of cephalochordates, and in the notochordal sheath of urochordates (Rychel *et al.*, 2006; Wada *et al.*, 2006; Zhang *et al.*, 2006). Vesicular cartilage has been identified in polychaete worms, horseshoe crabs, and mollusks (Cole and Hall, 2004b). To some degree, the vertebrate notochord can be considered a vesicular cell-rich cartilage, since notochordal cells are vacuolated and surrounded by cartilage-like extracellular matrix containing type II collage, type I collagen, type X collagen, aggrecan, and polysaccharides (Domowicz *et al.*, 1995; Eikenberry *et al.*, 1984; Linsenmayer *et al.*, 1986; Welsch *et al.*, 1991). How similar or different are the developmental processes and molecular mechanisms involved in chondrogenesis in invertebrates and vertebrates? The paucity of molecular and even embryological data on invertebrate cartilage development makes it difficult to answer this question. The structural similarities are striking, and given the conservatism of developmental evolution and the limited number of "toolkit genes," it would be surprising if entirely different mechanisms were utilized to build this tissue type in different lineages. Nonetheless, the possibility of convergence using different mechanisms remains, and comparative analyses of chondrogenesis will be required to resolve this evolutionary mystery.

6. Elaborating the Chondrogenetic Toolkit: Gene/Genome Duplication and the Origin of Collagenous Cartilage

Given the dominant role that fibrillar collagens play in constructing the matrices of a diverse array of vertebrate connective tissues, it seems likely that expansion of this gene family would have been a critical step toward the evolutionary diversification of skeletal tissues. Molecular phylogenetic analyses of deuterostome collagens indicate that a gene ancestral to the vertebrate clade A collagens had arisen in chordates before the origin of vertebrates, but the duplication and divergence of clade A collagens (*Col1α1*, *Col1α2*, *Col2α1*, *Col3α1*, and *Col5α2*) and clade B collagens (*Col5α1*, *Col5α3*, *Col11α1*, and *Col11α2*) occurred within the vertebrate lineage (Boot-Handford and Tuckwell, 2003; Zhang and Cohn, 2006, 2008). A number of findings point to deep conservation of chondrogenic mechanisms, such as the evidence that horseshoe crab cartilage contains chondroitin-6-sulfate (Sugahara *et al.*, 1996) and that squid and cuttlefish cartilages may contain collagen, although these appear to be different than type II collagen (Bairati and Gioria, 2004; Bairati *et al.*, 1999; Kimura and Karasawa, 1985; Kimura and Matsuura, 1974). Fibrillar collagens also have been identified in cartilage-like tissues of protostomes, including sponge, sea urchin, abalone, and hydra. Similarities have been described between the sea urchin α1 and vertebrate α2(I) chains, and between hydra Hcol1 and vertebrate collagen type I/II (Deutzmann *et al.*, 2000; Exposito *et al.*, 1992). Indeed, some invertebrate cartilage-like tissues crossreact with antibodies against vertebrate types II, V, and X collagen, and proteoglycans (Bairati *et al.*, 1999; Cole and Hall, 2004a,b; Sivakumar and Chandrakasan, 1998), although published phylogenies of the collagen family suggest it unlikely that these vertebrate antibodies are detecting strict orthologs of Col2, Col5, or Col10 in invertebrates (Boot-Handford and Tuckwell, 2003; Rychel *et al.*, 2006; Wada *et al.*, 2006; Zhang *et al.*, 2006, 2007). Nonetheless, the fundamental structure of fibrillar collagens was established early in metazoan evolution (Boot-Handford and Tuckwell, 2003). The phylogenetic distribution of cartilage and cartilage-like tissues suggests that this tissue type evolved independently and multiple times in metazoans (Cole and Hall, 2004a), and while the evidence for convergent evolution precludes structural homologies of invertebrate and vertebrate cartilages, it does not rule out the possibility of homologous developmental mechanisms. The striking structural and molecular similarities between invertebrate and vertebrate cartilages, such as utilization of fibrillar collagens and chondroitin-6-sulfate, suggests that a common suite of developmental tools was used repeatedly by metazoans to generate cartilaginous tissues, much like the deeply conserved eye development program involving Pax6 has been redeployed time and again to build eyes. The area of invertebrate cartilage biology is ripe for comparative studies using modern molecular developmental approaches.

REFERENCES

Akiyama, H., Lyons, J. P., Mori-Akiyama, Y., Yang, X., Zhang, R., Zhang, Z., Deng, J. M., Taketo, M. M., Nakamura, T., Behringer, R. R., McCrea, P. D., and de Crombrugghe, B. (2004). Interactions between Sox9 and beta-catenin control chondrocyte differentiation. *Genes Dev.* **18,** 1072–1087.

Akiyama, H., Kim, J. E., Nakashima, K., Balmes, G., Iwai, N., Deng, J. M., Zhang, Z., Martin, J. F., Behringer, R. R., Nakamura, T., and de Crombrugghe, B. (2005). Osteochondroprogenitor cells are derived from Sox9 expressing precursors. *Proc. Natl. Acad. Sci. USA* **102,** 14665–14670.

Aouacheria, A., Cluzel, C., Lethias, C., Gouy, M., Garrone, R., and Exposito, J. Y. (2004). Invertebrate data predict an early emergence of vertebrate fibrillar collagen clades and an anti-incest model. *J. Biol. Chem.* **279,** 47711–47719.

Bairati, A., and Gioria, M. (2004). Collagen fibrils of an invertebrate (Sepia officinalis) are heterotypic: Immunocytochemical demonstration. *J. Struct. Biol.* **147,** 159–165.

Bairati, A., Comazzi, M., Gioria, M., Hartmann, D. J., Leone, F., and Rigo, C. (1999). Immunohistochemical study of collagens of the extracellular matrix in cartilage of Sepia officinalis. *Eur. J. Histochem.* **43,** 211–225.

Benjamin, M. (1989). Hyaline-cell cartilage (chondroid) in the heads of teleosts. *Anat. Embryol. (Berl.)* **179,** 285–303.

Benjamin, M. (1990). The cranial cartilages of teleosts and their classification. *J. Anat.* **169,** 153–172.

Benjamin, M., and Evans, E. J. (1990). Fibrocartilage. *J. Anat.* **171,** 1–15.

Benjamin, M., and Ralphs, J. R. (1991). Extracellular matrix of connective tissues in the heads of teleosts. *J. Anat.* **179,** 137–148.

Benjamin, M., and Ralphs, J. R. (2004). Biology of fibrocartilage cells. *Int. Rev. Cytol.* **233,** 1–45.

Benjamin, M., Ralphs, J. R., and Eberewariye, O. S. (1992). Cartilage and related tissues in the trunk and fins of teleosts. *J. Anat.* **181**(Pt. 1), 113–118.

Bi, W., Deng, J. M., Zhang, Z., Behringer, R. R., and de Crombrugghe, B. (1999). Sox9 is required for cartilage formation. *Nat. Genet.* **22,** 85–89.

Bodine, P. V., Zhao, W., Kharode, Y. P., Bex, F. J., Lambert, A. J., Goad, M. B., Gaur, T., Stein, G. S., Lian, J. B., and Komm, B. S. (2004). The Wnt antagonist secreted frizzled-related protein-1 is a negative regulator of trabecular bone formation in adult mice. *Mol. Endocrinol.* **18,** 1222–1237.

Boot-Handford, R. P., and Tuckwell, D. S. (2003). Fibrillar collagen: The key to vertebrate evolution? A tale of molecular incest. *Bioessays* **25,** 142–151.

Bordat, C. (1987). Ultrastructural study of the vertebrae of the selachian *Scyliorhinus canicula*. *Can. J. Zool.* **65,** 1435–1444.

Boyden, L. M., Mao, J., Belsky, J., Mitzner, L., Farhi, A., Mitnick, M. A., Wu, D., Insogna, K., and Lifton, R. P. (2002). High bone density due to a mutation in LDL-receptor-related protein 5. *N. Engl. J. Med.* **346,** 1513–1521.

Bruckner, P., and van der Rest, M. (1994). Structure and function of cartilage collagens. *Microsc. Res. Tech.* **28,** 378–384.

Carroll, R. L. (1988). "Vertebrate Paleontology and Evolution." Freeman, New York.

Coates, M. I., and Sequeira, S. E. K. (2001). A new stethacanthid chondrichthyan from the Lower Carboniferous of Bearsden, Scotland. *J. Vertebr. Paleontol.* **21,** 438–459.

Coates, M. I., Sequeira, S. E. K., Sansom, I. J., and Smith, M. M. (1998). Spines and tissues of ancient sharks. *Nature* **396,** 729–730.

Cole, F. J. (1905). A monograph on the general morphology of the myxinoid fishes based on a study of Myxine. 1. The anatomy of the skeleton. *Trans. R. Soc. Edinburgh* **41,** 749–791.

Cole, A. G., and Hall, B. K. (2004a). Cartilage is a metazoan tissue; integrating data from nonvertebrate sources. *Acta Zool. (Stockholm)* **85,** 69–80.

Cole, A. G., and Hall, B. K. (2004b). The nature and significance of invertebrate cartilages revisited: Distribution and histology of cartilage and cartilage-like tissues within the Metazoa. *Zoology (Jena)* **107,** 261–273.

Colvin, J. S., Bohne, B. A., Harding, G. W., McEwen, D. G., and Ornitz, D. M. (1996). Skeletal overgrowth and deafness in mice lacking fibroblast growth factor receptor 3. *Nat. Genet.* **12,** 390–397.

Daniel, J. F. (1934). "The Elasmobranch Fishes." University of California Press, Berkeley.

Day, T. F., Guo, X., Garrett-Beal, L., and Yang, Y. (2005). Wnt/beta-catenin signaling in mesenchymal progenitors controls osteoblast and chondrocyte differentiation during vertebrate skeletogenesis. *Dev. Cell* **8,** 739–750.

Dean, M. N., and Summers, A. P. (2006). Mineralized cartilage in the skeleton of chondrichthyan fishes. *Zoology (Jena)* **109,** 164–168.

Deng, C., Wynshaw-Boris, A., Zhou, F., Kuo, A., and Leder, P. (1996). Fibroblast growth factor receptor 3 is a negative regulator of bone growth. *Cell* **84,** 911–921.

Deutzmann, R., Fowler, S., Zhang, X., Boone, K., Dexter, S., Boot-Handford, R. P., Rachel, R., and Sarras, M. P., Jr. (2000). Molecular, biochemical and functional analysis of a novel and developmentally important fibrillar collagen (Hcol-I) in hydra. *Development* **127,** 4669–4680.

Doege, K. J., Sasaki, M., Kimura, T., and Yamada, Y. (1991). Complete coding sequence and deduced primary structure of the human cartilage large aggregating proteoglycan, aggrecan. Human-specific repeats, and additional alternatively spliced forms. *J. Biol. Chem.* **266,** 894–902.

Domowicz, M., Li, H., Hennig, A., Henry, J., Vertel, B. M., and Schwartz, N. B. (1995). The biochemically and immunologically distinct CSPG of notochord is a product of the aggrecan gene. *Dev. Biol.* **171,** 655–664.

Donoghue, P. C., and Sansom, I. J. (2002). Origin and early evolution of vertebrate skeletonization. *Microsc. Res. Tech.* **59,** 352–372.

Donoghue, P. C. J., and Smith, M. P. (2001). The anatomy of *Turinia pagei* (Powrie) and the phylogenetic status of the Thelodonti. *Tran. R. Soc. Edinburgh (Earth Sci.)* **92,** 15–37.

Donoghue, P. C., Sansom, I. J., and Downs, J. P. (2006). Early evolution of vertebrate skeletal tissues and cellular interactions, and the canalization of skeletal development. *J. Exp. Zool. B Mol. Dev. Evol.* **306,** 278–294.

Ducy, P., Zhang, R., Geoffroy, V., Ridall, A. L., and Karsenty, G. (1997). Osf2/Cbfa1: A transcriptional activator of osteoblast differentiation. *Cell* **89,** 747–754.

Ducy, P., Starbuck, M., Priemel, M., Shen, J., Pinero, G., Geoffroy, V., Amling, M., and Karsenty, G. (1999). A Cbfa1-dependent genetic pathway controls bone formation beyond embryonic development. *Genes Dev.* **13,** 1025–1036.

Eames, B. F., and Helms, J. A. (2004). Conserved molecular program regulating cranial and appendicular skeletogenesis. *Dev. Dyn.* **231,** 4–13.

Eames, B. F., Sharpe, P. T., and Helms, J. A. (2004). Hierarchy revealed in the specification of three skeletal fates by Sox9 and Runx2. *Dev. Biol.* **274,** 188–200.

Eames, B. F., Allen, N., Young, J., Kaplan, A., Helms, J. A., and Schneider, R. A. (2007). Skeletogenesis in the swell shark *Cephaloscyllium ventriosum*. *J. Anat.* **210,** 542–554.

Eikenberry, E. F., Childs, B., Sheren, S. B., Parry, D. A., Craig, A. S., and Brodsky, B. (1984). Crystalline fibril structure of type II collagen in lamprey notochord sheath. *J. Mol. Biol.* **176,** 261–277.

Exposito, J. Y., D'Alessio, M., Solursh, M., and Ramirez, F. (1992). Sea urchin collagen evolutionarily homologous to vertebrate pro-alpha 2(I) collagen. *J. Biol. Chem.* **267,** 15559–15562.

Exposito, J. Y., Cluzel, C., Garrone, R., and Lethias, C. (2002). Evolution of collagens. *Anat. Rec.* **268,** 302–316.

Eyre, D. R., and Wu, J. J. (1983). Collagen of fibrocartilage: A distinctive molecular phenotype in bovine meniscus. *FEBS Lett.* **158,** 265–270.

Eyre, D. R., Pietka, T., Weis, M. A., and Wu, J. J. (2004). Covalent cross-linking of the NC1 domain of collagen type IX to collagen type II in cartilage. *J. Biol. Chem.* **279,** 2568–2574.

Fang, J., and Hall, B. K. (1997). Chondrogenic cell differentiation from membrane bone periostea. *Anat. Embryol. (Berl.)* **196,** 349–362.

Foster, J. W., Dominguez-Steglich, M. A., Guioli, S., Kowk, G., Weller, P. A., Stevanovic, M., Weissenbach, J., Mansour, S., Young, I. D., Goodfellow, P. N., Brook, J. D., and Schafer, A. J. (1994). Campomelic dysplasia and autosomal sex reversal caused by mutations in an SRY-related gene. *Nature* **372,** 525–530.

Franz-Odendaal, T. A., Ryan, K., and Hall, B. K. (2007). Developmental and morphological variation in the teleost craniofacial skeleton reveals an unusual mode of ossification. *J. Exp. Zool. B Mol. Dev. Evol.* **308,** 709–721.

Fukada, K., Shibata, S., Suzuki, S., Ohya, K., and Kuroda, T. (1999). *In situ* hybridisation study of type I, II, X collagens and aggrecan mRNAs in the developing condylar cartilage of fetal mouse mandible. *J. Anat.* **195**(Pt. 3), 321–329.

Fukuoka, H., Shibata, S., Suda, N., Yamashita, Y., and Komori, T. (2007). Bone morphogenetic protein rescues the lack of secondary cartilage in Runx2-deficient mice. *J. Anat.* **211,** 8–15.

Furumatsu, T., Tsuda, M., Taniguchi, N., Tajima, Y., and Asahara, H. (2005). Smad3 induces chondrogenesis through the activation of SOX9 via CREB-binding protein/p300 recruitment. *J. Biol. Chem.* **280,** 8343–8350.

Glass, D. A., II, Bialek, P., Ahn, J. D., Starbuck, M., Patel, M. S., Clevers, H., Taketo, M. M., Long, F., McMahon, A. P., Lang, R. A., and Karsenty, G. (2005). Canonical Wnt signaling in differentiated osteoblasts controls osteoclast differentiation. *Dev. Cell* **8,** 751–764.

Gong, Y., Slee, R. B., Fukai, N., Rawadi, G., Roman-Roman, S., Reginato, A. M., Wang, H., Cundy, T., Glorieux, F. H., Lev, D., Zacharin, M., Oexle, K., *et al.* (2001). LDL receptor-related protein 5 (LRP5) affects bone accrual and eye development. *Cell* **107,** 513–523.

Goodrich, E. S. (1930). "Studies on the Structure and Development of the Vertebrates." Macmillan, London.

Hall, B. K. (1975). Evolutionary consequences of skeletal differentiation. *Am. Zool.* **15,** 329–350.

Hall, B. K. (2005). "Bone and Cartilage: Development and Evolutionary Skeletal Biology." Elsevier Academic Press, San Diego.

Hardingham, T. E., and Fosang, A. J. (1992). Proteoglycans: Many forms and many functions. *FASEB J.* **6,** 861–870.

Hartmann, C., and Tabin, C. J. (2001). Wnt-14 plays a pivotal role in inducing synovial joint formation in the developing appendicular skeleton. *Cell* **104,** 341–351.

Healy, C., Uwanogho, D., and Sharpe, P. T. (1996). Expression of the chicken Sox9 gene marks the onset of cartilage differentiation. *Ann. N. Y. Acad. Sci.* **785,** 261–262.

Healy, C., Uwanogho, D., and Sharpe, P. T. (1999). Regulation and role of Sox9 in cartilage formation. *Dev. Dyn.* **215,** 69–78.

Hill, T. P., Spater, D., Taketo, M. M., Birchmeier, W., and Hartmann, C. (2005). Canonical Wnt/beta-catenin signaling prevents osteoblasts from differentiating into chondrocytes. *Dev. Cell* **8,** 727–738.

Hu, H., Hilton, M. J., Tu, X., Yu, K., Ornitz, D. M., and Long, F. (2005). Sequential roles of Hedgehog and Wnt signaling in osteoblast development. *Development* **132,** 49–60.

Hung, I. H., Yu, K., Lavine, K. J., and Ornitz, D. M. (2007). FGF9 regulates early hypertrophic chondrocyte differentiation and skeletal vascularization in the developing stylopod. *Dev. Biol.* **307,** 300–313.

Ikeda, T., Kamekura, S., Mabuchi, A., Kou, I., Seki, S., Takato, T., Nakamura, K., Kawaguchi, H., Ikegawa, S., and Chung, U. I. (2004). The combination of SOX5, SOX6, and SOX9 (the SOX trio) provides signals sufficient for induction of permanent cartilage. *Arthritis Rheum.* **50,** 3561–3573.

Inada, M., Yasui, T., Nomura, S., Miyake, S., Deguchi, K., Himeno, M., Sato, M., Yamagiwa, H., Kimura, T., Yasui, N., Ochi, T., Endo, N., *et al.* (1999). Maturational disturbance of chondrocytes in Cbfa1-deficient mice. *Dev. Dyn.* **214,** 279–290.

Iozzo, R. V. (1998). Matrix proteoglycans: From molecular design to cellular function. *Annu. Rev. Biochem.* **67,** 609–652.

Ishii, M., Suda, N., Tengan, T., Suzuki, S., and Kuroda, T. (1998). Immunohistochemical findings type I and type II collagen in prenatal mouse mandibular condylar cartilage compared with the tibial anlage. *Arch. Oral Biol.* **43,** 545–550.

Iwamoto, M., Kitagaki, J., Tamamura, Y., Gentili, C., Koyama, E., Enomoto, H., Komori, T., Pacifici, M., and Enomoto-Iwamoto, M. (2003). Runx2 expression and action in chondrocytes are regulated by retinoid signaling and parathyroid hormone-related peptide (PTHrP). *Osteoarthr. Cartil.* **11,** 6–15.

Jacob, A. L., Smith, C., Partanen, J., and Ornitz, D. M. (2006). Fibroblast growth factor receptor 1 signaling in the osteo-chondrogenic cell lineage regulates sequential steps of osteoblast maturation. *Dev. Biol.* **296,** 315–328.

Janvier, P. (1996). "Early Vertebrates." Oxford University Press, Oxford.

Janvier, P., and Arsenault, M. (2002). Palaeobiology: Calcification of early vertebrate cartilage. *Nature* **417,** 609.

Kadler, K. E., Holmes, D. F., Trotter, J. A., and Chapman, J. A. (1996). Collagen fibril formation. *Biochem. J.* **316**(Pt. 1), 1–11.

Karaplis, A. C., Luz, A., Glowacki, J., Bronson, R. T., Tybulewicz, V. L., Kronenberg, H. M., and Mulligan, R. C. (1994). Lethal skeletal dysplasia from targeted disruption of the parathyroid hormone-related peptide gene. *Genes Dev.* **8,** 277–289.

Karp, S. J., Schipani, E., St-Jacques, B., Hunzelman, J., Kronenberg, H., and McMahon, A. P. (2000). Indian hedgehog coordinates endochondral bone growth and morphogenesis via parathyroid hormone related-protein-dependent and -independent pathways. *Development* **127,** 543–548.

Karsenty, G. (2003). The complexities of skeletal biology. *Nature* **423,** 316–318.

Karsenty, G. (2008). Transcriptional control of skeletogenesis. *Annu. Rev. Genomics Hum. Genet.* **9,** 183–196.

Karsenty, G., and Wagner, E. F. (2002). Reaching a genetic and molecular understanding of skeletal development. *Dev. Cell* **2,** 389–406.

Kato, M., Patel, M. S., Levasseur, R., Lobov, I., Chang, B. H., Glass, D. A., II, Hartmann, C., Li, L., Hwang, T. H., Brayton, C. F., Lang, R. A., Karsenty, G., *et al.* (2002). Cbfa1-independent decrease in osteoblast proliferation, osteopenia, and persistent embryonic eye vascularization in mice deficient in Lrp5, a Wnt coreceptor. *J. Cell Biol.* **157,** 303–314.

Kemp, N. E., and Westrin, S. K. (1979). Ultrastructure of calcified cartilage in the endoskeletal tesserae of sharks. *J. Morphol.* **160,** 75–109.

Kim, I. S., Otto, F., Zabel, B., and Mundlos, S. (1999). Regulation of chondrocyte differentiation by Cbfa1. *Mech. Dev.* **80,** 159–170.

Kimura, S., and Karasawa, K. (1985). Squid cartilage collagen: Isolation of type I collagen rich in carbohydrate. *Comp. Biochem. Physiol. B* **81,** 361–365.

Kimura, S., and Matsuura, F. (1974). The chain compositions of several invertebrate collagens. *J. Biochem. (Tokyo)* **75,** 1231–1240.

Knudson, C. B., and Knudson, W. (2001). Cartilage proteoglycans. *Semin. Cell Dev. Biol.* **12,** 69–78.

Kobayashi, T., Lyons, K. M., McMahon, A. P., and Kronenberg, H. M. (2005). BMP signaling stimulates cellular differentiation at multiple steps during cartilage development. *Proc. Natl. Acad. Sci. USA* **102,** 18023–18027.

Komori, T., Yagi, H., Nomura, S., Yamaguchi, A., Sasaki, K., Deguchi, K., Shimizu, Y., Bronson, R. T., Gao, Y. H., Inada, M., Sato, M., Okamoto, R., *et al.* (1997). Targeted disruption of Cbfa1 results in a complete lack of bone formation owing to maturational arrest of osteoblasts. *Cell* **89,** 755–764.

Kronenberg, H. M. (2006). PTHrP and skeletal development. *Ann. N. Y. Acad. Sci.* **1068,** 1–13.

Kusserow, A., Pang, K., Sturm, C., Hrouda, M., Lentfer, J., Schmidt, H. A., Technau, U., von Haeseler, A., Hobmayer, B., Martindale, M. Q., and Holstein, T. W. (2005). Unexpected complexity of the Wnt gene family in a sea anemone. *Nature* **433,** 156–160.

Langille, R. M., and Hall, B. K. (1988). The organ culture and grafting of lamprey cartilage and teeth. *In Vitro Cell Dev. Biol.* **24,** 1–8.

Lanske, B., Karaplis, A. C., Lee, K., Luz, A., Vortkamp, A., Pirro, A., Karperien, M., Defize, L. H., Ho, C., Mulligan, R. C., Abou-Samra, A. B., Juppner, H., *et al.* (1996). PTH/PTHrP receptor in early development and Indian hedgehog-regulated bone growth. *Science* **273,** 663–666.

Lee, P. N., Pang, K., Matus, D. Q., and Martindale, M. Q. (2006). A WNT of things to come: Evolution of Wnt signaling and polarity in cnidarians. *Semin. Cell Dev. Biol.* **17,** 157–167.

Lefebvre, V., Behringer, R. R., and de Crombrugghe, B. (2001). L-Sox5, Sox6 and Sox9 control essential steps of the chondrocyte differentiation pathway. *Osteoarthr. Cartil.* **9** (Suppl. A), S69–S75.

Lefebvre, V., and de Crombrugghe, B. (1998). Toward understanding SOX9 function in chondrocyte differentiation. *Matrix Biol.* **16,** 529–540.

Lefebvre, V., Huang, W., Harley, V. R., Goodfellow, P. N., and de Crombrugghe, B. (1997). SOX9 is a potent activator of the chondrocyte-specific enhancer of the pro alpha1(II) collagen gene. *Mol. Cell Biol.* **17,** 2336–2346.

Lefebvre, V., Li, P., and de Crombrugghe, B. (1998). A new long form of Sox5 (L-Sox5), Sox6 and Sox9 are coexpressed in chondrogenesis and cooperatively activate the type II collagen gene. *EMBO J.* **17,** 5718–5733.

Levanon, D., Brenner, O., Negreanu, V., Bettoun, D., Woolf, E., Eilam, R., Lotem, J., Gat, U., Otto, F., Speck, N., and Groner, Y. (2001). Spatial and temporal expression pattern of Runx3 (Aml2) and Runx1 (Aml1) indicates non-redundant functions during mouse embryogenesis. *Mech. Dev.* **109,** 413–417.

Li, X., and Cao, X. (2006). BMP signaling and skeletogenesis. *Ann. N. Y. Acad. Sci.* **1068,** 26–40.

Li, Y., Lacerda, D. A., Warman, M. L., Beier, D. R., Yoshioka, H., Ninomiya, Y., Oxford, J. T., Morris, N. P., Andrikopoulos, K., Ramirez, F., *et al.* (1995). A fibrillar collagen gene, Col11a1, is essential for skeletal morphogenesis. *Cell* **80,** 423–430.

Li, T. F., Dong, Y., Ionescu, A. M., Rosier, R. N., Zuscik, M. J., Schwarz, E. M., O'Keefe, R. J., and Drissi, H. (2004). Parathyroid hormone-related peptide (PTHrP) inhibits Runx2 expression through the PKA signaling pathway. *Exp. Cell Res.* **299,** 128–136.

Lian, J. B., Balint, E., Javed, A., Drissi, H., Vitti, R., Quinlan, E. J., Zhang, L., Van Wijnen, A. J., Stein, J. L., Speck, N., and Stein, G. S. (2003). Runx1/AML1 hematopoietic transcription factor contributes to skeletal development *in vivo*. *J. Cell. Physiol.* **196,** 301–311.

Linsenmayer, T. F., Gibney, E., and Schmid, T. M. (1986). Segmental appearance of type X collagen in the developing avian notochord. *Dev. Biol.* **113,** 467–473.

Little, R. D., Carulli, J. P., Del Mastro, R. G., Dupuis, J., Osborne, M., Folz, C., Manning, S. P., Swain, P. M., Zhao, S. C., Eustace, B., Lappe, M. M., Spitzer, L., *et al.* (2002). A mutation in the LDL receptor-related protein 5 gene results in the autosomal dominant high-bone-mass trait. *Am. J. Hum. Genet.* **70,** 11–19.

Liu, Y., Li, H., Tanaka, K., Tsumaki, N., and Yamada, Y. (2000). Identification of an enhancer sequence within the first intron required for cartilage-specific transcription of the alpha2(XI) collagen gene. *J. Biol. Chem.* **275,** 12712–12718.

Logan, C. Y., and Nusse, R. (2004). The Wnt signaling pathway in development and disease. *Annu. Rev. Cell Dev. Biol.* **20,** 781–810.

Lyons, K. M., Hogan, B. L., and Robertson, E. J. (1995). Colocalization of BMP 7 and BMP 2 RNAs suggests that these factors cooperatively mediate tissue interactions during murine development. *Mech. Dev.* **50,** 71–83.

Maisey, J. G. (1988). Phylogeny of early vertebrate skeletal induction and ossification patterns. *In* "Evolutionary Biology" (M. K. Hecht, B. Wallace, and G. T. Prance, Eds.), pp. 1–36. Plenum Publishing Corporation, New York.

Mak, K. K., Kronenberg, H. M., Chuang, P. T., Mackem, S., and Yang, Y. (2008). Indian hedgehog signals independently of PTHrP to promote chondrocyte hypertrophy. *Development* **135,** 1947–1956.

Mallatt, J., and Chen, J. Y. (2003). Fossil sister group of craniates: Predicted and found. *J. Morphol.* **258,** 1–31.

Meulemans, D., and Bronner-Fraser, M. (2007). Insights from amphioxus into the evolution of vertebrate cartilage. *PLoS ONE* **2,** e787.

Miller, J. R. (2002). The Wnts. *Genome Biol.* **3,** REVIEWS3001.

Miller, E. J., and Matukas, V. J. (1969). Chick cartilage collagen: A new type of alpha 1 chain not present in bone or skin of the species. *Proc. Natl. Acad. Sci. USA* **64,** 1264–1268.

Minina, E., Wenzel, H. M., Kreschel, C., Karp, S., Gaffield, W., McMahon, A. P., and Vortkamp, A. (2001). BMP and Ihh/PTHrP signaling interact to coordinate chondrocyte proliferation and differentiation. *Development* **128,** 4523–4534.

Mizuta, S., Hwang, J.-H., and Yoshinaka, R. (2003). Molecular species of collagen in pectoral fin cartilage of skate (Raja Kenojei). *Food Chem.* **80,** 1–7.

Mori-Akiyama, Y., Akiyama, H., Rowitch, D. H., and de Crombrugghe, B. (2003). Sox9 is required for determination of the chondrogenic cell lineage in the cranial neural crest. *Proc. Natl. Acad. Sci. USA* **100,** 9360–9365.

Morrison, S. L., Campbell, C. K., and Wright, G. M. (2000). Chondrogenesis of the branchial skeleton in embryonic sea lamprey, *Petromyzon marinus*. *Anat. Rec.* **260,** 252–267.

Moss, M. L. (1970). Enamel and bone in shark teeth: With a note on fibrous enamel in fishes. *Acta Anat. (Basel)* **77,** 161–187.

Moss, M. L. (1977). Skeletal tissues in sharks. *Am. Zool.* 335–342.

Mundlos, S., and Olsen, B. R. (1997a). Heritable diseases of the skeleton. Part I. Molecular insights into skeletal development-transcription factors and signaling pathways. *FASEB J.* **11,** 125–132.

Mundlos, S., and Olsen, B. R. (1997b). Heritable diseases of the skeleton. Part II. Molecular insights into skeletal development-matrix components and their homeostasis. *FASEB J.* **11,** 227–233.

Mundlos, S., Huang, L. F., Selby, P., and Olsen, B. R. (1996). Cleidocranial dysplasia in mice. *Ann. N. Y. Acad. Sci.* **785,** 301–302.

Nakashima, K., Zhou, X., Kunkel, G., Zhang, Z., Deng, J. M., Behringer, R. R., and de Crombrugghe, B. (2002). The novel zinc finger-containing transcription factor osterix is required for osteoblast differentiation and bone formation. *Cell* **108,** 17–29.

Naumann, A., Dennis, J. E., Awadallah, A., Carrino, D. A., Mansour, J. M., Kastenbauer, E., and Caplan, A. I. (2002). Immunochemical and mechanical characterization of cartilage subtypes in rabbit. *J. Histochem. Cytochem.* **50,** 1049–1058.

Ng, L. J., Wheatley, S., Muscat, G. E., Conway-Campbell, J., Bowles, J., Wright, E., Bell, D. M., Tam, P. P., Cheah, K. S., and Koopman, P. (1997). SOX9 binds DNA, activates transcription, and coexpresses with type II collagen during chondrogenesis in the mouse. *Dev. Biol.* **183,** 108–121.

Ohtani, K., Yao, T., Kobayashi, M., Kusakabe, R., Kuratani, S., and Wada, H. (2008). Expression of Sox and fibrillar collagen genes in lamprey larval chondrogenesis with implications for the evolution of vertebrate cartilage. *J. Exp. Zool. B Mol. Dev. Evol.* **310,** 596–607.

Olsen, B. R., Reginato, A. M., and Wang, W. (2000). Bone development. *Annu. Rev. Cell Dev. Biol.* **16,** 191–220.

Ornitz, D. M., and Marie, P. J. (2002). FGF signaling pathways in endochondral and intramembranous bone development and human genetic disease. *Genes Dev.* **16,** 1446–1465.

Orvig, T. (1951). Histologic studies of Placoderms and fossil Elasmobranchs. I. The endoskeleton, with remarks on the hard tissues of lower vertebrates in general. *Ark. Zool.* **2,** 321–456.

Ostrander, G. K., Cheng, K. C., Wolf, J. C., and Wolfe, M. J. (2004). Shark cartilage, cancer and the growing threat of pseudoscience. *Cancer Res.* **64,** 8485–8491.

Otto, F., Thornell, A. P., Crompton, T., Denzel, A., Gilmour, K. C., Rosewell, I. R., Stamp, G. W., Beddington, R. S., Mundlos, S., Olsen, B. R., Selby, P. B., and Owen, M. J. (1997). Cbfa1, a candidate gene for cleidocranial dysplasia syndrome, is essential for osteoblast differentiation and bone development. *Cell* **89,** 765–771.

Parker, W. (1883). On the skeleton of the marsipobranch fishes. Part II. Petromyzon. *Philos. Trans. R. Soc. Lond. B Biol. Sci.* **174,** 411–457.

Patterson, C. M., Kruger, B. J., and Daley, T. J. (1977). Lipid and protein histochemistry of enamel—Effects of fluoride. *Calcif. Tissue Res.* **24,** 119–123.

Peignoux-Deville, J., Lallier, F., and Vidal, B. (1982). Evidence for the presence of osseous tissue in dogfish vertebrae. *Cell Tissue Res.* **222,** 605–614.

Person, P., and Mathews, M. B. (1967). Endoskeletal cartilage in a marine polychaete, *Eudistylia polymorpha. Biol. Bull.* **132,** 244–252.

Pogue, R., and Lyons, K. (2006). BMP signaling in the cartilage growth plate. *Curr. Top. Dev. Biol.* **76,** 1–48.

Pratta, M. A., Yao, W., Decicco, C., Tortorella, M. D., Liu, R. Q., Copeland, R. A., Magolda, R., Newton, R. C., Trzaskos, J. M., and Arner, E. C. (2003). Aggrecan protects cartilage collagen from proteolytic cleavage. *J. Biol. Chem.* **278,** 45539–45545.

Provot, S., Kempf, H., Murtaugh, L. C., Chung, U. I., Kim, D. W., Chyung, J., Kronenberg, H. M., and Lassar, A. B. (2006). Nkx3.2/Bapx1 acts as a negative regulator of chondrocyte maturation. *Development* **133,** 651–662.

Prud'homme, B., Lartillot, N., Balavoine, G., Adoutte, A., and Vervoort, M. (2002). Phylogenetic analysis of the Wnt gene family. Insights from lophotrochozoan members. *Curr. Biol.* **12,** 1395.

Rama, S., and Chandrakasan, G. (1984). Distribution of different molecular species of collagen in the vertebral cartilage of shark (*Carcharias acutus*). *Connect Tissue Res.* **12,** 111–118.

Rawadi, G., Vayssiere, B., Dunn, F., Baron, R., and Roman-Roman, S. (2003). BMP-2 controls alkaline phosphatase expression and osteoblast mineralization by a Wnt autocrine loop. *J. Bone Miner. Res.* **18,** 1842–1853.

Razzaque, M. S., Soegiarto, D. W., Chang, D., Long, F., and Lanske, B. (2005). Conditional deletion of Indian hedgehog from collagen type 2alpha1-expressing cells results in abnormal endochondral bone formation. *J. Pathol.* **207,** 453–461.

Reif, W. E. (1980). Development of dentition and dermal skeleton in embryonic *Scyliorhinus canicula*. *J. Morphol.* **166,** 275–288.

Rodda, S. J., and McMahon, A. P. (2006). Distinct roles for Hedgehog and canonical Wnt signaling in specification, differentiation and maintenance of osteoblast progenitors. *Development* **133,** 3231–3244.

Romer, A. S. (1985). The vertebrate body. *In* "Saunders Series in Organismic Biology." Saunders College Publishing, Philadelphia.

Rychel, A. L., and Swalla, B. J. (2007). Development and evolution of chordate cartilage. *J. Exp. Zool. B Mol. Dev. Evol.* **308,** 325–335.

Rychel, A. L., Smith, S. E., Shimamoto, H. T., and Swalla, B. J. (2006). Evolution and development of the chordates: Collagen and pharyngeal cartilage. *Mol. Biol. Evol.* **23,** 541–549.

Sahni, M., Ambrosetti, D. C., Mansukhani, A., Gertner, R., Levy, D., and Basilico, C. (1999). FGF signaling inhibits chondrocyte proliferation and regulates bone development through the STAT-1 pathway. *Genes Dev.* **13,** 1361–1366.

Saito, T., Ikeda, T., Nakamura, K., Chung, U. I., and Kawaguchi, H. (2007). S100A1 and S100B, transcriptional targets of SOX trio, inhibit terminal differentiation of chondrocytes. *EMBO Rep.* **8,** 504–509.

Sansom, I. J., Donoghue, P. C., and Albanesi, G. (2005). Histology and affinity of the earliest armoured vertebrate. *Biol. Lett.* **1,** 446–449.

Schipani, E., Kruse, K., and Juppner, H. (1995). A constitutively active mutant PTH–PTHrP receptor in Jansen-type metaphyseal chondrodysplasia. *Science* **268,** 98–100.

Schwartz, N. B., Pirok, E. W., III, Mensch, J. R., Jr., and Domowicz, M. S. (1999). Domain organization, genomic structure, evolution, and regulation of expression of the aggrecan gene family. *Prog. Nucleic Acid Res. Mol. Biol.* **62,** 177–225.

Seki, K., and Hata, A. (2004). Indian hedgehog gene is a target of the bone morphogenetic protein signaling pathway. *J. Biol. Chem.* **279,** 18544–18549.

Sire, J. Y., and Huysseune, A. (2003). Formation of dermal skeletal and dental tissues in fish: A comparative and evolutionary approach. *Biol. Rev. Camb. Philos. Soc.* **78,** 219–249.

Sivakumar, P., and Chandrakasan, G. (1998). Occurrence of a novel collagen with three distinct chains in the cranial cartilage of the squid Sepia officinalis: Comparison with shark cartilage collagen. *Biochim. Biophys. Acta* **1381,** 161–169.

Smith, M. M., and Hall, B. K. (1990). Development and evolutionary origins of vertebrate skeletogenic and odontogenic tissues. *Biol. Rev. Camb. Philos. Soc.* **65,** 277–373.

Smith, M. M., and Johanson, Z. (2003). Separate evolutionary origins of teeth from evidence in fossil jawed vertebrates. *Science* **299,** 1235–1236.

Smith, N., Dong, Y., Lian, J. B., Pratap, J., Kingsley, P. D., van Wijnen, A. J., Stein, J. L., Schwarz, E. M., O'Keefe, R. J., Stein, G. S., and Drissi, M. H. (2005). Overlapping expression of Runx1(Cbfa2) and Runx2(Cbfa1) transcription factors supports cooperative induction of skeletal development. *J. Cell. Physiol.* **203,** 133–143.

Smits, P., and Lefebvre, V. (2003). Sox5 and Sox6 are required for notochord extracellular matrix sheath formation, notochord cell survival and development of the nucleus pulposus of intervertebral discs. *Development* **130,** 1135–1148.

Soderhall, C., Marenholz, I., Kerscher, T., Ruschendorf, F., Esparza-Gordillo, J., Worm, M., Gruber, C., Mayr, G., Albrecht, M., Rohde, K., Schulz, H., Wahn, U., *et al.* (2007). Variants in a novel epidermal collagen gene (COL29A1) are associated with atopic dermatitis. *PLoS Biol.* **5,** e242.

Solloway, M. J., Dudley, A. T., Bikoff, E. K., Lyons, K. M., Hogan, B. L., and Robertson, E. J. (1998). Mice lacking Bmp6 function. *Dev. Genet.* **22,** 321–339.

Song, S. J., Cool, S. M., and Nurcombe, V. (2007). Regulated expression of syndecan-4 in rat calvaria osteoblasts induced by fibroblast growth factor-2. *J. Cell. Biochem.* **100,** 402–411.

St-Jacques, B., Hammerschmidt, M., and McMahon, A. P. (1999). Indian hedgehog signaling regulates proliferation and differentiation of chondrocytes and is essential for bone formation. *Genes Dev.* **13,** 2072–2086.

Stemple, D. L. (2004). The notochord. *Curr. Biol.* **14,** R873–R874.

Stensio, E. A. (1927). The Devonian and Downtonian vertebrates of Spitsbergen. Part I. Family Cephalaspidae. *Skr. Svalbard Ishavet* **12,** 1–391.

Stolt, C. C., Schlierf, A., Lommes, P., Hillgartner, S., Werner, T., Kosian, T., Sock, E., Kessaris, N., Richardson, W. D., Lefebvre, V., and Wegner, M. (2006). SoxD proteins influence multiple stages of oligodendrocyte development and modulate SoxE protein function. *Dev. Cell* **11,** 697–709.

Stricker, S., Fundele, R., Vortkamp, A., and Mundlos, S. (2002). Role of Runx genes in chondrocyte differentiation. *Dev. Biol.* **245,** 95–108.

Sugahara, K., Tanaka, Y., Yamada, S., Seno, N., Kitagawa, H., Haslam, S. M., Morris, H. R., and Dell, A. (1996). Novel sulfated oligosaccharides containing 3-O-sulfated glucuronic acid from king crab cartilage chondroitin sulfate K. Unexpected degradation by chondroitinase ABC. *J. Biol. Chem.* **271,** 26745–26754.

Takeda, S., Bonnamy, J. P., Owen, M. J., Ducy, P., and Karsenty, G. (2001). Continuous expression of Cbfa1 in nonhypertrophic chondrocytes uncovers its ability to induce hypertrophic chondrocyte differentiation and partially rescues Cbfa1-deficient mice. *Genes Dev.* **15,** 467–481.

Toma, C. D., Schaffer, J. L., Meazzini, M. C., Zurakowski, D., Nah, H. D., and Gerstenfeld, L. C. (1997). Developmental restriction of embryonic calvarial cell populations as characterized by their *in vitro* potential for chondrogenic differentiation. *J. Bone Miner. Res.* **12,** 2024–2039.

Tsuda, M., Takahashi, S., Takahashi, Y., and Asahara, H. (2003). Transcriptional co-activators CREB-binding protein and p300 regulate chondrocyte-specific gene expression via association with Sox9. *J. Biol. Chem.* **278,** 27224–27229.

van der Rest, M., and Garrone, R. (1991). Collagen family of proteins. *FASEB J.* **5,** 2814–2823.

van Wijnen, A. J., Stein, G. S., Gergen, J. P., Groner, Y., Hiebert, S. W., Ito, Y., Liu, P., Neil, J. C., Ohki, M., and Speck, N. (2004). Nomenclature for Runt-related (RUNX) proteins. *Oncogene* **23,** 4209–4210.

Vortkamp, A., Lee, K., Lanske, B., Segre, G. V., Kronenberg, H. M., and Tabin, C. J. (1996). Regulation of rate of cartilage differentiation by Indian hedgehog and PTH-related protein. *Science* **273,** 613–622.

Wada, H., Okuyama, M., Satoh, N., and Zhang, S. (2006). Molecular evolution of fibrillar collagen in chordates, with implications for the evolution of vertebrate skeletons and chordate phylogeny. *Evol. Dev.* **8,** 370–377.

Wagner, T., Wirth, J., Meyer, J., Zabel, B., Held, M., Zimmer, J., Pasantes, J., Bricarelli, F. D., Keutel, J., Hustert, E., Wolf, U., and Tommerup, N. (1994).

Autosomal sex reversal and campomelic dysplasia are caused by mutations in and around the SRY-related gene SOX9. *Cell* **79,** 1111–1120.
Wang, Y., Belflower, R. M., Dong, Y. F., Schwarz, E. M., O'Keefe, R. J., and Drissi, H. (2005). Runx1/AML1/Cbfa2 mediates onset of mesenchymal cell differentiation toward chondrogenesis. *J. Bone Miner. Res.* **20,** 1624–1636.
Welsch, U., Erlinger, R., and Potter, I. C. (1991). Proteoglycans in the notochord sheath of lampreys. *Acta Histochem.* **91,** 59–65.
Wong, M., and Tuan, R. S. (1995). Interactive cellular modulation of chondrogenic differentiation *in vitro* by subpopulations of chick embryonic calvarial cells. *Dev. Biol.* **167,** 130–147.
Wright, G. M., Keeley, F. W., Youson, J. H., and Babineau, D. L. (1984). Cartilage in the Atlantic hagfish, *Myxine glutinosa. Am. J. Anat.* **169,** 407–424.
Wright, E., Hargrave, M. R., Christiansen, J., Cooper, L., Kun, J., Evans, T., Gangadharan, U., Greenfield, A., and Koopman, P. (1995). The Sry-related gene Sox9 is expressed during chondrogenesis in mouse embryos. *Nat. Genet.* **9,** 15–20.
Wright, G. M., Keeley, F. W., and Robson, P. (2001). The unusual cartilaginous tissues of jawless craniates, cephalochordates and invertebrates. *Cell Tissue Res.* **304,** 165–174.
Yamashiro, T., Wang, X. P., Li, Z., Oya, S., Aberg, T., Fukunaga, T., Kamioka, H., Speck, N. A., Takano-Yamamoto, T., and Thesleff, I. (2004). Possible roles of Runx1 and Sox9 in incipient intramembranous ossification. *J. Bone Miner. Res.* **19,** 1671–1677.
Yang, X., and Karsenty, G. (2002). Transcription factors in bone: Developmental and pathological aspects. *Trends Mol. Med.* **8,** 340–345.
Yoshida, C. A., Yamamoto, H., Fujita, T., Furuichi, T., Ito, K., Inoue, K., Yamana, K., Zanma, A., Takada, K., Ito, Y., and Komori, T. (2004). Runx2 and Runx3 are essential for chondrocyte maturation, and Runx2 regulates limb growth through induction of Indian hedgehog. *Genes Dev.* **18,** 952–963.
Zangerl, R. (1966). A new shark in the family Edestidae, *Ornithoprion hertwigi* from the Pennsylvania Mecca and Logan Quarry Shales of Indiana. *Fieldiana Geol.* **16,** 1–43.
Zelzer, E., and Olsen, B. R. (2003). The genetic basis for skeletal diseases. *Nature* **423,** 343–348.
Zhang, G., and Cohn, M. J. (2006). Hagfish and lancelet fibrillar collagens reveal that type II collagen-based cartilage evolved in stem vertebrates. *Proc. Natl. Acad. Sci. USA* **103,** 16829–16833.
Zhang, G., and Cohn, M. J. (2008). Genome duplication and the origin of the vertebrate skeleton. *Curr. Opin. Genet. Dev.* **18,** 387–393.
Zhang, D., Schwarz, E. M., Rosier, R. N., Zuscik, M. J., Puzas, J. E., and O'Keefe, R. J. (2003a). ALK2 functions as a BMP type I receptor and induces Indian hedgehog in chondrocytes during skeletal development. *J. Bone Miner. Res.* **18,** 1593–1604.
Zhang, P., Jimenez, S. A., and Stokes, D. G. (2003b). Regulation of human COL9A1 gene expression. Activation of the proximal promoter region by SOX9. *J. Biol. Chem.* **278,** 117–123.
Zhang, G., Miyamoto, M. M., and Cohn, M. J. (2006). Lamprey type II collagen and Sox9 reveal an ancient origin of the vertebrate collagenous skeleton. *Proc. Natl. Acad. Sci. USA* **103,** 3180–3185.
Zhang, X., Boot-Handford, R. P., Huxley-Jones, J., Forse, L. N., Mould, A. P., Robertson, D. L., Lili, M., Athiyal, M., and Sarras, M. P., Jr. (2007). The collagens of hydra provide insight into the evolution of metazoan extracellular matrices. *J. Biol. Chem.* **282,** 6792–6802.
Zhou, G., Lefebvre, V., Zhang, Z., Eberspaecher, H., and de Crombrugghe, B. (1998). Three high mobility group-like sequences within a 48-base pair enhancer of the Col2a1 gene are required for cartilage-specific expression *in vivo. J. Biol. Chem.* **273,** 14989–14997.

Zhou, G., Zheng, Q., Engin, F., Munivez, E., Chen, Y., Sebald, E., Krakow, D., and Lee, B. (2006). Dominance of SOX9 function over RUNX2 during skeletogenesis. *Proc. Natl. Acad. Sci. USA* **103,** 19004–19009.

Zou, H., Wieser, R., Massague, J., and Niswander, L. (1997). Distinct roles of type I bone morphogenetic protein receptors in the formation and differentiation of cartilage. *Genes Dev.* **11,** 2191–2203.

CHAPTER THREE

Caenorhabditis Nematodes as a Model for the Adaptive Evolution of Germ Cells

Eric S. Haag

Contents

Abstract

A number of major adaptations in animals have been mediated by alteration of germ cells and their immediate derivatives, the gametes. Here, several such cases are discussed, including examples from echinoderms, vertebrates, insects, and nematodes. A feature of germ cells that make their development (and hence evolution) distinct from the soma is the prominent role played by posttranscriptional controls of mRNA translation in the regulation of proliferation and differentiation. This presents a number of special challenges for investigation of the evolution of germline development. *Caenorhabditis* nematodes represent a particularly favorable system for addressing these challenges, both because of technical advantages and (most importantly) because of natural variation in mating system that is rooted in alterations of germline sex determination. Recent studies that employ comparative genetic methods in this rapidly maturing system are discussed, and likely areas for future progress are identified.

Department of Biology, University of Maryland, College Park, Maryland, USA

Current Topics in Developmental Biology, Volume 86
ISSN 0070-2153, DOI: 10.1016/S0070-2153(09)01003-5

1. Introduction

Beginning in the early 1980s, developmental biology was transformed by two nearly simultaneous revolutions, namely the advent of molecular-level developmental genetics and the rebirth of evolutionary developmental biology. These two revolutions were linked from the beginning, and were often furthered by the same researchers (e.g., see Bonner, 1981). Since its early days, a central goal of evolutionary developmental biology has been to understand how development is modified to enable major adaptations. However, the bulk of the animal adaptations that have been scrutinized developmentally are somatic attributes of larvae or adults, such as pigmentation, skeletal and exoskeletal morphologies, etc. This chapter is generally concerned with a less-appreciated type of developmental evolution, in which reproductive adaptations are mediated wholly or in part by changes in germ cells and their derivatives, the gametes. After an overview, recent studies using the model nematode genus *Caenorhabditis* are reviewed and synthesized.

2. Germ Cell Adaptation and the Evolution of New Life Histories

Examples of germline-mediated adaptations of great ecological significance include both everyday and more obscure organisms. Every time someone cracks a hen's egg into a bowl, they are holding in their hands one of the most spectacular of these adaptations: the amniote oocyte and surrounding albumen (the "white") and shell (formed around the oocyte by the shell gland). The amniote lineage has been so successful in large part because of the derived properties of this egg and its associated coverings (Packard and Seymour, 1997; Stewart, 1997). First, the desiccation-resistant shell allowed them to commit to a fully terrestrial life cycle, while the extraembryonic membranes evolved to facilitate gas exchange, waste sequestration, and (in the archosaurs) calcium absorption required for direct development of a bony skeleton. Second, the enormous yolk reserves of the oocyte proper allow direct development of the embryo into a miniature adult, eliminating the larval phase of amphibian tetrapods. These traits likely first appeared in the Pennsylvanian epoch of the Carboniferous era, roughly 300 million years ago, although they are inferred indirectly from the features of fossil adults (Clack, 2002).

Perhaps less familiar to many are examples of the relatively large eggs of some animals, generally associated with major shifts in lifestyle and reproductive strategy. One example is direct development in anuran amphibians, which is associated with terrestrial or arboreal life. A well-studied case is *Eleutherodactylus coqui*, native to Puerto Rico. Although embryonic development is radically altered to allow the development of a miniature frog at hatching, it all begins with a large (3.5-mm diameter) egg (e.g., Callery *et al.*, 2001; Elinson and Beckham, 2002). *E. coqui* eggs also differ from those of tadpole-forming species in tolerating some polyspermy (Elinson, 1987).

Another extreme case of egg enlargement is associated with the evolution of direct development in some echinoderms. A model system here is the Australian echinoid *Heliocidaris erythrogramma*, whose eggs are 100 times the volume of their indirect-developing sister species *H. tuberculata* (Fig. 3.1A). In terms of bulk constituents, a significant portion of the increase in cytoplasmic volume can be attributed to large lipid droplets, initially described by Williams and Anderson (1975) as "vesicular yolk." These droplets are later secreted into the blastocoel to form an acellular nutritive deposit (Henry *et al.*, 1991) that persists through metamorphosis (Haag *et al.*, 1999). Similar lipid droplets have evolved independently in different lineages with large eggs and direct development (Villinski *et al.*, 2002). They were inferred by thin layer chromatography to be composed largely of waxy esters by Villinski *et al.* (2002), but other studies using other methods have failed to support this diagnosis (Byrne *et al.*, 2008; Prowse *et al.*, 2008). Whatever their precise composition, the droplets are deposited during oogenesis after a conserved, early phase of yolk production is completed (Byrne *et al.*, 1999). This post-vitellogenic phase thus represents a novel aspect of direct-developing oogenesis. Surprisingly, this lipid is not necessary for the completion of metamorphosis, but is required for survival of the nonfeeding juvenile stage that occurs between metamorphosis and eruption of the mouth (Emlet and Hoegh-Guldberg, 1996; Williams and Anderson, 1975). With respect to developmental patterning, embryological experiments indicate that the egg of the direct-developing *H. erythrogramma* incorporates axial patterning cues that are specified only after first cleavage in its indirect-developing relatives (Henry and Raff, 1990; Henry *et al.*, 1990).

In addition to size, germ cells also mediate extreme shifts in reproductive mode by facilitating the loss of obligate mating. An example from the vertebrates is parthenogenesis, seen both in lizards and in salamanders. This is invariably associated with hybrid species with cytogenetically distinguishable karyotypes, and includes both allodiploids and allotriploids, the latter presumably formed by fertilization of a diploid oocyte of one species by a sperm of another (Uzzell, 1970). Parthenogenic species may, in principle, use several genetic mechanisms to produce oocytes with the same ploidy as their somatic cells (Uzzell, 1970). In both the related salamanders *Ambystoma platineum* and *Ambystoma jeffersonianum* (Macgregor and Uzzell, 1964), and in the whiptail lizard *Cnemidophorus uniparens* (Cuellar, 1971; Fig. 3.1B), it appears that triploid primary oocytes undergo one round of mitosis without cytokinesis. This allows formation of a set of pseudobivalents composed of

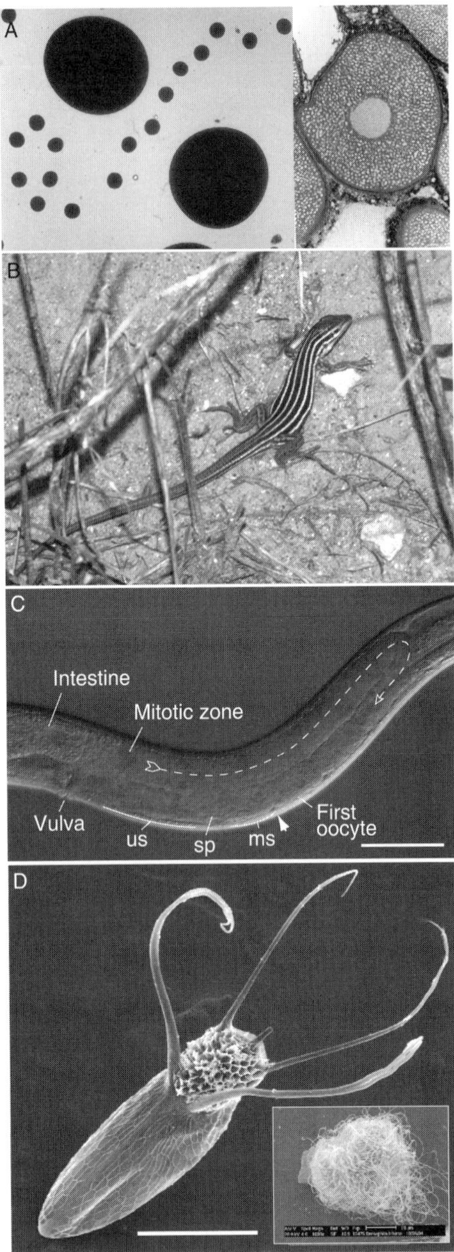

Figure 3.1 Examples of extreme germ cell adaptation in four phyla. (A) The left panel is a micrograph of a mixture of spawned, mature eggs from the Australian congeneric sea urchins *Heliocidaris tuberculata* (ca. 95-μm diameter) and *H. erythrogramma* (ca. 420-μm diameter). *H. tuberculata* is a typical indirect developer with a feeding pluteus larva, while *H. erythrogramma* is a lecithotrophic direct developer. The right panel shows a paraffin

pairs of newly replicated daughter chromosomes. As pairing is always between chromosomes of the same hybrid parent species, the two karyotypes present in the hybrid genome undergo no effective recombination. It is unclear how rapidly this sort of pseudomeiosis can evolve, but it is possible that it represents a latent capacity of oocytes that cannot produce interspecies bivalents. If so, this trait would appear as soon as a hybrid lineage forms, though selection may further enhance its reliability.

A second example of germ cell changes that decouple mating and reproduction is the evolution of self-fertile hermaphroditism in nematodes of the family Rhabditidae, including the model species *Caenorhabditis elegans* and its close relative, *C. briggsae*. This trait has evolved repeatedly from the ancestral male/female (gonochoristic) condition in soil nematodes, and even the two *Caenorhabditis* cases are likely cases of convergent evolution (Kiontke and Fitch, 2005; Kiontke and Sudhaus, 2006; Kiontke *et al.*, 2004). In all cases, self-fertility is mediated by the evolution of limited spermatogenesis in the XX (female) sex (Fig. 3.1C). This situation may evolve repeatedly instead of the sperm-swapping hermaphroditism seen in other protostomes (e.g., gastropod molluscs; Jarne, 2006) because of the extreme sexual dimorphism and the associated internal fertilization of the gonochoristic ancestors. More specifically, because the entire posterior of the male is specialized for copulation and sperm transfer, an entirely new mechanism would be required to allow hermaphrodites to accomplish the same task.

Though the above examples are all instances where germ cell attributes have been the result of natural selection acting on life history, germ cells are also involved in extreme examples of sexual selection. For example, in the

section through a maturing *H. erythrogramma* oocyte in the ovary, which reveals abundant cytoplasmic lipid droplets. (B) Adult of the all-female parthenogenetic whiptail lizard, *Cnemidophorus uniparens*, in its natural setting in Arizona. (C) Differential interference contrast micrograph of the posterior gonad arm of a young adult hermaphrodite *Caenorhabditis briggsae* nematode, showing the completion of spermatogenesis and initiation of oogenesis in the same germ cell population. In females of gonochoristic *Caenorhabditis* species, spermatocytes are absent. Germ cells move from the distal, mitotic stem cell niche half the length of the gonad arm, at which point the arm reflexes and converges on the uterus and spermatheca (us). The first hundred or so germ cells that differentiate produce sperm. Both meiotic spermatocytes (ms) and mature (but inactive) spermatids (sp) can be seen here. Immediately behind the spermatocytes, the first oocytes are starting to differentiate, with an abrupt transition between them (arrowhead). Scale bar is 50 μm. (D) Giant sperm in the dipteran insect *Drosophila bifurca*, as seen in scanning electron micrographs. The oocyte, with its elaborate chorion, is shown in the main panel, while a single spermatozoan with its extensively coiled axoneme is shown in the inset panel. Scale bar in the main panel is 200 μm; the sperm image is magnified 2.5$\times$ relative to the egg. Image credits: A (left) by Jeff Villinski (courtesy of Rudolf Raff) and A (right) by Maria Byrne; B by Twan Leenders; C by the author; and D by Romano Dallai (courtesy of Scott Pitnick).

dipteran insect *Drosophila bifurca*, the sperm are many times longer than the adult male that produces them, and require elaborate coiling in order to fit into the female reproductive tract (Fig. 3.1D). These giant sperm are likely to have evolved from runaway postcopulatory sexual selection imposed by elongation of the female's seminal receptacles (Miller and Pitnick, 2002), and are now so big that in many respects the species is effectively isogamous (Bjork and Pitnick, 2006). All of the above examples make the point that, far from being adaptively inert custodians of the genome, germ cells themselves can rapidly respond to selection to enable important adaptations or extreme sexual traits.

3. Germ Cell Adaptation: Evo-Devo Meets RNA

Because it is a premier model species for developmental genetics, and even more so because germline sex determination has a long history of genetic and molecular research (Ellis and Schedl, 2007), *C. elegans* and its close relatives make a powerful system for addressing the evolution of hermaphrodite development. The author, along with his students and colleagues, has spent much of the last decade developing tools for non-*elegans* species of *Caenorhabditis* that will enable the realization of the great potential in this area. But to understand our approach, it is important to first understand some of the ways in which germ cell development is different from that of somatic tissues.

While the question of whether transcriptional regulation or coding sequence changes contribute more to phenotypic evolution has received much recent attention (Carroll, 2008; Hoekstra and Coyne, 2007; Prud'homme *et al.*, 2007), evidence from *C. elegans* and other systems suggests that germ cells often use a third type of regulation to control cell cycle progression and differentiation, that of posttranscriptional control of mRNA translation. Why germ cells rely so heavily upon RNA-level regulation is still unclear, but one idea (Kimble and Crittenden, 2007; Seydoux and Braun, 2006) is that they are poised on the cusp of initiating embryonic differentiation via their diverse maternal mRNAs, but are restrained from doing so prematurely by translational repression via various RNP complexes. We might term this the "frozen almost-embryo" hypothesis, and there is a large body of data supporting it (Evans and Hunter, 2005). An alternative idea is that meiosis, which is unique to germ cells, may impose special requirements on gene expression. As the vast majority of adult *C. elegans* germ cells are in various stages of meiosis, which is marked by condensed chromatin, it may be that differentiation and cell cycle control must be handled to a large extent in the absence of new transcription. We could term this the "meiotic transcriptional block" hypothesis, and a number of studies in *C. elegans* have indeed suggested that transcription in germ cells is generally repressed by chromatin modifications (Kelly and Fire, 1998; Schaner and Kelly, 2006). Of course, these two hypotheses are not mutually exclusive, and both are probably relevant.

Though the evolution of RNA-level controls is only just beginning to be investigated, they do bear some similarity to the more familiar *cis*-regulatory control of transcription by DNA-binding transcription factors. For example, they are generally mediated by *cis*-elements, typically in 3′-untranslated regions (UTR) of mRNA. These UTR elements serve as specific docking sites for various RNA-binding proteins (RBPs) that, like transcription factors, are often combinatorial in their effects on a single target and highly pleiotropic in that they bind many different mRNAs (Jin *et al.*, 2001; Lee and Schedl, 2001; Luitjens *et al.*, 2000; Piqué *et al.*, 2008; Standart and Minshall, 2008; Wickens *et al.*, 2002). To understand how adaptive evolution works in germ cells, it is important to develop methods that allow the discovery and functional perturbation of potentially complex regulatory networks in multiple species. Such studies would necessarily address both target mRNAs and the RBPs that regulate them. As the most obvious adaptation in hermaphroditic *Caenorhabditis* is a change in sexual fate of germ cells from oocytes to sperm, the target mRNAs we focus upon are those encoding components of the sex-determination pathway.

4. Overview of *C. elegans* Sex Determination

In *C. elegans*, germ cell sex is controlled by same pathway of negative regulation that governs sex in the rest of the body (the core pathway; Fig. 3.2). At the simplest level, this pathway links the ratio of X chromosomes to autosomes to the activity state of the terminal global regulator, the transcription factor TRA-1 (Zarkower and Hodgkin, 1992). TRA-1 exists at high levels in XX hermaphrodites as a proteolytically processed form (TRA-1^{100}) that represses male development (Schvarzstein and Spence, 2006). An unprocessed form of TRA-1 is present at much lower levels in both sexes (Schvarzstein and Spence, 2006). As complete loss of TRA-1 via mutations converts XX animals into near-perfect males that can sire progeny (Hodgkin, 1987; Hodgkin and Brenner, 1977), most of *tra-1*'s activity can be ascribed to repression of male fates by TRA-1^{100}.

We can examine the molecular logic that underlies the diagram shown in Fig. 3.2 by backing up from *tra-1*. The sex difference in TRA-1^{100} abundance is due to male-specific ubiquitination and proteolysis, which is mediated by the three cytoplasmic FEM proteins acting in a complex (Chin-Sang and Spence, 1996; Starostina *et al.*, 2007; Tan *et al.*, 2001). In XX animals, the FEM proteins are prevented from targeting TRA-1 for

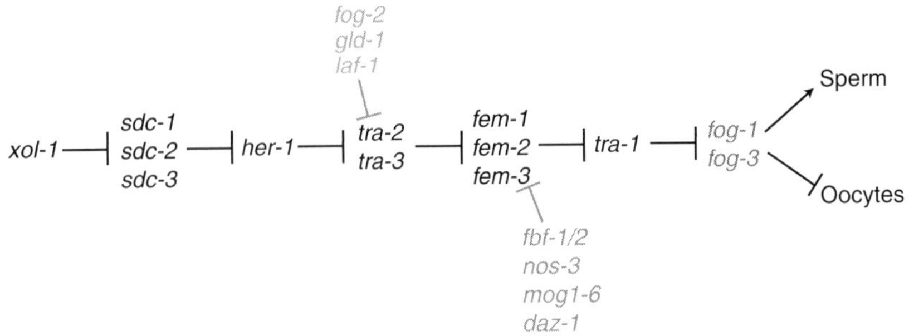

Figure 3.2 The *C. elegans* sex-determination pathway and its germline-specific modifiers. The "core pathway" acting in all cells is depicted in black typeface, which germline-specific genes are in gray. Germline genes required for the onset of XX spermatogenesis are shown above the horizontal midline, and affect *tra-2*. Genes required for the sperm–oocyte switch are shown below the midline, and affect *fem-3*.

degradation by an interaction between FEM-3 and the membrane protein TRA-2 (Mehra *et al.*, 1999). *tra-2* function also requires that TRA-2 be cleaved by the calpain protease TRA-3 (Hodgkin and Brenner, 1977; Sokol and Kuwabara, 2000), indicating that repression of the FEM proteins by TRA-2 may actually be accomplished by a cytoplasmic C-terminal fragment rather than the intact transmembrane protein. The TRA-2–FEM interaction, in turn, is prevented in XO males by the secreted protein HER-1, which interacts with the extracellular domain of TRA-2 (Hamaoka *et al.*, 2004). In keeping with the cell nonautonomy implied by HER-1 secretion, germ cell sex can be influenced by surrounding somatic tissues (Cho *et al.*, 2007; Hunter and Wood, 1992; McCarter *et al.*, 1997).

Continuing upstream, HER-1 levels are regulated at the transcription level by the SDC proteins, which also mediate dosage compensation of the X chromosomes (Chu *et al.*, 2002). This dual function of the SDC proteins ensures that transcription of both *her-1* and most X-linked genes are repressed in XX cells. Finally, the *sdc* genes are regulated by *xol-1*, which sits atop the signaling cascade and whose transcription is directly controlled by the relative levels of X-linked and autosomal factors (Meyer, 2005).

With the above pathway in mind, we now return to the subject of the derived germ cell differentiation of hermaphrodites. It is crucial to note that although hermaphroditic *Caenorhabditis*, such as *C. elegans* and *C. briggsae*, make sperm, they do so without expressing HER-1 (Trent *et al.*, 1991). Therefore, they must set the downstream part of the sex-determination pathway in male mode without HER-1, and only in the germline. A large body of genetic and molecular work has revealed that this feat requires the activity of a number of germline-specific factors. Two, the cytoplasmic polyadenylation element-binding (CPEB) protein homologue FOG-1 and the TOB domain protein FOG-3, act downstream of TRA-1, with *fog-3* being a direct transcriptional target (Barton and Kimble, 1990; Chen and Ellis, 2000; Ellis and Kimble, 1995). Another group of RBPs affect sex-determination upstream of *tra-1*, and several have been shown to directly regulate sex-determination mRNAs. In particular, the KH domain RBP GLD-1 (Francis *et al.*, 1995a,b; Jones and Schedl, 1995), its cofactor, FOG-2 (Clifford *et al.*, 2000; Schedl and Kimble, 1988), and the RNA helicase LAF-1 (Goodwin *et al.*, 1997; A. Hubert, submitted for publication) are all required to allow initiation of XX spermatogenesis. All of these factors are directly or indirectly involved in regulating the translation of *tra-2* mRNA, which harbors an essential GLD-1-binding site in its 3′-UTR (Goodwin *et al.*, 1993; Jan *et al.*, 1999; Lee and Schedl, 2001). This has led to model in which XX spermatogenesis requires, and may be specifically activated by, repression of *tra-2* translation, which mimics HER-1 inhibition of TRA-2 activity in the XO male (Fig. 3.2).

Cessation of spermatogenesis, the "sperm-to-oocyte switch," is also a crucial step in hermaphrodite development. Again, a large body of work has implicated RBPs in the translational control of a second sex-determination gene, the male-promoting *fem-3*. As with *tra-2, fem-3* contains a crucial binding site for an RBP complex (Ahringer and Kimble, 1991; Barton *et al.*, 1987), which is composed of the PUF family members FBF-1 and FBF-2 and their cofactor, the *Nanos* homologue NOS-3 (Kraemer *et al.*, 1999; Zhang *et al.*, 1997). The translational repression of *fem-3* also requires the six *mog* genes (Gallegos *et al.*, 1998), at least three of which encode homologues of mRNA splicing factors and as well as a cyclophilin-related protein (Belfiore *et al.*, 2004; Puoti and Kimble, 1999, 2000). Finally, the RBP DAZ-1 appears to promote the sperm–oocyte switch by stimulating translation of the *fbf-1* and *fbf-2* mRNA (Otori *et al.*, 2006).

While the above two paragraphs catalog an impressive array of discoveries in the area of germline sex determination, a cautionary note is appropriate. While many factors are necessary for proper execution of the sperm-then-oocyte pattern of hermaphrodite germline development, the identity of the sex-determination pathway component(s) whose activity is differentially modulated under natural physiological conditions to effect the switch represented by the arrowhead in Fig. 3.1C is still not known. To underscore this point, when the *tra-2* and *fem-3* translational controls described above are both abrogated through mutations that eliminate their translational control elements, self-fertile hermaphrodites are produced at high frequencies (Barton *et al.*, 1987; Schedl and Kimble, 1988). Whichever factor serves as the natural switch element, the distal expression of *rme-2* mRNA (encoding an egg-specific yolk receptor) in the last larval (L4) stage implies that oocyte fate is specified in, or soon after cells exit from, the distal mitotic stem cell zone (Ellis and Schedl, 2007).

There are additional complications that make germline sex determination different from that seen in the soma. One is that while XX *tra-1* loss-of-function mutants are transformed into mating males, they usually have intersexual germline development, rather than the full maleness seen in the soma (Hodgkin, 1987). This suggests that, unlike in the soma, the repression of maleness is not TRA-1's only function in germ cells. As XO *tra-1(lf)* mutants also suffer germline feminization, it is likely that this phenotype results from a germline-specific requirement for the full-length (unprocessed) form of TRA-1 in reliable specification of the sperm fate. Thus, *tra-1* may have both repressive and activating roles in male development, which would be reminiscent of the similar dual roles of its homologues, the *hedgehog* pathway effectors *Cubitus interruptus* (in *Drosophila*) and Gli (in vertebrates; reviewed by Østerlund and Kogerman, 2006).

A second complication comes from double mutant analyses. The core sex-determination pathway shown in Fig. 3.2 indicates that the sole purpose of the FEM proteins is to regulate TRA-1 activity. In the soma this seems to hold up well, as the three possible *fem; tra-1* double mutants all have the same completely male anatomy and behavior found in true XO males (Doniach and Hodgkin, 1984; Hodgkin, 1986). However, the germline phenotype of these double mutants is complete feminization. This unexpected result suggests that the FEM proteins may promote sperm fate independently of their action on TRA-1, such that the already partially feminized germline of *tra-1(lf)* mutants is pushed into completely female territory when they are compromised. A more specific variation on this is that TRA-1 transcriptionally represses the *fem* genes in the germline as part of its general male-repressing function. Under this model, loss of *tra-1* produces a partly masculinized germline because of upregulation of *fem* transcription, which in turn promotes spermatogenesis. Mutations in *fem* genes thus reverse this phenotype by preventing them from responding to reduced TRA-1.

5. *Caenorhabditis*: A Window on the World of Germline Adaptation

As we have seen, germ cell biology is marked by a strong reliance upon RNA–protein complexes, many of which serve to regulate mRNA translation, and sex determination in *C. elegans* is no exception. Germ cell translational control is mediated by a number of widely conserved, often germline-specific proteins. Since choosing between oocyte and spermatocyte fate is the main task that a nematode germ cell must accomplish prior to fertilization, perturbations of many translational regulators produce sexual phenotypes. This may be further exaggerated by the existence of reinforcement, feedback, and threshold controls that are normally in place to prevent intersexuality. Such controls would be expected to create sharp phenotypic

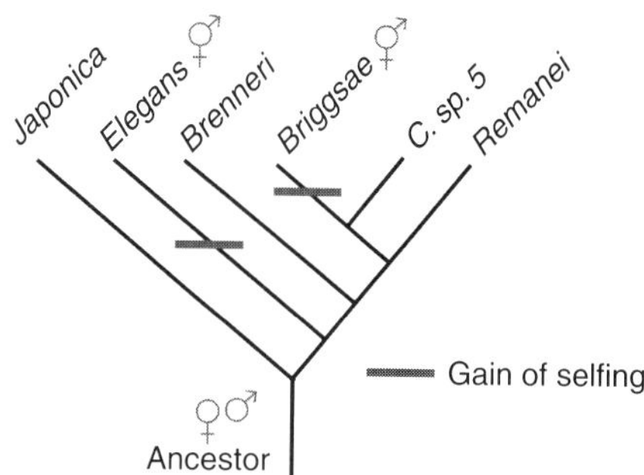

Figure 3.3 The current phylogenetic hypothesis for the relationships among *Caenorhabditis* species, with the most parsimonious reconstruction of mating system evolution mapped upon it. This figure synthesizes results of Braendle and Felix (2006), Cho *et al.* (2004), Hill *et al.* (2006), Kiontke *et al.* (2004), Nayak *et al.* (2005), and Sudhaus and Kiontke (2007).

transitions upon experimental perturbation. *Caenorhabditis* gives us a system to explore how these posttranscriptional controls are modified to produce an ecologically important adaptation—XX spermatogeneis. Two main approaches we have used are:

1. Evaluation of candidate translational controls in gonochoristic species
2. Genetic and molecular comparison of sex determination in convergently evolved hermaphrodites (Fig. 3.3)

Below, recent results from both areas are summarized.

6. What Makes a Female Different from a Hermaphrodite?

In the simplest possible model, the translational controls that regulate *tra-2* and *fem-3* levels in *C. elegans* are the essence of hermaphrodite development, and evolved specifically for this purpose. Motivated by this hypothesis, Haag and Kimble (2000) characterized the first sex-determination gene from a gonochoristic nematode, the ortholog of *tra-2* in *C. remanei*. RNAi interference experiments showed that TRA-2 promotes female fates in both the soma and germline, as in *C. elegans*. The study also revealed two surprising aspects of *tra-2* evolution. First, though *C. remanei* females never initiate spermatogenesis, the 3′-UTR of *Cr-tra-2* nevertheless bound a factor in extracts that had properties similar to DRF, the GLD-1-containing translational repressor. This suggested that it was not the evolution of translational control *per se* that enabled hermaphrodite spermatogenesis, and that perhaps more subtle modulation of preexisting controls was closer to the truth. Second, though TRA-2 was overall rather divergent, as expected from earlier work on the *C. briggsae* homologue (Kuwabara, 1996), the C-terminal cytoplasmic domain shown to bind FEM-3 was hypervariable—so much so that there are essentially no conserved residues in a three-way alignment. Given the essential nature of the TRA-2–FEM-3 interaction, this lack of sequence constraint was wholly unexpected.

Given the results for *Cr-tra-2*, it became important to also examine *Cr-fem-3*. Previous attempts to clone homologues of *fem-3* from other *Caenorhabditis* species by low-stringency nucleic acid hybridization failed, presumably due to unusually low sequence conservation (J. Kimble, personal communication). Using the synteny-based strategy pioneered by Kuwabara and Shah (1994), Haag *et al.* (2002) identified phage and fosmid clones from *C. remanei* and *C. briggsae*, respectively, that contained both the conserved copine gene used to identify the clones as well as highly diverged orthologs of *fem-3*. As with the domain of TRA-2 with which it interacts, conservation of FEM-3 as a whole is remarkably poor, with pairwise identities ranging from 31% to 38% and only very short motifs conserved in all three homologues. Despite this rapid sequence evolution, however, in all three species the C-terminus of TRA-2 interacts strongly with the conspecific FEM-3 homologue in yeast two-hybrid assays (Haag *et al.*, 2002). That none of the mixed-species pairings did suggested that rapid coevolution was occurring, prompting the author to examine both the theoretical and empirical population genetics of this phenomenon (Haag, 2007; Haag and Ackerman, 2005; Haag and Molla, 2005).

Functional assays also support a conserved interaction between *fem*-3 and *tra-2* products. *fem-3(RNAi)* feminizes the soma of XO animals of both *C. remanei* and *C. briggsae*. Importantly, knocking down both *Cr-fem-3* and *Cr-tra-2* reversed the somatic masculinization of *Cr-tra-2(RNAi)* alone, indicating that despite their molecular divergence they perform similar roles and have similar epistatic relationships that are independent of reproductive mode. However, the one tissue in XO males that was not feminized by *Cr-fem-3(RNAi)* was the germline. Further, *Cr-fem-3(RNAi)* could not suppress the masculinized germline of XX *Cr-tra-2(RNAi)* animals, even though it did reverse somatic phenotypes. Taken together, these results indicated that *C. remanei fem-3* is important for male somatic development, but is not used to regulate germ cell fates.

Although the above results might suggest that *fem-3* translational control would not occur in *C. remanei*, the 3′-UTR of *Cr-fem-3* nevertheless contains a well-conserved point mutation element (Haag *et al.*, 2002), the short sequence known to bind the FBF-1 and FBF-2 proteins in *C. elegans* (Ahringer and Kimble, 1991; Zhang *et al.*, 1997). Similar to the case with *Cr-tra-2*, we see that translational controls *per se* probably preceded the evolution of self-fertility, though they may have been modified in hermaphrodite lineages. As translational control of both *tra-2* and *fem-3* occurs in the *C. elegans* soma (Gallegos *et al.*, 1998; Jan *et al.*, 1997), this may be their original site of action in gonochoristic species.

So, what evidence is there that hermaphrodites do have unique translational controls that act on sex-determination genes? The most compelling so far is the case of *fog-2*. Mutant *C. elegans* hermaphrodites lacking *fog-2* activity are converted into true females, yet homozygous males make copious sperm (Schedl and Kimble, 1988). *fog-2* was cloned when its F-box protein product was found as an interactor of the RBP GLD-1 (Clifford *et al.*, 2000). GLD-1 had previously been identified as a major component of DRF, the repressor of *tra-2* translation (Jan *et al.*, 1999). Interestingly, *fog-2* is the recent product of recent tandem duplications. Nayak *et al.* (2005) expanded on this initial observation by showing that FOG-2 is part of a large family of F-box-containing proteins, and that the entire *C. elegans* gene family coalesces to a common ancestral gene that is younger than the time at which *C. elegans* split from the lineage it shared with *C. briggsae*. Further, Nayak *et al.* demonstrated that only FOG-2, and not its paralogs, has the C-terminal sequences necessary to mediate an interaction with GLD-1. Taken together, *fog-2* is a lineage-specific gene with a new function in germline sex that is required to make a hermaphrodite a hermaphrodite. It is therefore likely that the evolution of *fog-2* was a key step in the evolution of XX spermatogenesis in the *C. elegans* lineage.

7. Are There Really 50 Ways to Leave Your Lover?

Another asset of the *Caenorhabditis* system is the existence of at least two outwardly similar hermaphroditic species, *C. elegans* and *C. briggsae*, which are inferred from phylogenies to be independently evolved (Cho *et al.*, 2004; Kiontke and Fitch, 2005; Kiontke *et al.*, 2004; Fig. 3.3). This enables us to examine how reproducible the evolution of XX spermatogenesis is at the level of developmental genetics. Although the convergent acquisition of selfing was not known at the time, some of the earliest gene homologues to be characterized in non-*elegans Caenorhabditis* species were components of the *C. briggsae* sex-determination pathway (Chen *et al.*, 2001; de Bono and Hodgkin, 1996; Haag *et al.*, 2002; Hansen and Pilgrim, 1998; Kuwabara, 1996; Streit *et al.*, 1999). These studies found that sequence conservation was generally lower than for typical orthologous pairs (Stein *et al.*, 2003), ranging from roughly one- to two-thirds amino acid identity (summarized by Haag, 2005; Nayak *et al.*, 2005). Nevertheless, using cross-species transgenic rescue assays and RNA interference methods, these studies generally found that sex-determination functions were conserved. A notable exception, however, was seen in the *Cb-fem-2* and *Cb-fem-3* genes, which could not be implicated in germline sex determination using these assays (see also Stothard *et al.*, 2002). These results are reminiscent of those for *C. remanei* described above, in that the germline function of the *fem* genes emerges as an exception to more general conservation.

Though considered cutting edge at the time, neither RNAi nor cross-species transgenes produce completely penetrant phenotypes. As a result, doubt remained whether the unexpected results for the *C. briggsae fem* homologues were due to true functional differences or to technical limitations of the method. To provide the same standard of proof used in *C. elegans*, the author and his coworkers have developed mutational methods in *C. briggsae* (see Table 3.1 for summary). We began by following the historically successful approach (Hodgkin and Brenner, 1977) of screening for masculinized (Tra) mutants among mutagenized *C. briggsae* animals. This work identified multiple mutant alleles of the homologues of the three known *tra* loci, *Cb-tra-1, Cb-tra-2*, and *Cb-tra-3*, including conditional alleles of the latter two (Kelleher *et al.*, 2008). The phenotypes of these mutants are generally congruent with those of their *C. elegans* equivalents, and specifically they cause complete germline masculinization.

As noted above, the *Cb-fem* genes were the ones that showed unexpected germline phenotypes in knockdown and rescue experiments. To identify true mutations in these genes, we took two approaches. One was to screen for suppressors of the Tra phenotype of *Cb-tra-2(ts)* and *Cb-tra-3(ts)* at nonpermissive temperature, similar to earlier work in *C. elegans* (Hodgkin, 1986). Using two different alleles of *Cb-tra-2*, 75 different alleles were isolated that reversed the somatic masculinization of XX *Cb-tra-2(ts)* mutations (Hill *et al.*, 2006). Interestingly, none of these mutations produced true females, as their *C. elegans* equivalents would, but instead converted the Tra pseudomales into self-fertile hermaphrodites. However, as provocative as these results were, the identities of the suppressors and the nature of their molecular lesions remained unknown.

In a more direct approach, PCR-based screens for deletion mutations were used to isolate null alleles of *Cb-fem-2* and *Cb-fem-3* (Hill *et al.*, 2006). Confirming previous RNAi studies, both of these mutations had no effect on XX hermaphrodites. Further, XO homozygotes are converted into self-fertile hermaphrodites. In contrast, in *C. elegans* both XX and XO *fem* homozygotes are converted into true females. Thus, while both males and hermaphrodites require the *fem* genes for spermatogenesis in *C. elegans*, in *C. briggsae* the only germline function of the *fem* genes appears to be to prevent males from switching to oogenesis. Overall, the ability to produce mutations in *C. briggsae* sex-determination genes delivers a new level of precision to the analysis of developmental evolution *Caenorhabditis*. They enable us to infer with considerable confidence that the genetic control of

Table 3.1 Summary of functional characterization of *C. briggsae* sex-determination genes

Gene	*C. elegans* mutant phenotype (*lf*)	*C. briggsae* RNAi phenotype	*C. briggsae* transgene in *C. elegans*	*C. briggsae* mutant phenotype	References
her-1	XO: Her XX: no effect	XO: weak Her XX: no effect	[*Punc-54::Cb-*HER-1] XX: Tra XO: ND	ND	Hodgkin (1980) and Streit *et al.* (1999)
tra-2	XO: no effect XX: imperfect Tra	XO: ND XX: weak Tra	ND	XO: no effect XX: imperfect Tra	Hodgkin and Brenner (1977), Kelleher *et al.* (2008), and Kuwabara (1996)
tra-3	XO: no effect XX: imperfect Tra, maternally rescued	XO: ND XX: no effect	ND	XO: no effect XX: imperfect Tra, maternally rescued	Hodgkin and Brenner (1977) and Kelleher *et al.* (2008)
fem-2	XO: Fem XX: Fem	XO: germline feminized, soma intersex XX: no effect	Somatic rescue of Fem phenotype in XO *fem-2(lf)*, no rescue of germline Fem phenotype in XX or XO	XO: Her XX: no effect	Hansen and Pilgrim (1998), Hill *et al.* (2006), Kimble *et al.* (1984), and Stothard *et al.* (2002)
fem-3	XO: Fem XX: Fem	XO: weak Fem XX: no effect	ND	XO: Her XX: no effect	Haag *et al.* (2002) and Hodgkin (1986)

(continued)

Table 3.1 (*continued*)

Gene	*C. elegans* mutant phenotype (*lf*)	*C. briggsae* RNAi phenotype	*C. briggsae* transgene in *C. elegans*	*C. briggsae* mutant phenotype	References
tra-1	XX: Tra soma, intersexual germline XO: male soma, intersexual germline	XO: germline feminization XX: intersex	Rescues nongonadal soma of XX *tra-1* mutants; feminizes wild-type XO animals	XO: intersexual germline XX: Tra soma, intersexual germline	de Bono and Hodgkin (1996), Hodgkin and Brenner (1977), and Kelleher *et al.* (2008)
fog-3	XO: Fog XX: Fog	XO: Fog XX: Fog	Rescues Fog	ND	Chen *et al.* (2001) and Ellis and Kimble (1995)
gld-1	XO: no effect XX: Fog, tumorous	XO: ND XX: Mog	ND	XO: no effect XX: Mog, tumorous	A. Doty and E.S. Haag (unpublished data), Francis *et al.* (1995a), and Nayak *et al.* (2005)

hermaphrodite germline development is fundamentally different in *C. elegans* and *C. briggsae*, and more specifically that the locus of regulation of XX spermatogenesis in *C. briggsae* probably lies downstream of the *Cb-fem* genes. In combination with the parsimonious reading of current phylogenies (Cho *et al.*, 2004; Kiontke *et al.*, 2004; Fig. 3.3), these results further indicate that nearly identical germline phenotypes have evolved using distinct genetic paths. The general lesson here is that within the general constraints imposed by the sex-determination pathway, considerable flexibility exists in how adaptation can occur.

8. Evolutionary Dynamics of Germline RNA-Binding Proteins

The above synopsis makes clear that the global *Caenorhabditis* sex pathway, while subject to rapid sequence evolution, is generally intact in all species examined thus far. With the exception of *fog-2*, however, little has been said about the germline-specific regulators shown in Fig. 3.2. Although less is known here, it already appears that germline-specific sex-determination genes are often well conserved at the protein level, yet have evolutionary dynamics that go beyond point mutation, including duplication, divergence in functional domains, and cooption into new roles. FOG-2 has all of these attributes, and an RNAi study of its binding partner, GLD-1, suggests that it too presents surprises (Nayak *et al.*, 2005). While *C. briggsae* GLD-1 is very similar at the amino acid level to its *C. elegans* homologue (Haag, 2005), *Cb-gld-1(RNAi)* has a phenotype that is opposite. Specifically, while reduction in *C. elegans gld-1* causes loss of XX spermatogenesis (presumably because of failure to translationally repress *tra-2* translation), *Cb-gld-1(RNAi)* causes germline masculinization (Nayak *et al.*, 2005). Aided by this result, two strong loss-of-function alleles of *Cb-gld-1* have been identified in forward screens for *C. briggsae* Mog mutants (A. Doty and E.S. Haag, unpublished data). This confirms the different roles of *gld-1* in germline sex determination of *C. elegans* and *C. briggsae*.

fog-2 is not the only germline sex-determination gene that is the product of lineage-specific gene duplication. In *C. elegans*, FBF is encoded by two nearly identical genes that are the product of a recent duplication (Zhang *et al.*, 1997). In *C. briggsae*, the closest PUF family relatives of FBF are encoded by a three-member clade of similarly duplicated genes (Lamont *et al.*, 2004; discussed in Haag, 2005). Recent work in the author's lab suggests that these genes also have unexpected functions in germ cell sex determination, as well as in other processes (Q. Liu and E. S. Haag, unpublished data).

9. Challenges and Future Directions

This chapter has demonstrated that the bulk of functional divergence in the *Caenorhabditis* sex-determination pathway lies in the germline. As this is the same tissue that undergoes the most dramatic phenotypic evolution, this is perhaps not surprising. However, much of this divergence may be due to inherently dynamic evolution of germline regulators, and not be specifically related to adaptive shifts in phenotype (see True and Haag, 2001 for further discussion). To identify the subset of changes responsible for germline sex-determination adaptation, we must first recognize several challenges. First, RBPs are often pleiotropic and have many targets, with the result that their loss-of-function phenotypes are often complex. For example, most of the germline-specific sex regulators discussed in this chapter have other phenotypes when inactivated, such as cell cycle defects and embryonic lethality (e.g., Crittenden *et al.*, 2002; Francis *et al.*, 1995a; Graham *et al.*, 1993). For translation-regulating RBPs, this may be the manifestation of a large number of target mRNAs. Second, these sexual regulators are often encoded by members of gene families, in which members may have either similar or dissimilar functions. Therefore, both redundancy and unexpectedly paralog-specific phenotypes could emerge, and we see evidence for both in our ongoing studies. Third, *in vivo* assays for translational control are technically more difficult than those for transcriptional control, and are even harder in the germ line due to transgene silencing.

While the above challenges are indeed rather daunting, we can still make progress. For example, it is likely that the different phenotypes of otherwise conserved RBPs are due to evolutionary changes in target mRNAs. By extending the same sort of systematic characterization of RBP target mRNAs that has been done in *C. elegans* (e.g., Lee and Schedl, 2001) to other species, species-specific targets could be discovered. With respect to redundancy, we do appear to be fortunate in that RNAi by injection produces fairly reliable germline phenotypes in *C. briggsae* (e.g., compare the results of Haag *et al.*, 2002; Stothard *et al.*, 2002 with those of Hill *et al.*, 2006). This allows rapid searches for synthetic phenotypes via double RNAi experiments. Another key method will be production of transgenes that express well in germ cells. The most reliable method currently in *C. elegans* is based on particle bombardment of DNA constructs into an *unc-119* mutant strain (Praitis *et al.*, 2001), and we have recently identified the equivalent mutant in *C. briggsae* (C. Thomas and E. S. Haag, unpublished data). Finally, the ongoing discovery of new *Caenorhabditis* species, in particular by M. A. Félix (Institut Jacques Monod, Paris), is opening up the possibility of using hybrids between hermaphroditic and gonochoristic species as a new route to understanding how XX spermatogenesis evolves

(M.A. Félix, G. Woodruff, and E.S. Haag, unpublished data). Overall, it is fair to say that *Caenorhabditis* is maturing into a sophisticated model metasystem for probing the genetic basic of germ cell adaptations.

ACKNOWLEDGMENTS

The author thanks those who contributed images and unpublished results to this chapter. He also thanks members of his laboratory, R. Ellis, and T. Schedl for useful discussions about some of the ideas presented here. Research in the author's lab is supported by the generous support of the National Institute of General Medical Sciences (1R01GM079414).

REFERENCES

Ahringer, J., and Kimble, J. (1991). Control of the sperm–oocyte switch in *Caenorhabditis elegans* hermaphrodites by the *fem-3* 3′ untranslated region. *Nature* **349,** 346–348.

Barton, M., and Kimble, J. (1990). *fog-1*, a regulatory gene required for specification of spermatogenesis in the germ line of *Caenorhabditis elegans*. *Genetics* **125,** 29–39.

Barton, M. K., Schedl, T. B., and Kimble, J. (1987). Gain-of-function mutations of *fem-3*, a sex-determination gene in *Caenorhabditis elegans*. *Genetics* **115,** 107–119.

Belfiore, M., Pugnale, P., Saudan, Z., and Puoti, A. (2004). Roles of the *C. elegans* cyclophilin-like protein MOG-6 in MEP-1 binding and germline fates. *Development* **131,** 2935–2945.

Bjork, A., and Pitnick, S. (2006). Intensity of sexual selection along the anisogamy–isogamy continuum. *Nature* **441,** 742–745.

Bonner, J. (Ed.), (1981). Evolution and development (Report of the Dahlem Workshop). *In* S. Bernhard, (Series Ed.), "Life Sciences Research Reports". Springer–Verlag, Berlin.

Braendle, C., and Felix, M. A. (2006). Sex determination: Ways to evolve a hermaphrodite. *Curr. Biol.* **16,** R468–R471.

Byrne, M., Villinski, J. T., Cisternas, P., Siegel, R. K., Popodi, E., and Raff, R. A. (1999). Maternal factors and the evolution of developmental mode: Evolution of oogenesis in *Heliocidaris erythrogramma*. *Dev. Genes Evol.* **209,** 275–283.

Byrne, M., Prowse, T., Sewell, M., Dworjanyn, S., and Williamson, J. (2008). Maternal provisioning for larvae and larval provisioning for juveniles in the toxopneustid sea urchin *Tripneustes gratilla*. *Mar. Biol.* **155,** 473–482.

Callery, E. M., Fang, H., and Elinson, R. P. (2001). Frogs without polliwogs: Evolution of anuran direct development. *Bioessays* **23,** 233–241.

Carroll, S. B. (2008). Evo-devo and an expanding evolutionary synthesis: A genetic theory of morphological evolution. *Cell* **134,** 25–36.

Chen, P., and Ellis, R. E. (2000). TRA-1A regulates transcription of fog-3, which controls germ cell fate in *C. elegans*. *Development* **127,** 3119–3129.

Chen, P., Cho, S., Jin, S., and Ellis, R. (2001). Specification of germ cell fates by FOG-3 has been conserved during nematode evolution. *Genetics* **158,** 1513–1525.

Chin-Sang, I. D., and Spence, A. M. (1996). *Caenorhabditis elegans* sex-determining protein FEM-2 is a protein phosphatase that promotes male development and interacts directly with FEM-3. *Genes Dev.* **10,** 2314–2325.

Cho, S., Jin, S. W., Cohen, A., and Ellis, R. E. (2004). A phylogeny of *Caenorhabditis* reveals frequent loss of introns during nematode evolution. *Genome Res.* **14,** 1207–1220.

Cho, S., Rogers, K. W., and Fay, D. S. (2007). The *C. elegans* glycopeptide hormone receptor ortholog, FSHR-1, regulates germline differentiation and survival. *Curr. Biol.* **17,** 203–212.

Chu, D. S., Dawes, H. E., Lieb, J. D., Chan, R. C., Kuo, A. F., and Meyer, B. J. (2002). A molecular link between gene-specific and chromosome-wide transcriptional repression. *Genes Dev.* **16,** 796–805.

Clack, J. (2002). "Gaining Ground: The Origin and Evolution of Tetrapods." Indiana University Press, Bloomington, IN.

Clifford, R., Lee, M., Nayak, S., Ohmachi, M., Giorgini, F., and Schedl, T. (2000). FOG-2, a novel F-box-containing protein, associates with the GLD-1 RNA-binding protein and directs male sex determination in the *C. elegans* hermaphrodite germline. *Development* **127,** 5265–5276.

Crittenden, S. L., Bernstein, D. S., Bachorik, J. L., Thompson, B. E., Gallegos, M., Petcherski, A. G., Moulder, G., Barstead, R., Wickens, M., and Kimble, J. (2002). A conserved RNA-binding protein controls germline stem cells in *Caenorhabditis elegans*. *Nature* **417,** 660–663.

Cuellar, O. (1971). Reproduction and the mechanism of meiotic restitution in the parthenogenetic lizard *Cnemidophorus uniparens*. *J. Morphol.* **133,** 139–165.

de Bono, M., and Hodgkin, J. (1996). Evolution of sex determination in *Caenorhabditis*: Unusually high divergence of *tra-1* and its functional consequences. *Genetics* **144,** 587–595.

Doniach, T., and Hodgkin, J. (1984). A sex-determining gene, fem-1, required for both male and hermaphrodite development in *Caenorhabditis elegans*. *Dev. Biol.* **106,** 223–235.

Elinson, R. P. (1987). Fertilization and aqueous development of the Puerto-Rican terrestrial-breeding frog, *Eleutherodactylus coqui*. *J. Morphol.* **193,** 217–224.

Elinson, R. P., and Beckham, Y. (2002). Development in frogs with large eggs and the origin of amniotes. *Zoology (Jena)* **105,** 105–117.

Ellis, R., and Kimble, J. (1995). The *fog-3* gene and regulation of cell fate in the germ line of *Caenorhabditis elegans*. *Genetics* **139,** 561–577.

Ellis, R., and Schedl, T. (2007). Sex determination in the germ line (March 5, 2007). *In* "WormBook" (The *C. elegans* Research Community, Ed.), pp. 1-13.

Emlet, R., and Hoegh-Guldberg, O. (1996). Effects of egg size on postlarval performance: Experimental evidence from a Sea Urchin. *Evolution* **51,** 141–152.

Evans, T., and Hunter, C. (2005). "Translational Control of Maternal RNAs" (10 November 2005).

Francis, R., Barton, M. K., Kimble, J., and Schedl, T. (1995a). *gld-1,* a tumor suppressor gene required for oocyte development in *Caenorhabditis elegans*. *Genetics* **139,** 579–606.

Francis, R., Maine, E., and Schedl, T. (1995b). Analysis of the multiple roles of *gld-1* in germline development: Interactions with the sex determination cascade and the *glp-1* signaling pathway. *Genetics* **139,** 607–630.

Gallegos, M., Ahringer, J., Crittenden, S., and Kimble, J. (1998). Repression by the 3′UTR of *fem-3*, a sex-determining gene, relies on a ubiquitous *mog*-dependent control in *Caenorhabditis elegans*. *EMBO J.* **17,** 6337–6347.

Goodwin, E. B., Okkema, P. G., Evans, T. C., and Kimble, J. (1993). Translational regulation of *tra-2* by its 3′ untranslated region controls sexual identity in *C. elegans*. *Cell* **75,** 329–339.

Goodwin, E. B., Hofstra, K., Hurney, C. A., Mango, S., and Kimble, J. (1997). A genetic pathway for regulation of *tra-2* translation. *Development* **124,** 749–758.

Graham, P. L., Schedl, T., and Kimble, J. (1993). More *mog* genes that influence the switch from spermatogenesis to oogenesis in the hermaphrodite germ line of *Caenorhabditis elegans*. *Dev. Genet.* **14,** 471–484.

Haag, E. (2005). The evolution of nematode sex determination: *C. elegans* as a reference point for comparative biology. *In* "WormBook" (The *C. elegans* Research Community, Ed.).

Haag, E. S. (2007). Compensatory vs. pseudocompensatory evolution in molecular and developmental interactions. *Genetica* **129,** 45–55.

Haag, E. S., and Ackerman, A. D. (2005). Intraspecific variation in fem-3 and tra-2, two rapidly coevolving nematode sex-determining genes. *Gene* **349,** 35–42.

Haag, E. S., and Molla, M. N. (2005). Compensatory evolution of interacting gene products through multifunctional intermediates. *Evolution* **59,** 1620–1632.

Haag, E. S., Sly, B. J., Andrews, M. E., and Raff, R. A. (1999). Apextrin, a novel extracellular protein associated with larval ectoderm evolution in *Heliocidaris erythrogramma. Dev. Biol.* **211,** 77–87.

Haag, E., and Kimble, J. (2000). Regulatory elements required for development of *C. elegans* hermaphrodites are conserved in the *tra-2* homologue of *C. remanei*, a male/female sister species. *Genetics* **155,** 105-116.

Haag, E. S., Wang, S., and Kimble, J. (2002). Rapid coevolution of the nematode sex-determining genes *fem-3* and *tra-2. Curr. Biol.* **12,** 2035–2041.

Hamaoka, B. Y., Dann, C. E. III, Geisbrecht, B. V., and Leahy, D. J. (2004). Crystal structure of *Caenorhabditis elegans* HER-1 and characterization of the interaction between HER-1 and TRA-2A. *Proc. Natl. Acad. Sci. USA* **101,** 11673–11678.

Hansen, D., and Pilgrim, D. (1998). Molecular evolution of a sex determination protein. FEM-2 (pp2c) in *Caenorhabditis. Genetics* **149,** 1353–1362.

Henry, J. J., and Raff, R. A. (1990). Evolutionary change in the process of dorsoventral axis determination in the direct developing sea urchin, *Heliocidaris erythrogramma. Dev. Biol.* **141,** 55–69.

Henry, J. J., Wray, G. A., and Raff, R. A. (1990). The dorsoventral axis is specified prior to first cleavage in the direct developing sea urchin *Heliocidaris erythrogramma. Development* **110,** 875–884.

Henry, J. J., Wray, G. A., and Raff, R. A. (1991). Mechanism of an alternate type of echinoderm blastula formation: The wrinkled blastula of the sea urchin *Heliocidaris erythrogramma. Dev. Growth Differ.* **33,** 317–328.

Hill, R. C., de Carvalho, C. E., Salogiannis, J., Schlager, B., Pilgrim, D., and Haag, E. S. (2006). Genetic flexibility in the convergent evolution of hermaphroditism in *Caenorhabditis* nematodes. *Dev. Cell* **10,** 531–538.

Hodgkin, J. (1980). More sex-determination mutants of *Caenorhabditis elegans. Genetics* **96,** 649–664.

Hodgkin, J. (1986). Sex determination in the nematode *C. elegans*: Analysis of *tra-3* suppressors and characterization of *fem* genes. *Genetics* **114,** 15–52.

Hodgkin, J. (1987). A genetic analysis of the sex-determining gene, *tra-1*, in the nematode *Caenorhabditis elegans. Genes Dev.* **1,** 731–745.

Hodgkin, J. A., and Brenner, S. (1977). Mutations causing transformation of sexual phenotype in the nematode *Caenorhabditis elegans. Genetics* **86,** 275–287.

Hoekstra, H. E., and Coyne, J. A. (2007). The locus of evolution: Evo devo and the genetics of adaptation. *Evolution* **61,** 995–1016.

Hunter, C. P., and Wood, W. B. (1992). Evidence from mosaic analysis of the masculinizing gene her-1 for cell interactions in *C. elegans* sex determination. *Nature* **355,** 551–555.

Jan, E., Yoon, J. W., Walterhouse, D., Iannaccone, P., and Goodwin, E. B. (1997). Conservation of the *C. elegans tra-2* 3′YTP translational control. *EMBO J.* **16,** 6301–6313.

Jan, E., Motzny, C. K., Graves, L. E., and Goodwin, E. B. (1999). The STAR protein, GLD-1, is a translational regulator of sexual identity in *Caenorhabditis elegans. EMBO J.* **18,** 258–269.

Jarne, P., and Auld, J. R. (2006). Animals mix it up too: The distribution of self-fertilization among hermaphroditic animals. *Evolution* **60,** 1816–1824.

Jin, S. W., Kimble, J., and Ellis, R. E. (2001). Regulation of cell fate in *Caenorhabditis elegans* by a novel cytoplasmic polyadenylation element binding protein. *Dev. Biol.* **229,** 537–553.

Jones, A. R., and Schedl, T. (1995). Mutations in *gld-1*, a female germ cell-specific tumor suppressor gene in *Caenorhabditis elegans*, affect a conserved domain also found in Src-associated protein Sam68. *Genes Dev.* **9,** 1491–1504.

Kelleher, D. F., de Carvalho, C. E., Doty, A. V., Layton, M., Cheng, A. T., Mathies, L. D., Pilgrim, D., and Haag, E. S. (2008). Comparative genetics of sex determination: Masculinizing mutations in *Caenorhabditis briggsae*. *Genetics* **178,** 1415–1429.

Kelly, W. G., and Fire, A. (1998). Chromatin silencing and the maintenance of a functional germline in *Caenorhabditis elegans*. *Development* **125,** 2451–2456.

Kimble, J., and Crittenden, S. L. (2007). Controls of germline stem cells, entry into meiosis, and the sperm/oocyte decision in *Caenorhabditis elegans*. *Annu. Rev. Cell Dev. Biol.* **23,** 405–433.

Kimble, J., Edgar, L., and Hirsh, D. (1984). Specification of male development in *Caenorhabditis elegans*: The *fem* genes. *Dev. Biol.* **105,** 234–239.

Kiontke, K., and Fitch, D. (2005). The phylogenetic relationships of *Caenorhabditis* and other rhabditids. *In* "WormBook: The Online Review of *C. elegans* Biology" (The *C. elegans* Research Community. Ed.).

Kiontke, K., and Sudhaus, W. (2006). Ecology of *Caenorhabditis* species (January 2006). *In* "WormBook" (The *C. elegans* Research Community. Ed.).

Kiontke, K., Gavin, N. P., Raynes, Y., Roehrig, C., Piano, F., and Fitch, D. H. (2004). *Caenorhabditis* phylogeny predicts convergence of hermaphroditism and extensive intron loss. *Proc. Natl. Acad. Sci. USA* **101,** 9003–9008.

Kraemer, B., Crittenden, S., Gallegos, M., Moulder, G., Barstead, R., Kimble, J., and Wickens, M. (1999). NANOS-3 and FBF proteins physically interact to control the sperm–oocyte switch in *Caenorhabditis elegans*. *Curr. Biol.* **9,** 1009–1018.

Kuwabara, P. E. (1996). Interspecies comparison reveals evolution of control regions in the nematode sex-determining gene *tra-2*. *Genetics* **144,** 597–607.

Kuwabara, P. E., and Shah, S. (1994). Cloning by synteny: Identifying *C. briggsae* homologues of *C. elegans* genes. *Nucleic Acids Res.* **22,** 4414–4418.

Lamont, L. B., Crittenden, S. L., Bernstein, D., Wickens, M., and Kimble, J. (2004). FBF-1 and FBF-2 regulate the size of the mitotic region in the *C. elegans* germline. *Dev. Cell* **7,** 697–707.

Lee, M. H., and Schedl, T. (2001). Identification of *in vivo* mRNA targets of GLD-1, a maxi-KH motif containing protein required for *C. elegans* germ cell development. *Genes Dev.* **15,** 2408–2420.

Luitjens, C., Gallegos, M., Kraemer, B., Kimble, J., and Wickens, M. (2000). CPEB proteins control two key steps in spermatogenesis in *C. elegans*. *Genes Dev.* **14,** 2596–2609.

Macgregor, H. C., and Uzzell, T. M. Jr., (1964). Gynogenesis in salamanders related to *Ambystoma jeffersonianum*. *Science* **143,** 1043–1045.

McCarter, J., Bartlett, B., Dang, T., and Schedl, T. (1997). Soma-germ cell interactions in *Caenorhabditis elegans*: Multiple events of hermaphrodite germline development require the somatic sheath and spermathecal lineages. *Dev. Biol.* **181,** 121–143.

Mehra, A., Gaudet, J., Heck, L., Kuwabara, P. E., and Spence, A. M. (1999). Negative regulation of male development in *Caenorhabditis elegans* by a protein–protein interaction between TRA-2A and FEM-3. *Genes Dev.* **13,** 1453–1463.

Meyer, B. (2005). X-chromosome dosage compensation (25 June 2005). *In* "WormBook" (The *C. elegans* Research Community. Ed.).

Miller, G. T., and Pitnick, S. (2002). Sperm–female coevolution in *Drosophila*. *Science* **298,** 1230–1233.

Nayak, S., Goree, J., and Schedl, T. (2005). *fog-2* and the evolution of self-fertile hermaphroditism in *Caenorhabditis*. *PLoS Biol.* **3,** e6.

Østerlund, T., and Kogerman, P. (2006). Hedgehog signalling: How to get from Smo to Ci and Gli. *Trends Cell Biol.* **16,** 176–180.

Otori, M., Karashima, T., and Yamamoto, M. (2006). The *Caenorhabditis elegans* homologue of deleted in azoospermia is involved in the sperm/oocyte switch. *Mol. Biol. Cell* **17,** 3147–3155.

Packard, M., and Seymour, R. (1997). Evolution of the amniote egg. *In* "Amniote Origins: Completing the Transition to Land" (S. Sumida and K. Martin, Eds.). Academic Press, San Diego, CA.

Piqué, M., Lopez, J. M., Foissac, S., Guigo, R., and Méndez, R. (2008). A combinatorial code for CPE-mediated translational control. *Cell* **132,** 434–448.

Praitis, V., Casey, E., Collar, D., and Austin, J. (2001). Creation of low-copy integrated transgenic lines in *Caenorhabditis elegans*. *Genetics* **157,** 1217–1226.

Prowse, T., Sewell, M., and Byrne, M. (2008). Fuels for development: Evolution of maternal provisioning in asterinid sea stars. *Mar. Biol.* **153,** 337–349.

Prud'homme, B., Gompel, N., and Carroll, S. B. (2007). Emerging principles of regulatory evolution. *Proc. Natl. Acad. Sci. USA* **104**(Suppl. 1), 8605–8612.

Puoti, A., and Kimble, J. (1999). The *Caenorhabditis elegans* sex determination gene *mog-1* encodes a member of the DEAH-Box protein family. *Mol. Cell. Biol.* **19,** 2189–2197.

Puoti, A., and Kimble, J. (2000). The hermaphrodite sperm/oocyte switch requires the *Caenorhabditis elegans* homologs of PRP2 and PRP22. *Proc. Natl. Acad. Sci. USA* **97,** 3276–3281.

Schaner, C. E., and Kelly, W. G. (2006). Germline chromatin. *In* "WormBook" (The *C. elegans* Research Community, Ed.), pp. 1–14.

Schedl, T., and Kimble, J. (1988). *fog-2*, a germ-line-specific sex determination gene required for hermaphrodite spermatogenesis in *Caenorhabditis elegans*. *Genetics* **119,** 43–61.

Schvarzstein, M., and Spence, A. M. (2006). The *C. elegans* sex-determining GLI protein TRA-1A is regulated by sex-specific proteolysis. *Dev. Cell* **11,** 733–740.

Seydoux, G., and Braun, R. E. (2006). Pathway to totipotency: Lessons from germ cells. *Cell* **127,** 891–904.

Sokol, S., and Kuwabara, P. (2000). Proteolysis in *Caenorhabditis elegans* sex determination: Cleavage of TRA-2A by TRA-3. *Genes Dev.* **14,** 901–906.

Standart, N., and Minshall, N. (2008). Translational control in early development: CPEB, P-bodies and germinal granules. *Biochem. Soc. Trans.* **36,** 671–676.

Starostina, N. G., Lim, J. M., Schvarzstein, M., Wells, L., Spence, A. M., and Kipreos, E. T. (2007). A CUL-2 ubiquitin ligase containing three FEM proteins degrades TRA-1 to regulate *C. elegans* sex determination. *Dev. Cell* **13,** 127–139.

Stein, L. D., Bao, Z., Blasiar, D., Blumenthal, T., Brent, M. R., Chen, N., Chinwalla, A., Clarke, L., Clee, C., Coghlan, A., Coulson, A., D'Eustachio, P., *et al.* (2003). The genome sequence of *Caenorhabditis briggsae*: A platform for comparative genomics. *PLoS Biol.* **1,** 166–192.

Stewart, J. (1997). Morphology and evolution of the egg of oviparous amniotes. *In* "Amniote Origins: Completing the Transition to Land" (S. Sumida and K. Martin, Eds.), Academic Press, San Diego, CA.

Stothard, P., Hansen, D., and Pilgrim, D. (2002). Evolution of the PP2C family in *Caenorhabditis*: Rapid divergence of the sex-determining protein FEM-2. *J. Mol. Evol.* **54,** 267–282.

Streit, A., Li, W., Robertson, B., Schein, J., Kamal, I., Marra, M., and Wood, W. (1999). Homologs of the *Caenorhabditis elegans* masculinizing gene *her-1* in *C. briggsae* and the filarial parasite *Brugia malayi*. *Genetics* **152,** 1573–1584.

Sudhaus, W., and Kiontke, K. (2007). Comparison of the cryptic nematode species *Caenorhabditis brenneri* sp. n. and *C. remanei* (Nematoda: Rhabditidae) with the stem species pattern of the *Caenorhabditis elegans* group. *Zootaxa* **1456,** 45–62.

Tan, K. M., Chan, S. L., Tan, K. O., and Yu, V. C. (2001). The *Caenorhabditis elegans* sex-determining protein FEM-2 and its human homologue, hFEM-2, are Ca^{2+}/calmodulin-dependent protein kinase phosphatases that promote apoptosis. *J. Biol. Chem.* **276,** 44193–44202.

Trent, C., Purnell, B., Gavinski, S., Hageman, J., Chamblin, C., and Wood, W. B. (1991). Sex-specific transcriptional regulation of the *C. elegans* sex-determining gene her-1. *Mech. Dev.* **34,** 43–55.

True, J. R., and Haag, E. S. (2001). Developmental system drift and flexibility in evolutionary trajectories. *Evol. Dev.* **3,** 109–119.

Uzzell, T. (1970). Meiotic mechanisms of naturally occurring unisexual vertebrates. *Am. Nat.* **104,** 433–445.

Villinski, J. T., Villinski, J. C., Byrne, M., and Raff, R. A. (2002). Convergent maternal provisioning and life-history evolution in echinoderms. *Evolution* **56,** 1764–1775.

Wickens, M., Bernstein, D. S., Kimble, J., and Parker, R. (2002). A PUF family portrait: 3′YTP regulation as a way of life. *Trends Genet.* **18,** 150–157.

Williams, D., and Anderson, D. (1975). The reproductive system, embryonic development, larval development, and metamorphosis of the sea urchin *Heliocidaris erythrogramma* (Val.) (Echinoidea: Echinometridae). *Aust. J. Zool.* **23,** 371–403.

Zarkower, D., and Hodgkin, J. (1992). Molecular analysis of the *C. elegans* sex-determining gene tra-1: A gene encoding two zinc finger proteins. *Cell* **70,** 237–249.

Zhang, B., Gallegos, M., Puoti, A., Durkin, A., Fields, S., Kimble, J., and Wickens, M. P. (1997). A conserved RNA binding protein that regulates sexual fates in the *C. elegans* hermaphrodite germ line. *Nature* **390,** 477–484.

CHAPTER FOUR

New Model Systems for the Study of Developmental Evolution in Plants

Elena M. Kramer

Contents

Abstract

The number of genetically tractable plant model systems is rapidly increasing, thanks to the decreasing cost of sequencing and the wide amenability of plants to stable transformation and other functional approaches. In this chapter, I discuss emerging model systems from throughout the land plant phylogeny and consider how their unique attributes are contributing to our understanding

Department of Organismic and Evolutionary Biology, Harvard University, Cambridge, Massachusetts, USA

Current Topics in Developmental Biology, Volume 86
ISSN 0070-2153, DOI: 10.1016/S0070-2153(09)01004-7

of development, evolution, and ecology. These new models are being developed using two distinct strategies: in some cases, they are selected because of their close relationship to the established models, while in others, they are chosen with the explicit intention of exploring distantly related plant lineages. Such complementary approaches are yielding exciting new results that shed light on both micro- and macroevolutionary processes in the context of developmental evolution.

1. Introduction

Developing a new genetic model system is not a trivial process and typically involves the collaborative efforts of multiple laboratories. It has become much easier, however, due to a number of technological advances. Perhaps, the most important is the relative speed and decreasing cost of DNA sequencing, which greatly facilitates the generation of important resources such as EST databases and whole-genome sequences (Table 4.1). The ready availability of large amounts of DNA sequence, in turn, makes the generation of polymorphism data much easier and provides extensive information for genetic mapping (Borevitz *et al.*, 2003). Another important aspect of any model system is the capability to do functional tests, whether using stable transformation or transient siRNA. The fact that so many plants are amenable to one or both of these techniques makes them particularly well suited for genetic research (Robertson, 2004; Veluthambi *et al.*, 2003). It remains true, however, that every plant would not make a good model system. Whether due to a large genome size, long generation time or restrictive growth conditions, some plants are simply not good candidates. For these reasons, not to mention limited financial resources, choices need to be made with care so that new models will be as powerful as possible and encompass a wide range of potential research questions.

Following these considerations, a significant number of new genetic models have been drawn from across the entire phylogeny of land plants (Fig. 4.1; Table 4.1), although it is still true that most are concentrated in the flowering plants or angiosperms. Many morphological innovations can be studied using these models but one of the most distinct in comparison to animals is the alternation of generations that is common to all land plants (Fig. 4.2). Plants alternate between diploid (sporophyte) and haploid (gametophyte) generations over the course of their life cycles. These two phases are radically different in their morphology, meaning that the same genome can produce entirely distinct body plants depending on whether it is haploid or diploid. One of the major trends in land plant evolution is the stepwise reduction of the gametophyte phase from the persistent, dominant generation in nonvascular plants to no more than a few cells in the angiosperms.

Table 4.1 Tools and resources

Organism	Resources
Physcomitrella patens	http://www.cosmoss.org/
Selaginella moellendorffii	http://selaginella.genomics.purdue.edu/
Arabidopsis thaliana	http://www.arabidopsis.org/
Legumes	http://www.comparative-legumes.org/
	http://www.bio.indiana.edu/~nsflegume/
	http://www.lotusjaponicus.org/
Populus trichocarpa	http://www.ornl.gov/sci/ipgc/about_the_consortium.htm
	http://www.populus.db.umu.se/
	http://www.populusgenome.info/
Mimulus	http://www.mimulusevolution.org/
	http://openwetware.org/wiki/Mimulus_Community
Solanaceae	http://www.nhm.ac.uk/research-curation/research/projects/solanaceaesource/
	http://www.sgn.cornell.edu/
Asteraceae	http://compgenomics.ucdavis.edu/
Aquilegia	https://www.genome.clemson.edu/activities/projects/aquilegia/
Poaceae	http://www.gramene.org/
Gene indices	http://compbio.dfci.harvard.edu/tgi/plant.html
	http://pgn.cornell.edu/

During the course of land plant diversification, we also see the evolution of numerous features, including multicellular meristems, complex forms of branching, true leaves, seeds, and flowers. The continued development and utilization of diverse plant model systems is helping us to understand the origins and diversification of these morphological features.

2. Lower Land Plants

The early branches of the plant phylogeny (Fig. 4.1)—including mosses, lycophytes, and ferns—are of considerable interest since they hold the potential to help us understand how plants moved onto land. Many morphological innovations are associated with this transition, including the evolution of a multicellular sporophyte, a three-dimensional gametophyte, a cuticle, stomata, and, eventually, true leaves and roots. Studies of the

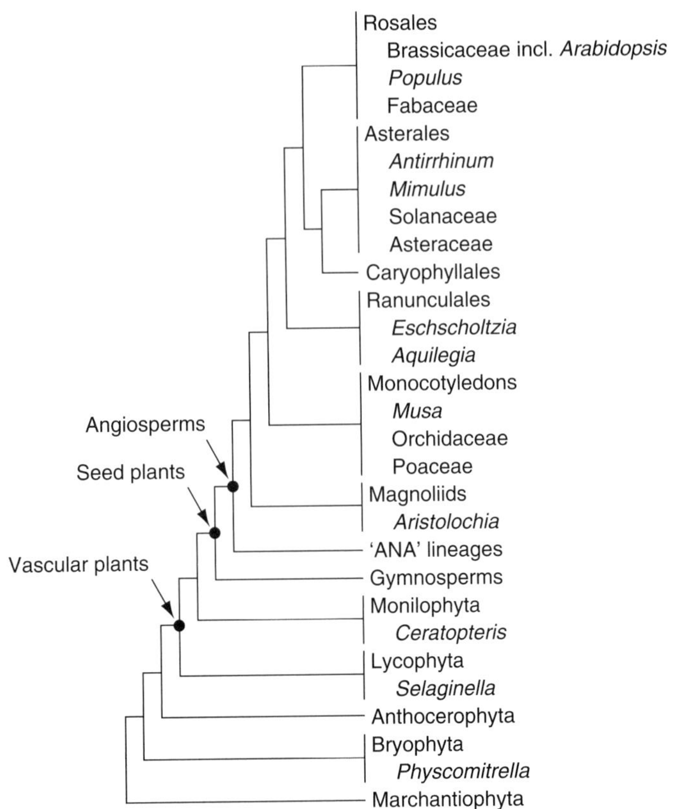

Figure 4.1 Simplified phylogeny of the land plants based on Moore *et al.* (2007) and Qiu *et al.* (2007). Major model systems associated with the various land plant lineages are listed.

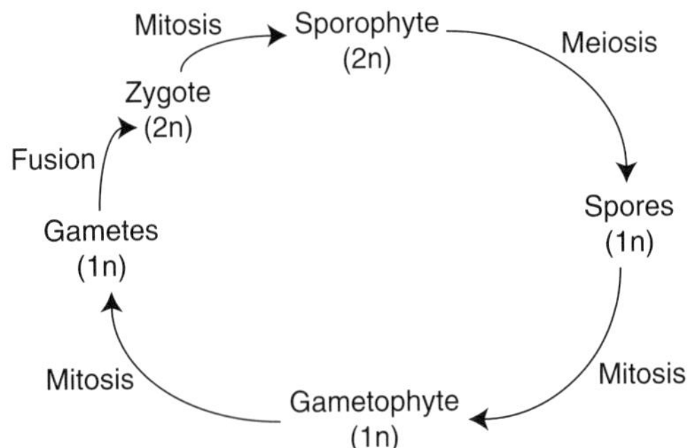

Figure 4.2 A diagram showing the alternation of generations in land plants. The multicellular, diploid sporophyte generation gives rise to haploid spores via meiosis, which germinate and divide to form the multicellular, haploid gametophyte generation. The gametophyte then produces haploid gametes via mitosis that can fuse, reform the diploid zygote and sporophyte.

recently completed genome sequences for *Physcomitrella patens* (a moss) and *Selaginella moellendorffii* (a lycophyte) reveal deep conservation in many developmentally important gene lineages (Floyd and Bowman, 2007; Rensing *et al.*, 2008). Some notable exceptions are transcription factors families such as the YABBYs, which are not found outside of seed plants, and the ARPs, which are limited to vascular plants (Floyd and Bowman, 2007). Both the YABBY and ARP gene lineages play roles in the establishment of organ polarity and the repression of the meristem identity *KNOTTED*-like homeobox or KNOX genes (reviewed in Bowman and Floyd, 2008). Their absence from the early lineages of land plants indicates that these regulatory interactions have evolved more recently, perhaps as a feature of true leaves in the sporophyte. Given the overall genetic conservation across land plants, the question becomes How has gene function evolved in conjunction with morphological innovations? Powerful functional tools in lower land plant models are now helping to answer this question.

2.1. *Physcomitrella*

The incredible utility of *P. patens* for developmental research lies in part in its relative ease of culture and available genome sequence, but it is particularly notable for its amenability to transformation (Cove, 2005; Quatrano *et al.*, 2007). Not only is it possible to perform stable transformation, *P. patens* shows high levels of homologous recombination, similar to yeast systems. This capacity is unique among land plant models to date and provides powerful tools for targeted gene knockout or modification. Current research in *P. patens* has focused on several areas including hormone response pathways and the function of homologs of important seed plant developmental loci. Studying these genetic programs in moss has a unique aspect since work in seed plants typically focuses on the sporophyte while the dominant life stage of the moss is the gametophyte. In terms of hormones, the auxin, abscisic acid (ABA), and gibberellin pathways are being investigated in *P. patens* (reviewed in Cove *et al.*, 2006). Auxin is notable due to its pleiotropic and critical role in the establishment of cell and organismal polarity in higher plants (Friml *et al.*, 2006). In *P. patens*, auxin has been shown to regulate aspects of gametophyte development but this may not involve the same kinds of polar auxin transport that is seen in higher plant sporophytes (Fujita *et al.*, 2008). The moss sporophytes, however, do exhibit polar transport and appear to use auxin to establish their apical–basal polarity during embryogenesis, highlighting a major difference between these life stages. Research on ABA response has focused on the role of this hormone in desiccation tolerance, which is particularly important for mosses, being nonvascular. Moreover, the evolution of extreme desiccation tolerance may have been a critical step in the early movement onto land.

Experimental studies have found that, similar to vascular plants, application of ABA to mosses enhances tolerance to environmental stresses including freezing and rapid desiccation (Cove, 2005; Minami *et al.*, 2003). Genomic and candidate genes approaches are being used to investigate the genetic basis for these ABA responses. Recently, it was found that the *P. patens* homolog of the *Arabidopsis* gene *ABSCISIC ACID INSENSITIVE3* (*ABI3*) is quite conserved, both in terms of its biochemical and genetic roles in ABA response (Marella *et al.*, 2006). In contrast, the gibberellin response pathway, which is dependent on the GID1 and DELLA proteins in vascular plants, is not found in *Physcomitrella*, representing a major exception to the overall trend of conservation.

Another developmental pathway that has received considerable attention is the *LEAFY* (*LFY*) regulatory network. In seed plants, *LFY* homologs are expressed in reproductive meristems and are generally essential to their identity (Sablowski, 2007). They also sometimes play roles in leaf development and/or phase change (Champagne *et al.*, 2007; Hofer *et al.*, 1997). The *P. patens LFY* homolog, *PpLFY*, is expressed in both gametophyte and sporophyte life stages but deletion mutants only show a phenotype in the sporophyte (Tanahashi *et al.*, 2005). Rather dramatically, the developing homozygous mutant zygote arrests during the first cell division, suggesting a critical role in diploid development. It has further been shown that PpLFY cannot activate any of LFY's normal targets in *Arabidopsis*, which is apparently due to changes in DNA binding specificity (Maizel *et al.*, 2005). These studies indicate that while many gene lineages are found throughout the land plants, the degree of conservation in biochemical and developmental aspects of gene function are likely to vary considerably.

2.2. *Selaginella*

The lycophytes represent an important transition point in the diversification of land plants—they are the earliest branch of the vascular plants and still retain free-living gametophytes, although these are often highly reduced. *Selaginella* offers the ability to investigate the genetic basis of primitive vascular systems and their associated microphylls as well as an independent derivation of heterospory, which also evolved in the seed plants. *S. moellendorffii* was selected for genome sequencing due to its very small genome, approximately 100 Mbp (Wang *et al.*, 2005). Although this genome is still undergoing annotation, it has already yielded significant information about the conservation of important gene lineages (Floyd and Bowman, 2007; Hirano *et al.*, 2007). As of yet, transformation protocols have not been developed for *Selaginella* but work with another species, *S. kraussiana*, highlights the genus' potential for developmental research. Harrison *et al.* (2007) used X-ray irradiation to conduct clonal analysis of the activity of the *Selaginella* meristem, which is structurally much simpler than those of seed plants. This study demonstrated that *S. kraussiana* has a meristem composed of two primary initials that turn over during the course of development. In terms of leaf development, it was determined that these simple leaves, or microphylls, are derived from only two progenitor cells, far fewer than in seed plants, and initiate their inner layers through later patterns of cell division rather than direct recruitment. At the molecular level, both similarities and differences with seed plants have been observed in expression studies. Comparative analysis of the class III HD-Zips suggests that the mechanism of vascularization is distinct in microphylls relative to the true leaves, or megaphylls, of seed plants (Floyd and Bowman, 2006). When other genetic pathways are examined, however, particularly the KNOX/ARP program, it appears that aspects of their genetic interaction are conserved. The difficulty with these comparisons, of course, lies in the large phylogenetic distance that separates early vascular plants like *Selaginella* from the seed plant models. Some observed discrepancies may be simply indicative of developmental system drift (True and Haag, 2001) while others could reflect fundamental differences in the evolutionary history of the organs. What is clear is that we need functional tools in *Selaginella* to better understand the genetic function underlying observed expression patterns.

2.3. Ceratopteris

The monilophytes—which include whisk ferns (*Psilotum*), the horsetails (e.g., *Equisetum*), and both homosporous and heterosporous ferns—are of considerable interest due to their phylogenetic position as well as their diversity. One significant challenge to working with them, from a genetic standpoint, is that they are often polyploid with large genomes (see Nakazato *et al.*, 2006 and references therein). One tractable system is the homosporous fern *Ceratopteris richardii*, which has been the subject of significant research into gametophyte sex determination (Banks, 1999). Like other early land plants with free-living gametophytes, one advantage to this model is the ability to conduct mutagenesis of single cell haploid spores, and many mutants have been recovered for the pheromone-based gametophyte sex determination pathway (reviewed in Tanurdzic and Banks, 2004). This pathway is distinct from what has been observed in many animal sex determination mechanisms in that it is not linear and involves antagonistic male- and hermaphrodite-determining factors (Tanurdzic and Banks, 2004). These genes respond to the environmental concentration of the pheromone antheridiogen, which results in density-dependent variation in the gender ratio. Genetic and functional resources for *Ceratopteris* are growing and now include ESTs and an RNAi-based reverse genetic technique (Rutherford *et al.*, 2004; Stout *et al.*, 2003). Further development of all of these nonseed plant models will significantly increase our understanding of the genetic control of gametophyte development as well as the early evolution of the land plants.

3. Angiosperms: The Core Eudicots

One may note that there is a rather large phylogenetic gap between the models of the lower land plants and the angiosperms, in large part represented by the diversity of extant gymnosperms (Fig. 4.1). Unfortunately, gymnosperms are almost exclusively woody, long-lived plants with moderate sized genomes (Soltis *et al.*, 2003). Although significant DNA sequence is available and transformation is possible in several gymnosperms, especially the economically important members of the Pinaceae, their limitations as fully functional genetic model systems remain rather daunting (Pavy *et al.*, 2007; Tang *et al.*, 2007b). In contrast, the angiosperms offer an enormous wealth of morphological diversity and include many rapid cycling, herbaceous plants that are well suited for genetic analysis. As shown in Fig. 4.1, the evolutionary history of the angiosperms is one of successive rounds of radiation: first the so-called "ANA" lineages including *Amborella* and the water lilies, then the magnoliid dicots and monocots, and finally, the eudicots, encompassing 75% of all angiosperm species (Magallon *et al.*, 1999; Moore *et al.*, 2007). The eudicots are further divided into the lower eudicots, a grade that includes the Ranunculales, and the core eudicots, which represents the bulk of the clade as well as many established model systems. While the evolution and diversification of floral morphology often receives disproportionate attention, other traits such as leaf or fruit morphology are even more evolutionarily labile and are now being dissected at the genetic level.

3.1. *Arabidopsis* and the Brassicaceae

It is only right to start any consideration of angiosperm model species with the dominant plant genetic model, *Arabidopsis thaliana*. Many recent studies have demonstrated the utility of *Arabidopsis* for investigating questions related to evolutionary and ecological genetics (e.g., Bomblies and Weigel, 2007; Mitchell-Olds and Schmitt, 2006; Tang *et al.*, 2007a) but, even within *A. thaliana*, there are opportunities for evo-devo research. For example, Mouchel *et al.* (2006) surveyed morphological variation in natural ecotypes and discovered a new gene family controlling aspects of root development. Also, *A. thaliana* has proven useful for studying the effects of gene duplication, such as the paralogous MYB genes *WEREWOLF* (*WER*) and *GLABROUS1* (*GL1*). These loci have been shown to be biochemically equivalent despite their respective roles in trichome and root hair development (Lee and Schiefelbein, 2001). This initial study is now being extended to additional duplicated members of the genetic module, revealing that other paralog pairs have experienced biochemical as well as regulatory divergence (Simon *et al.*, 2007).

Given the extensive genetic and genomic resources available for *Arabidopsis*, developing model systems among its close relatives would seem like a natural next step. One stumbling block, however, was the fact that the established taxonomic relationships of the family Brassicaceae appeared to be highly homoplastic (Koch *et al.*, 1999). Luckily, significant progress has been made in understanding relationships among the Brassicaceae and we now have a reasonable phylogenetic framework in which to evaluate morphological evolution (Al-Shehbaz *et al.*, 2006; Beilstein *et al.*, 2006). New model systems under development include *Arabidopsis lyrata*, *Capsella rubella*, *Boechera stricta*, multiple species of *Brassica*, and *Thellungiella halophila* (Schranz *et al.*, 2007). One model that holds particular interest from an evo-devo perspective is *Cardamine hirsuta*, which is a useful system to study leaf diversification. Across the angiosperms, leaf morphology is often among the most variable traits, even between closely related species, and this lability is well represented in the Brassicaceae (Fig. 4.3). To understand the genetic basis of this variation, Tsiantis and coworkers have developed significant genetic resources for *Cardamine*, including stable transformation protocols and mutagenized populations (Hay and Tsiantis, 2006). Hay and Tsiantis (2006) elegantly applied these tools to demonstrate that while the basic regulatory mechanisms of leaf formation are conserved between *Arabidopsis* and *Cardamine*, changes in the upstream regulatory regions of two *Cardamine* KNOX genes causes them to be expressed in developing leaves. This, in turn, promotes the development of compound or dissected leaves (Fig. 4.2F). More recently, *Cardamine* has been used to develop a more detailed model of leaflet formation. Perhaps not surprisingly, it appears that the basic molecular program for leaf initiation, which involves the PIN1-dependent establishment of auxin maxima, also underlies the positioning and formation of leaflets on the flanks of developing compound leaves (Barkoulas *et al.*, 2008). These studies demonstrate the considerable utility of *Cardamine* for investigating leaf developmental evolution but there are additional traits that can be investigated using the system, including floral and trichome development.

Other Brassicaceous taxa are being used to study a wide range of features. Although we may think of floral morphology in the family as being quite highly conserved (Fig. 4.4A), there is some observed variation in organ number and floral symmetry. The latter is being studied in the lab of Sabine Zachgo, who has shown that the bilaterally symmetric corolla of *Iberis amara* is due to late, differential expression of a homolog of the TCP gene *CYCLOIDEA* (*CYC*) (Busch and Zachgo, 2007), which was first found to be involved in floral symmetry in the asterid *Antirrhinum*

Figure 4.3 Variation in leaf morphology across the Brassicaceae. (A) Three leaf types found on a single plant of *Cakile lanceolata*. (B) *Brassica oleracea*. (C) and (D) Two different species of *Erucaria*. (E) *Moricanda* sp. (F) *Cardamine hirsuta*.

(Luo *et al.*, 1996). Interestingly, members of this gene lineage appear to have been recruited on many separate occasions to promote bilateral floral symmetry, termed zygomorphy (see below). Another characteristic that varies across the family is inflorescence structure. While most taxa produce indeterminate racemes, some take an alternative strategy of making solitary flowers in their basal rosette (Shu *et al.*, 2000). Since it is well established that the functions of the floral meristem identity gene *LEAFY* (*LFY*) are highly conserved across angiosperms, this developmental shift is likely to be due to changes in the expression pattern of *LFY*, a hypothesis that has been confirmed in several of the rosette-flowering taxa (Bosch *et al.*, 2008; Shu *et al.*, 2000). Further studies of heterologous promoter expression suggest that both *cis*- and *trans*-regulatory changes at the *LFY* locus underlie the conversion from raceme to rosette-flowering (Yoon and Baum, 2004). One trait that is almost as diverse across the Brassicaceae as leaf form is fruit

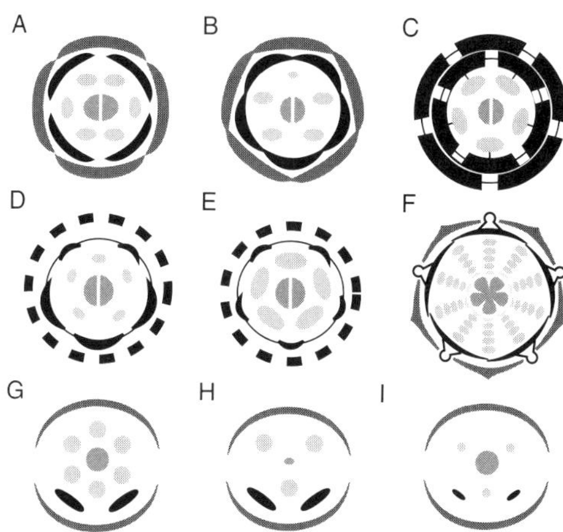

Figure 4.4 Floral diagrams of selected angiosperm model systems. (A) *Arabidopsis thaliana*, with four sepals, four petals, six stamens, and two carpels. (B) *Antirrhinum majus*, with five sepals, five petals, five stamens (four fertile, the dorsal stamen reduced as staminode), and two carpels. (C) *Petunia hybrida*, with five sepals, five petals, five stamens, and two carpels. There is within and between whorl fusion among the floral organs, indicated by thin lines. (D) Ray flower of *Gerbera hybrida*, with multiple pappus bristles in the first whorl, followed by five petals (upper two reduced), five reduced stamens, and two carpels. (E) Disk flower of *Gerbera hybrida*, with multiple pappus bristles, five reduced petals, five stamens, and two carpels. (F) *Aquilegia vulgaris*, with five sepals, five spurred petals, multiple whorls of ten stamens, one whorl of ten staminodia, and five carpels. (G) *Oryza sativa*, one palea (upper first whorl organ), one lemma (lower first whorl organ), two lodicules in the second whorl, six stamens, and one fertile carpel. (H) Male flower of *Zea mays*, one palea (upper first whorl organ), one lemma (lower first whorl organ), two lodicules in the second whorl, and three stamens. The central carpel is aborted. (I) Female flower of *Zea mays*, one palea (upper first whorl organ), one lemma (lower first whorl organ), two aborted lodicules in the second whorl, three aborted stamens, and one fertile carpel.

morphology (Beilstein *et al.*, 2006). Although we commonly think of the silique as being the diagnostic fruit type of the family, genuinely bizarre variations can be found, especially in the tribe Brassiceae. In the genus *Cakile*, for instance, the fruit is indehiscent and breaks transversely to produce two sealed propagules (Donohue, 1998). This unusual dispersal strategy is made possible by profound alterations in carpel development that are now being investigated at the molecular level (Hall *et al.*, 2006). Moving outside the Brassicaceae into other members of the Brassicales, genetic and genomic tools are being developed for members of the family Cleomaceae (Schranz and Mitchell-Olds, 2006; Schranz *et al.*, 2007) as well as the crop plant *Carica* (papaya) (Ming *et al.*, 2008). These new models will allow the investigation of diverse features including floral gynophores (a stalked gynoecium), independently derived instances of zygomorphy, and novel forms of sex determination.

3.2. Legumes

The legumes, more technically known as the family Fabaceae, are a diverse and economically important family that includes soy bean (*Glycine max*), pea (*Pisum sativa*), alfalfa (*Medicago sativa*), and trefoil (*Lotus japonicus*). In addition to agriculturally relevant characteristics such as seed biology and nitrogen fixation, these models are being used to investigate developmental evolution, particularly of leaf and floral morphology. As mentioned above, leaf characteristics tend to be highly variable, even among closely related taxa. The Fabaceae are similarly diverse but many members have compound or dissected leaves, which are independently derived relative to the case of *Cardamine* discussed above (Sinha, 1997). Broader studies across the vascular plants have shown a close association between the production of compound leaves and the expression of KNOX genes in the developing leaf primordia, which is usually not observed for simple leaves (Bharathan *et al.*, 2002). There is a notable exception to this pattern, however, in the legume model *Pisum*. There it was discovered that the mutant *unifoliata* (*uni*) actually encodes a homolog of the floral meristem identity gene *LFY* (Hofer *et al.*, 1997), which is often expressed at low levels in leaves but does not generally have a function in these organs (Sablowski, 2007). Moreover, the leaves of *Pisum* do not express KNOX genes (Bharathan *et al.*, 2002), suggesting that at least some legumes have evolved a novel genetic pathway for producing compound leaves. This finding has been further investigated by Champagne *et al.* (2007), who demonstrated that the transition from a KNOX-based compound leaf program to one using *LFY* homologs actually occurred within the legumes, at the base of the so-called the inverted repeat-lacking clade (IRLC) that includes *Wisteria*, *Medicago*, *Pisum*, and *Vicia* (fava bean). They further found that in legumes outside of this clade, *LFY* homologs play a weak role in compound leaf development, possibly representing a transitional state from the KNOX- to *LFY*-based mechanism. This work indicates that while recruitment of KNOX genes to control compound leaf development is a common occurrence, there is more than one to solve the genetic problem of promoting leaf indeterminacy and, in some cases, wholly novel genetic mechanisms may be employed.

Legume flowers are also of considerable interest due to their complex zygomorphic morphology. The genetic basis of zygomorphy was first dissected using the model system *Antirrhinum* (see below) and was found to involve a pair of recent paralogs of the TCP gene family called *CYC* and *DICHOTOMA* (*DICH*) (Luo *et al.*, 1996, 1999). Members of this gene lineage have been identified in many core eudicots and there had been suggestions that the loci might have been repeatedly, independently recruited to play roles in zygomorphy (Citerne *et al.*, 2000, 2003; Cubas, 2002). Hard functional evidence of the hypothesis was lacking, however, until a set of exceptional experiments in *L. japonicus* (Feng *et al.*, 2006). This study used reverse and forward genetic techniques to demonstrate that a *CYC* homolog in *Lotus*, *LjCYC2*, is essential to the establishment of dorsal organ identity, much like *CYC/DICH* in *Antirrhinum*. Further work has demonstrated that a second locus, *Keeled Wings1* (*KEW1*), actually encodes another TCP paralog, *LjCYC3* (Wang *et al.*, 2008). Double mutants of these loci in *Lotus* or their orthologs in *Pisum* show strongly ventralized phenotypes, similar to *cyc–dich* double mutants in *Antirrhinum* (Luo *et al.*, 1996, 1999; Wang *et al.*, 2008). It is important to note, however, that the *CYC/DICH* duplication event was completely independent from that which produced *LjCYC2/3* (Citerne *et al.*, 2003).

Given the fact that *CYC* homologs are often expressed on the dorsal side of even radially symmetric floral meristems (Cubas *et al.*, 2001), it may be that these loci have a deeply conserved function in determining dorsal–ventral polarity of axillary meristems (Cubas, 2002; Feng *et al.*, 2006). This underlying function could then have been recruited and elaborated many times independently to yield a homoplastic pattern of zygomorphy across the angiosperms. It also seems that the expression of *CYC* genes can be modified to yield reversions to radial symmetry (actinomorphy). Although it had been suggested that such events might be due to loss of function in *CYC*-like genes (Coen and Nugent, 1994), morphological evidence indicates the contrary that these cases were more likely to be due to dorsalization rather than the ventralization that is seen in *cyc* mutants (Donoghue *et al.*, 1998). This hypothesis has been confirmed in the actinomorphic legume *Cadia*, which has expanded expression of its *CYC* homolog such that it encompasses the entire corolla (Citerne *et al.*, 2006). Overall, it has become clear that legume models represent exceptionally useful systems for the genetic dissection of floral zygomorphy. Work in *Pisum* has even identified a novel type of symmetry mutant called *SYMMETRIC PETALS1* (*SYM1*) that promotes the internal asymmetry of individual organs (Wang *et al.*, 2008). These lines of research will help us to understand the similarities and differences between convergent genetic pathways.

3.3. *Populus*

The development of wood, known as secondary growth, has evolved several times during the course of land plant evolution (Stewart and Rothwell, 1993). As mentioned above, woodiness is often a barrier to genetic studies since woody plants tend to grow slowly. There are also the simple, physical difficulties of working with woody tissue, particularly for histology and traditional gene expression techniques. That being said, secondary growth is a fascinating and exquisitely coordinated developmental process. Understanding its genetic basis is important from economic, evolutionary, and developmental standpoints. Given that *Arabidopsis* has only limited secondary growth and the monocot grasses have none whatsoever, it has been necessary to develop a new model system for research into tree biology. The eudicot genus *Populus* represents a good candidate because of its rapid growth, ecological importance, and potential application for biofuels (Groover, 2007; Jansson and Douglas, 2007). A wide range of tools are available now including a genome sequence in *Populus trichocarpa*, stable transformation, inducible expression, RNAi techniques, and insertional mutagenesis (reviewed in Jansson and Douglas, 2007).

Several recent studies have begun to investigate the control of developmental dynamics and cell fate in the vascular cambium—a special type of meristem that simultaneously gives rise to new xylem on the inside of the stem and new phloem at the periphery. This bifacial, unicellular meristem forms a cylindrical sheath inside the stem of every woody plant. The rate and orientation of cell divisions within the cambium must be carefully coordinated to give the right balance of transport and support cells as well as the correct physical properties of the wood (Kramer, 2006). Perhaps not surprisingly, it has been found that some of the same genes that control development in apical meristems also control aspects of cambium activity and organization. In particular, overexpression of a KNOX homolog delays the differentiation of cambial derivatives and results in uncoordinated patterns of cell division (Groover *et al.*, 2006). Experiments also suggest that the critical plant hormone auxin plays an important role in the organization and activity of the cambium, similar to the apical meristems (Nilsson *et al.*, 2008; Schrader *et al.*, 2003). An important challenge for the future of this work is to move beyond traditional candidate genes. This can be accomplished through forward genetics, particularly insertional mutagenesis and enhancer trap screens (Busov *et al.*, 2005; Groover *et al.*, 2004), as well as genomic approaches (Schrader *et al.*, 2004). One recent study has extended microarray analyses to specific stem cell initials, some of which give rise to conductive tissue while others produce metabolically active parenchymal cells (Goue *et al.*, 2008). This work will hopefully provide microscale insight into the genetic differentiation of these unique cell types.

Another aspect of Poplar's biology that is intimately tied to its perennial, woody habit is deciduousness. Trees in temperate or tropical seasonal forests go through periods of developmental quiescence in response to environmental cues such as shortened day length or reduced precipitation. In preparation for this dormancy, the apical meristems alter their developmental behavior. First, they produce a set of modified protective leaves termed bud scales or cataphylls. These will protect the resting meristem and associated structures during the winter or dry season. At the same time, the internodes that separate successive leaves no longer elongate, creating a tightly packed bud. After the cataphylls are produced, additional foliage leaves and, often, floral meristems develop, but these structures do not expand and mature. Instead, they arrest at early developmental stages to wait out the quiescent period. Understanding the genetic basis for the developmental and physiological changes associated with seasonal dormancy is simply not possible using the main model systems (e.g., *Arabidopsis*) due to the very reasons that these taxa were selected as models—all are rapidly cycling annuals. Therefore, Poplar provides a unique opportunity to study seasonal dormancy. The first genetic investigation of this process began as an analysis of flowering time, involving a homolog of the *Arabidopsis* gene *FLOWERING LOCUS T* (*FT*). In *Arabidopsis*, *FT* is transcribed under long day (LD) and promotes the transition to flowering (Kardailsky *et al.*, 1999). The Poplar homolog, *PtFT1*, also promotes flowering but, in addition, overexpression prevents the normal transition to dormancy under short days (SD) while *PtFT1* RNAi plants exhibit hypersensitivity to SD and increased dormancy (Bohlenius *et al.*, 2006). Similar to *FT*, *PtFT1* is expressed in LD and quickly declines in SD, but in this case *PtFT1*'s function is expanded to include promotion of active vegetative growth and suppression of the dormancy developmental program. It is perhaps not surprising that this genetic module, which serves as a readout of day length to promote flowering in a number of different taxa (Simpson, 2003), has been recruited to control other developmental responses to day-length change. Perhaps most interesting though, the researchers found that natural variation in *PtFT1*'s transcriptional response to day length is correlated with a latitudinal cline in day-length thresholds that trigger growth cessation (Bohlenius *et al.*, 2006). This finding underscores another of Poplar's advantages as a model system. Its broad natural range, high genetic diversity, and tendency to grow in clonal clumps create an excellent opportunity for association mapping of economically and evolutionarily important traits (Hall *et al.*, 2007; Ingvarsson *et al.*, 2006).

3.4. *Antirrhinum* and relatives

Antirrhinum majus has more than a century of history as a genetic model but came to prominence more recently thanks in large part to the application of its transposon mutagenesis system to the subject of floral development (Carpenter and Coen, 1990). Along with *Arabidopsis*, work done in *Antirrhinum* helped to craft the well-known ABC model of floral organ identity (Coen and Meyerowitz, 1991). This model holds that there are three classes of gene activity functioning in the floral meristem that are expressed in overlapping domains: A in the first and second whorls, B in the second and third whorls, and C in the third and fourth (Fig. 4.5A). This creates a combinatorial code that determines the identity of primordia arising in each whorl: A = sepals, A + B = petals, B + C = stamens, and C = carpels. It must be noted, however, that the so-called "A" function has never been detected in *Antirrhinum* (Coen *et al.*, 1991; Davies *et al.*, 2006),

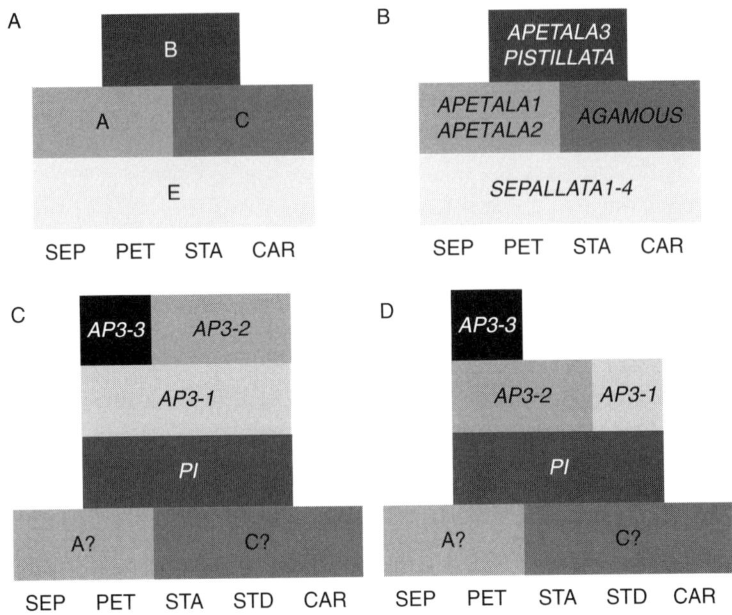

Figure 4.5 (A) The classic ABC model with the addition of the E function. (B) The corresponding ABCE genes from *Arabidopsis*. (C) and (D) The modified ABC model of *Aquilegia* based on expression studies of the B gene homologs. (C) corresponds to early developmental stages while (D) reflects expression after carpel initiation.

consistent with evidence from other taxa indicating that the functions of these genes are not well conserved (as well as being subject to alternative interpretations) (Litt, 2007). In contrast, C function is generally conserved across the angiosperms but gene duplications within the C gene lineage have led to independent patterns of subfunctionalization among paralogs (Kramer *et al.*, 2004). For instance, the primary C function gene in *Arabidopsis* is the MADS-box gene *AGAMOUS* (*AG*; Fig. 4.5B) (Bowman *et al.*, 1989; Yanofsky *et al.*, 1990). This locus is actually the product of a duplication that occurred at the base of the core eudicots, the other lineage being represented in *Arabidopsis* by the more recent paralogs *SHATTERPROOF1* and *2* (*SHP1/2*) (Kramer *et al.*, 2004; Liljegren *et al.*, 2000). These genes play specific functions in fruit and ovule development but appear to be biochemically equivalent to AG (Ferrandiz *et al.*, 2000; Liljegren *et al.*, 2000; Pinyopich *et al.*, 2003). Somewhat surprisingly, phylogenetic and synteny analyses have demonstrated that the *Antirrhinum* ortholog of *AG* is a gene called *FARINELLI*, which contributes specifically to stamen development (Causier *et al.*, 2005; Davies *et al.*, 1999; Kramer *et al.*, 2004). The primary C function gene is instead *PLENA* (*PLE*), the ortholog of *SHP1/2* (Bradley *et al.*, 1993; Causier *et al.*, 2005; Davies *et al.*,

1999). In this case, however, FAR and PLE have diverged biochemically as well as in their expression patterns and developmental functions (Causier *et al.*, 2005). This work demonstrates that while paralogs may be retained due to subfunctionalization, the process does not occur along the same paths in different organisms and may remain labile through long evolutionary periods.

Antirrhinum has also been used as a model for floral symmetry, floral color, epidermal cell development, and leaf shape (reviewed in Davies *et al.*, 2006; Schwarz-Sommer *et al.*, 2003). As discussed above, the identification of the TCP genes *CYC* and *DICH* laid the foundation for analyses of the genetic basis of zygomorphy across the angiosperms. In *Antirrhinum*, these recent paralogs are expressed on the dorsal side of the floral meristem and are responsible for the abortion of the dorsal stamen as well as the development of specific dorsal and lateral petal types (Fig. 4.4B) (Luo *et al.*, 1996, 1999). Evidence for their functional conservation was discovered soon after these genes were described. In particular, it was found that a radially symmetric mutant of the closely related genus *Linaria* is due to an epigenetically silenced allele of *CYC* (Cubas *et al.*, 1999), an intriguing example of how epigenetic modification could produce evolutionarily relevant genetic change. Other comparative studies of *CYC/DICH* homologs have taken advantage of the diversity in floral morphology that exists within the close phylogenetic vicinity of *Antirrhinum*. For instance, the genus *Mohavea* actually nests within a clade of North American *Antirrhinum* species but has traditionally been segregated as a separate genus due to its open floral morphology and the abortion of three stamens rather than the one that is typical for *Antirrhinum* (Oyama and Baum, 2004). It has been shown that these shifts in floral morphology are correlated with alteration in the *CYC/DICH* expression domains, including expansion into the aborted lateral stamens (Hileman *et al.*, 2003). Overall, the *Antirrhinum* species complex is proving to be exceptionally useful, especially for the study of flower color. Several studies have found that natural variation in flower color and pattern is due to genetic variation in loci that were first characterized in *Antirrhinum* (Schwinn *et al.*, 2006; Whibley *et al.*, 2006). These analyses have been particularly facilitated by interfertility among *Antirrhinum* species that allows interspecies complementation tests. This cross fertility has also been used to understand the evolutionary genetics of leaf size and shape (Langlade *et al.*, 2005). Another major model for evolutionary genetics that is more distantly related to *Antirrhinum* within the order Lamiales is the genus *Mimulus*, a classic model for adaptive radiation and speciation genetics (Wu *et al.*, 2008). Although *Mimulus* research has largely focused on QTL analysis, it is clear that the genus will be useful to investigate aspects of developmental evolution such as distinct floral forms (Bradshaw *et al.*, 1998; Fishman *et al.*, 2002).

3.5. Solanaceae

The family Solanaceae is notable for the presence of many economically and/or horticulturally important genera that are also genetically tractable, particularly *Nicotiana* (tobacco), *Solanum* (tomato, potato), and *Petunia*. All of these systems are easy to transform as well as being highly susceptible to RNAi techniques such as virus-induced gene silencing (VIGS) (Brigneti *et al.*, 2004; Burch-Smith *et al.*, 2004). The flowers of the Solanaceae share a fairly conserved morphology where the sepals and petals are fused within their whorls to form tube-like structures (Fig. 4.4C). The stamens are then fused at their bases to the inside of the petal tube. Floral development has been a major focus of research in *Petunia*, first using transgenics to knock down or overexpress homologs of the *Arabidopsis* ABC genes and, more recently, using an elegant transposon system to identify loss-of-function mutants in these loci (van der Krol and Chua, 1993; Vandenbussche *et al.*, 2003b). These studies have uncovered several interesting results, particularly in regards to evolution of paralogous gene functions. Just as most core eudicots have two representatives of the C class gene lineage (e.g., *AG* and *SHP1/2* in *Arabidopsis*), many have two representatives of the *APETALA3* (*AP3*) B class gene lineage (Kramer *et al.*, 1998). These two paralogous lineages are termed eu*AP3* and *TM6* and, unlike the case with *AG*, the paralogs are distinct in both their sequence and expression patterns (Kramer and Irish, 2000; Kramer *et al.*, 1998, 2006; Vandenbussche *et al.*, 2003a; Zahn *et al.*, 2005). While *Arabidopsis* has lost its *TM6* ortholog (Fig. 4.5B), *Petunia* has orthologs of both eu*AP3* and *TM6*, which allows their separate developmental functions to be studied. Analyses of insertional mutants of *PhDEF* (the eu*AP3* ortholog) and *PhTM6* have shown that the former has a typical B gene function promoting both petal and stamen identity, but the latter is only expressed in stamens and is sufficient for their identity (Rijpkema *et al.*, 2006). Furthermore, *Petunia* has two copies of the other B gene lineage, the *PISTILLATA* (*PI*) homologs, although these are much more recently derived than the *AP3* paralogs (Kramer *et al.*, 1998). These two PI proteins, termed PhGLO1 and PhGLO2, have become biochemically specialized (Vandenbussche *et al.*, 2004). It is typical for AP3 and PI proteins to function as obligate heterodimers (Riechmann *et al.*, 1996) but in *Petunia*, these interactions have become specific such that PhGLO2 primarily interacts with PhTM6 (Vandenbussche *et al.*, 2004). This interaction has been confirmed both by yeast two-hybrid and genetically. Given that the *PhDEF/PhTM6* duplication predates the *PhGLO1/GLO2* event by tens of millions of years (Kramer *et al.*, 1998), it would appear that the subfunctionalization of the earlier *AP3* paralog pair played a role in the subfunctionalization later arising *PI* pair. Characterization of some of the classic *Petunia* floral mutants has also proven useful for new gene discovery. For example, the mutant *blind*, which resembles a traditional A class mutation in having petal to stamen transformations, turns out to encode a miRNA and reveals a novel regulatory pathway for controlling C gene expression (Cartolano *et al.*, 2007).

Domesticated tomato, *Solanum lycopersicon*, has been particularly useful in the investigation of compound leaf development and the genetic control of fruit shape. It was work done in tomato that first demonstrated the role for KNOX homeodomain genes in the development of compound leaves (Janssen *et al.*, 1998). This initial finding has been elegantly elaborated upon to further investigate how leaflet number and position is determined. In particular, the *Antirrhinum* gene *PHANTASTICA* (*PHAN*) normally contributes to the dorsal/ventral patterning of leaf primordia (Waites *et al.*, 1998) but in tomato, its ortholog *LePHAN* has an additional function in positioning leaflets (Kim *et al.*, 2003b). It had already been shown that the juxtaposition of dorsal and ventral identity is necessary for laminar expansion in simple leaves (Waites and Hudson, 1995). The *LePHAN* study further showed that leaflet initiation, which is in some ways analogous to laminar expansion, requires a similar establishment of polarity. In order for this to occur, however, there must be simultaneous expression of *PHAN* and KNOX homologs in the leaf, a situation that is not normally seen in simple leaved taxa and must require alterations in their typically antagonistic regulatory interactions (Kim *et al.*, 2003b). Just as KNOX genes have been repeatedly recruited for a role in compound leaf production (Bharathan *et al.*, 2002), it also appears that this role of *PHAN* in leaflet positioning has evolved in many different instances (Kim *et al.*, 2003a). Natural variation in tomato leaf morphology has also proven to be very useful for dissecting the genetics of compound leaf development. A naturally occurring morphological variant is found in the Galapagos island where *S. galapagensis* has increased complexity in leaf dissection, a phenotype noted by Charles Darwin (Kimura *et al.*, 2008). Identification of the semidominant locus responsible for the trait uncovered a single base-pair deletion in the promoter of a locus called *TOMATO KNOX-LIKE HOMEODOMAIN PROTEIN 1* (*TKD1*) (Kimura *et al.*, 2008). This mutation results in overexpression of *TKD1*, which encodes a novel KNOX gene that has a MEINOX protein–protein interaction domain but lacks the homeodomain itself. The TKD1 protein can still interact with other proteins that act to attenuate KNOX function. Thus, the normal function of TKD1 is to act in a concentration-dependent manner to bind up specific inhibitors of KNOX proteins. In the variant, TKD1 is overexpressed, resulting in overactivity of KNOX proteins and increased leaf dissection. This mechanism was also found to be functioning in *Arabidopsis* (Magnani and Hake, 2008) and may represent an evolutionary rheostat for degrees of leaf dissection. Of course, the domestication of tomato has also provided enormous morphological variation for genetic analysis. One subject of intense research has been variation in tomato shape, which can vary in overall size and along multiple axes. These studies, using both QTL analyses and identification of Mendelian traits, have uncovered some previously known candidate genes as well as novel loci (Cong *et al.*, 2008; Liu *et al.*, 2002; van der Knaap and Tanksley, 2003; Xiao *et al.*, 2008). This work may lay the foundation for further studies across the family Solanaceae where fruit type (e.g., dry vs fleshy) can vary profoundly among closely related taxa (Knapp, 2002; Knapp *et al.*, 2004).

3.6. Asteraceae

The Asteraceae or Compositae are one of the largest families of flowering plants. Their diagnostic feature is their inflorescence type, the capitulum. Think of a typical daisy—a yellow central disk surrounded by white petals. This is not, in fact, a single flower but a composite inflorescence containing hundreds of individual florets that vary in their morphology across the inflorescence. These florets are produced by a flattened disk-shaped inflorescence meristem that gives rise to individual floral meristems in a spiral series. In the case of the daisy, the first meristems produced, which end up at the periphery of the mature inflorescence, are the ray florets and correspond to the "petals" of the daisy. In each floret, the true petals are fused to form a basal tube (corolla) but on the lower side of the tube, three of the corolla lobes are greatly elongated (Fig. 4.4D). Additionally, the ray florets are functionally female due to aborted stamens. By contrast, the florets produced in the center of the inflorescence, called disk florets, have reduced corollas with no elongated lobes and are fully hermaphroditic (Fig. 4.4E). Both of these floret types lack traditionally defined sepals but are surrounded by fine, hair-like structures called pappus, which assist in seed dispersal later in development. In some taxa, such as *Gerbera*, there are intermediate floret forms produced in zones between the ray and disk florets (Teeri *et al.*, 2006b). There are also members of the family that produce either only ray florets or only disk florets. The first step to understanding the genetic basis of these novel floral and inflorescence forms was determining the conservation of the ABC program in the Asteraceae. The primary model for this process has been *Gerbera hybrida*, which is amenable to stable transformation and has good genomic resources (Teeri *et al.*, 2006a). These studies determined that the ABC class genes were functioning in an analogous fashion in *Gerbera* and established that the pappus bristles are equivalent to sepals (Yu *et al.*, 1999). At the same time, intriguing data suggested that some of the ABC gene homologs function at the level of the inflorescence to influence the pattern of ray and disk florets. This is particularly true of homologs of a fourth class of genes, termed the E class for historical reasons, that is represented by the *SEPALLATA1–4* (*SEP1–4*) loci in *Arabidopsis* (Fig. 4.5A and B) (Ditta *et al.*, 2004; Pelaz *et al.*, 2000). Repression of the *SEP* homolog *GRCD2* results in indeterminacy of both the floral and inflorescence meristems, indicating a novel function for this locus (Uimari *et al.*, 2004). Similarly, microarray-based studies of the ray versus disk florets found evidence that ABC gene paralogs are often differentially expressed between the floret types (Laitinen *et al.*, 2006). These findings have led to the hypothesis that aspects of the ABC program normally used to pattern the floral meristem have been co-opted at a higher developmental level to pattern the inflorescence itself (Teeri *et al.*, 2006a).

Even if this is the case, however, it has become clear that other genetic pathways also contribute to the differentiation of ray versus disk florets. Studies of three different Asteraceae systems—*Gerbera*, *Helianthus*, and *Senecio*—have all implicated homologs of the *CYC* genes in the establishment of floral zygomorphy in the florets (Broholm *et al.*, 2008; Chapman *et al.*, 2008; Kim *et al.*, 2008). This represents yet another independent recruitment of the TCP genes to function in establishment of floral zygomorphy, but this case appears to be even more complex. Phylogenetic studies demonstrate that the *CYC* homologs of the Asteraceae have undergone a series of gene duplication events to give rise to at least four distinct lineages within the clade most closely related to *CYC* itself (Broholm *et al.*, 2008; Chapman *et al.*, 2008; Kim *et al.*, 2008). Moreover, several of these paralogs are differentially expressed, appear to have experienced diversifying selection, and may differ in their ability to promote or repress cell division (Broholm *et al.*, 2008; Chapman *et al.*, 2008; Kim *et al.*, 2008). Further study of *CYC* homologs in this system is, therefore, likely to reveal a complex interplay of sub- and neofunctionalization that may have been critical to the evolution of the distinct floret types of the capitulum inflorescence. What is particularly exciting about Asteraceae is that several new model systems appear to be poised for concurrent development, with extensive genetic and genomic resources available for the important evolutionary and ecological model *Helianthus* (Rieseberg *et al.*, 2003) and stable transformation being tractable in *Senecio* (Kim *et al.*, 2008).

4. Angiosperms: Lower Eudicots

The grade of lineages that branch off from early nodes in the eudicot clade are collectively termed the lower eudicots. These include morphologically diverse taxa ranging from tiny flowered sycamore trees (*Platanus*) to the showy, aquatic sacred lotus (*Nelumbo*). The fact that these two taxa are actually quite closely related underscores the lability in floral morphology that is seen in the lower eudicots, which do not show the fixed floral plan that characterizes the core eudicot radiation (Drinnan *et al.*, 1994; Hoot *et al.*, 1999; Magallon *et al.*, 1999). Unfortunately, many lower eudicot taxa are woody, slowly growing and/or full of latexes that can complicate histology and nucleic acid preparation. This has led a number of different research groups to focus on the first branch of the lower eudicots, represented by the order Ranunculales. The Ranunculales comprises seven families, several of which are primarily herbaceous (Hoot and Crane, 1995; Loconte *et al.*, 1995). The two families that have been the target of genetic research are the Papaveraceae or poppy family and the Ranunculaceae or buttercup family. Studies of these families have the added benefit of providing a phylogenetically intermediate, third data point for deep comparisons between the core eudicot and grass model systems (Fig. 4.1). In addition, they have the potential to shed light on the ancestral genetic toolkit that was present before the major radiation of the core eudicots.

4.1. *Papaver* and *Eschscholzia*

The poppy family represents one of the first branches of the Ranunculales clade and includes the long-standing self-incompatibility model system *Papaver rhoeas* as well as the alkaloid production model *Papaver somniferum* (Facchini and De Luca, 2008; Franklin-Tong, 2007). Stable transformation has been used in both of these but the development of RNAi-based techniques, specifically VIGS using the tobacco rattle virus platform, hold great promise (Chitty *et al.*, 2006; Hileman *et al.*, 2005). For example, this approach was successfully used to test the function of B gene homologs in *P. somniferum* (Drea *et al.*, 2007). In that case, there are two *AP3* paralogs that were derived independently of the core eudicot *AP3* duplication describe above, as well as two *PI* paralogs (Kramer *et al.*, 1998). Previous studies had found that *Papaver AP3-1* paralog was expressed in both petals and stamens while the other, *AP3-2*, was largely stamen specific (Kramer and Irish, 1999). Consistent with this, functional analyses in *P. somniferum* demonstrated that the paralogs contribute differentially to petal and stamen identity (Drea *et al.*, 2007). Knockdown of *AP3-1* function resulted in transformation of petals into sepals while *AP3-2* silencing produced stamen-to-carpel transformation. The two *PI* paralogs are more recently derived than the two *AP3*s and *PI-2* appears to be expressed at relatively low levels (Drea *et al.*, 2007; Kramer and Irish, 1999; Kramer *et al.*, 1998). Accordingly, *PI-1* silencing produces a strong phenotype while *PI-2* silencing alone has no effect, although silencing of both loci has an even stronger phenotype than *PI-1* alone (Drea *et al.*, 2007). In core eudicot models, AP3 and PI proteins function as obligate heterodimers (Riechmann *et al.*, 1996), which explains their equivalent mutant phenotypes (Bowman *et al.*, 1989). In *Papaver*, PI-1 dimerizes with both AP3-1 and AP3-2, suggesting that the distinct functions of the two AP3s is a product of biochemical differences between these two paralogs (Drea *et al.*, 2007). This exciting work in *Papaver* is complemented by genetic and genomic studies that are ongoing in *Eschscholzia*, the California poppy. Available resources in this system include an EST database, stable transformation, and VIGS protocols (Carlson *et al.*, 2006; MacLeod and Facchini, 2006; Wege *et al.*, 2007). In addition to questions related to floral organ identity, *Eschscholzia* is being used to investigate a wide range of morphological questions including leaf and carpel/fruit development (Becker *et al.*, 2005; Gleissberg, 2004).

4.2. *Aquilegia*

The Ranunculaceae or buttercup family is of particular evolutionary interest due to its diverse forms of sepals and petals. Across the angiosperms, botanists regularly invoke the idea that petaloid organs have evolved many times independently, but the Ranunculaceae are perhaps the best example of this idea (Kosuge, 1994; Tamura, 1965; Worsdell, 1903). Specifically, it has been hypothesized that the diverse petal forms found in this family were derived from outer stamens on multiple occasions. In addition to this issue, the sepals in many Ranunculaceae are petaloid. Normally, we think of B gene expression promoting petaloidy, but in models such as *Arabidopsis* and *Antirrhinum*, there is only one petal identity program, that is, expression of B genes can only produce one kind of petaloid organ (Krizek and Meyerowitz, 1996). So, how do these genera make their sepals petaloid and showy but still maintain morphological differentiation of the petals in the second whorl? Additional morphological questions arise in specific genera, including the emerging model system *Aquilegia* (reviewed in Hodges and Kramer, 2007). *Aquilegia*, commonly known as columbine, has nectar spurs on its petals, a recently evolved trait that was critical to the recent adaptive radiation of the genus (Hodges, 1997; Whittall and Hodges, 2007). The flowers also contain five distinct types of organs. In addition to the typical sepals, petals, stamens, and carpels, there is a novel form of sterile organ, termed the staminodium, inserted between the stamens and carpels (Fig. 4.4F). Thus, the flowers of *Aquilegia* encompass several questions related to the evolution of floral organ identity. How can you make two different types of petaloid organs in the same flower? How is the identity of the fifth organ type established? What is the genetic basis for spur development? On a broader level, we can ask whether the petal identity program that is functioning in *Aquilegia* appears to be conserved across the family or there is actually evidence for independent derivations of petals in different genera. The process of addressing these questions has been facilitated by the development of several genetic and genomic tools for *Aquilegia*, including a large EST database, BAC libraries, a physical map, a microarray platform, and VIGS techniques (Gould and Kramer, 2007; Hodges and Kramer, 2007; Kramer, 2009). The focus of early studies has been the ABC gene homologs, particularly the B genes. As has been seen in other taxa, there have been gene duplications in the Ranunculaceae *AP3* lineage, in this case three paralogs termed *AP3-1*, *AP3-2*, and *AP3-3* (Kramer *et al.*, 2003).

The expression of these genes in *Aquilegia* suggests temporal and spatial subfunctionalization: *AP3-1* is expressed at early stages in petals, stamens, and staminodia but quickly becomes restricted to the staminodia; *AP3-2* is expressed in early stamens and staminodia but, in a complementary pattern to *AP3-1*, loses expression in the staminodia and comes on in the petals; and *AP3-3* is strictly expressed in petals (Kramer *et al.*, 2007). The differential expression of *AP3-1* and *AP3-2* between the staminodia and stamens, respectively, may indicate neofunctionalization if *AP3-1* is actually required for the identity of this novel organ. The single *PI* ortholog is expressed continually throughout the petals, stamens, and staminodia, consistent with the fact that the PI protein forms heterodimers with each of the three AP3s (Kramer *et al.*, 2007). VIGS was used to knock down *PI* function in *Aquilegia* and flowers were recovered with petal-to-sepal, stamen-to-carpel, and staminodium-to-carpel transformations (Kramer *et al.*, 2007). This is consistent with a traditional B class mutant phenotype, with the novel addition of a role in staminodium identity. These findings suggest that staminodium identity is controlled by B function and is perhaps derived from the pre-existing stamen identity program. The next step in this analysis will be testing the functions of each *AP3* paralog, which will hopefully provide insight into their differential functions. Perhaps, the most surprising finding was that *PI* silencing had little effect on the gross or micromorphology of the sepals (Kramer *et al.*, 2007). This was consistent with the fact that none of the B gene homologs are expressed in the sepals at early stages when identity is normally established. This indicates that petaloidy of the sepals is not genetically controlled by the B gene homologs and that, therefore, alternative genetic mechanisms to promote petaloidy must have evolved. Now, these findings can be used as a basis for broader studies across the Ranunculaceae. In particular, the petal specific expression of the *AP3-3* paralog in *Aquilegia* creates a useful marker for conservation of the petal identity program. It turns out that orthologs of *AP3-3* are petal specific in their expression across widely divergence taxa in both the Ranunculaceae and Berberidaceae (Rasmussen *et al.*, 2009). Moreover, they are typically not expressed in flowers that lack petals (Kramer *et al.*, 2003; Rasmussen *et al.*, 2009). These data indicate that the genetic program in diverse petal types is likely to be homologous and does not support a model where petals were truly independently derived on many occasions.

Aside from these questions related to floral organ identity, *Aquilegia* is of considerable interest due to its recent, rapid adaptive radiation (Hodges and Arnold, 1994). This diversification appears to have been largely driven by pollinator shifts as the plants moved into new environments, which is correlated with changes in floral color, floral orientation, and spur length/shape (Whittall and Hodges, 2007). Therefore, establishing the genetic basis of spur development will be useful in terms of dissecting a morphological innovation as well as providing candidate genes for the diversification of spur morphology. Further, the high degree of interfertility among *Aquilegia* species combined with their low genetic divergence (Hodges and Arnold, 1994; Hodges *et al.*, 2002; Prazmo, 1965) means that genetic resources developed for one species will be useful for QTL studies between many different species pairs. These tools are already being used to dissect the genetic basis for independent shifts from red to white flowers. In this case, it has been shown that loss of anthocyanin production is usually correlated with changes across the entire genetic pathway, suggesting regulatory changes at a high genetic level (Whittall *et al.*, 2006). Additional characteristics of interest in *Aquilegia* are its compound leaves, flowering time control, and fruit development.

5. Angiosperms: Monocots and Magnoliids

The monocots are the second largest monophyletic clade of flowering plants, containing ~20% of angiosperm diversity. They also include taxa that provide the bulk of the world's caloric intake, namely grasses such as rice (*Oryza*) and maize (*Zea*). Of course, there are many other economically important monocots, such as bananas, gingers, and palms. While the magnoliid and so-called "ANA" lineages (the orders Amborellales, Nymphaeales, and Austrobaileyales) are not large in numbers, they are of considerable interest from an evolutionary standpoint since these early diverging angiosperms exhibit much more variation in certain aspects of floral morphology and often possess what are considered primitive traits (Doyle and Endress, 2000; Endress, 1994). Identifying good model systems from these lineages has been challenging but some progress is being made.

5.1. The Poaceae

The use of *Zea mays* as a genetic and developmental model dates back to the mid-twentieth century (Candela and Hake, 2008; Peterson, 2005). Similarly, extensive resources are available for *Oryza sativa*, which is more amenable to transformation than *Zea*. Early work on these models was subsequently leveraged to produce a huge amount of genomic information that spans the grass family, allowing detailed genetic studies to be conducted across this morphologically diverse and economically critical group (Liang *et al.*, 2008). The depth and breadth of work on the grasses is too great to attempt to summarize in this context but a few studies hold particular interest for developmental evolution. For example, the grass flower is highly modified relative to those in dicot model systems, largely because of their transition to wind pollination. In place of the sepals and petals, these flowers have two outer sterile organs termed the palea and lemma that surround a second type of sterile organ called the lodicule (Fig. 4.4G–I). The homology of these structures relative to the sepals and petals of dicots has been a matter of controversy for some time (Dahlgren *et al.*, 1984). In particular, it was unclear whether the lodicules were derived from preexisting stamens or petals, or were perhaps an entirely novel structure (Clifford, 1987). Molecular studies of the B gene homologs in *Zea* and *Oryza* have shown that lodicule identity is dependent on the function of *AP3/PI* homologs (Ambrose *et al.*, 2000; Kang *et al.*, 1998; Lee *et al.*, 2003; Nagasawa *et al.*, 2003), but this does leave open the possibility that the organs could be derived from either petals or stamens. Further comparative studies across the grasses have supported the hypothesis that lodicules represent modified petals (Whipple *et al.*, 2007). This suggests that the derivation of lodicules from petals is analogous to the evolution of halteres from hind wings in dipterans—the highest level of the identity program is conserved but the downstream components are clearly divergent (Warren *et al.*, 1994; Weatherbee *et al.*, 1998). Another unique feature of the grasses is their extremely complex and varied inflorescence structure, which has also been important as a major target for modification during domestication (Bommert *et al.*, 2005; Kellogg, 2000). Several loci that may have played roles in these evolutionary changes have been identified through a combination of forward and reverse genetic approaches along with QTL studies (Bomblies *et al.*, 2003; Doust *et al.*, 2004; McSteen, 2006; Vollbrecht *et al.*, 2005). Some aspects of this morphological diversification may have been driven by gene and genome duplication (Yu *et al.*, 2005). For example, the E class homologs of the grasses have a very complex evolutionary history and comparative gene expression studies indicate that their functions are highly labile, often correlating with differences in inflorescence structure (Malcomber and Kellogg, 2004, 2005; Malcomber *et al.*, 2006). Further functional studies are necessary to test this hypothesis but the general concept, that floral organ identity genes may have been co-opted to function in aspects of inflorescence structure, is reminiscent of what has been found in the Asteraceae. Lastly, the leaves of monocots are often highly modified relative to those of dicots, with clasping leaf bases that encircle the stem and support strap-like laminae. The development of these leaves is also distinct from that of dicot leaves from the earliest stages of primordium initiation (Sylvester *et al.*, 1990). Several studies have found evidence for both conservation and divergence in the genetic program that underlies grass versus dicot leaf development (Chitwood *et al.*, 2007; Tsiantis *et al.*, 1999). One notable example of divergence is that the YABBY gene family, which is expressed on the lower or abaxial side of leaves in *Arabidopsis* (Siegfried *et al.*, 1999), functions on the upper or adaxial side of leaves in maize (Juarez *et al.*, 2004). The advantage to the grass system is that these types of differences can be fully explored at the functional level, allowing us to move beyond comparative expression data.

5.2. Other potential monocot and magnoliid models

As noted above, the monocots include a considerable amount of diversity and several other families are of particular interest for development as model systems. One important case is the Orchidaceae, the largest family of flowering plants. This family is particularly well known for the beauty and complexity of its flowers, which are being targeted for research into the evolution of floral organ identity as well as floral symmetry. One major feature of orchid flowers is that, similar to *Aquilegia*, they produce more than one type of petaloid organ. In this case, however, they produce three types: petaloid sepals in the first whorl, two true petals in the second whorl, and a morphologically distinct "lip" petal, also in the second whorl. This lip petal is thought to represent a fusion between a second whorl petal and third whorl stamen (Mondragon-Palomino and Theissen, 2008; Tsai *et al.*, 2008). Again like *Aquilegia*, there is evidence that duplications in the *AP3* homologs may be important to the differentiation of these three petaloid organ types (Mondragon-Palomino and Theissen, 2008; Tsai *et al.*, 2004, 2008). It remains to be tested whether *CYC* homologs have also been recruited to promote zygomorphy in the orchids. Other monocot lineages that hold potential interest include the gingers, which show fascinating trends in stamen to petal transformations over the course of their evolution (Kress *et al.*, 2002). Outside of the monocots, it would be enormously valuable to have one or more models in the magnoliid dicot or ANA lineages. Unfortunately, many of these taxa are woody, slow-growing and/or difficult to grow. One possible exception is the species *Aristolochia fimbriata*, which is herbaceous, fast-growing and self-fertile with a relatively small genome (M.A. Jaramillo, A. Litt, and C. dePamphilis, personal communication). *Aristolochia* is characterized by unusual floral morphology—the petals have been lost and the first whorl sepals are fused to form a highly modified floral tube (Gonzalez and Stevenson, 2000). Studies of the *Aristolochia* B gene homologs determined that, although this structure is arguably petaloid, homologs of *AP3* and *PI* are not expressed in a manner that would suggest a role in organ identity or promoting petaloidy (Jaramillo and Kramer, 2004). This finding is consistent with what has now been observed in several taxa with petaloid first whorl organs, indicating that even if the *AP3/PI* genetic program is deeply conserved, there are other genetic mechanisms for promoting petaloidy (Jaramillo and Kramer, 2007; Kramer and Jaramillo, 2005).

6. Conclusions

Considering the wide breadth of new plant model systems, there are several themes that rise to the fore. First, there are major advantages to developing clusters of model systems, whether within the close vicinity of an established model such as *Arabidopsis*, *Antirrhinum*, or *Oryza*, or entirely new clusters such as the species of *Aquilegia*. Second, greater focus needs to be placed on underrepresented lineages of the plant tree of life. In depth analyses of models such as *Physcomitrella* are just beginning to take advantage of that model's enormous resources and are likely to be among the most exciting and intriguing advances in plant evo devo over the coming years. In terms of the results themselves, we see the repeated theme of gene duplication as a mechanism for functional and morphological evolution. Also, convergent genetic co-option events, as exemplified by the *CYC* homologs of the core eudicots, are particularly striking. Plants appear to be especially good models to study these types of phenomena and will provide ever more detailed examples as the plant evo-devo field takes makes full use of these new model systems.

ACKNOWLEDGMENTS

Work described here was funded by NSF-BE grant #0412727 and NSF-IBN grants #0319103 and #0720240 to E. M. Kramer.

REFERENCES

Al-Shehbaz, I. A., Beilstein, M. A., and Kellogg, E. A. (2006). Systematics and phylogeny of the Brassicaceae (Cruciferae): An overview. *Plant Syst. Evol.* **259,** 89–120.

Ambrose, B. A., Lerner, D. R., Ciceri, P., Padilla, C. M., Yanofsky, M. F., and Schmidt, R. J. (2000). Molecular and genetic analyses of the *Silky1* gene reveal conservation in floral organ specification between eudicots and monocots. *Mol. Cell* **5,** 569–579.

Banks, J. A. (1999). Gametophyte development in ferns. *Ann. Rev. Plant Phys. Plant Mol. Biol.* **50,** 163–186.

Barkoulas, M., Hay, A., Kougioumoutzi, E., and Tsiantis, M. (2008). A developmental framework for dissected leaf formation in the *Arabidopsis* relative *Cardamine hirsuta*. *Nature Gen.* **40,** 1136–1141.

Becker, A., Gleissberg, S., and Smyth, D. R. (2005). Floral and vegetative morphogenesis in California poppy (Eschscholzia californica Cham.). *Int. J. Plant Sci.* **166,** 537–555.

Beilstein, M. A., Al-Shehbaz, I. A., and Kellogg, E. A. (2006). Brassicaceae phylogeny and trichome evolution. *Am. J. Bot.* **93,** 607–619.

Bharathan, G., Goliber, T. E., Moore, C., Kessler, S., Pham, T., and Sinha, N. R. (2002). Homologies in leaf form inferred from KNOXI gene expression during development. *Science* **296,** 1858–1860.

Bohlenius, H., Huang, T., Charbonnel-Campaa, L., Brunner, A. M., Jansson, S., Strauss, S. H., and Nilsson, O. (2006). CO/FT Regulatory Module Controls Timing of Flowering an Seasonal Growth Cessation in Trees. *Science* **312,** 1040–1043.

Bomblies, K., Wang, R.-L., Ambrose, B. A., Schmidt, R., Meeley, R. B., and Doebley, J. (2003). Duplicate FLORICAULA/LEAFY homologs zfl1 and zfl2 control inflorescence architecture and flower patterning in maize. *Development* **130,** 2385–2395.

Bomblies, K., and Weigel, D. (2007). Arabidopsis - a model genus for speciation. *Curr. Opin. Gen. Dev.* **17,** 500–504.

Bommert, P., Satoh-Nagasawa, N., Jackson, D., and Hirano, H. Y. (2005). Genetics and evolution of inflorescence and flower development in grasses. *Plant Cell Phys.* **46,** 69–78.

Borevitz, J. O., Liang, D., Plouffe, D., Chang, H. S., Zhu, T., Weigel, D., Berry, C. C., Winzeler, E., and Chory, J. (2003). Large-scale identification of single-feature polymorphisms in complex genomes. *Genome Res.* **13,** 513–523.

Bosch, J. A., Heo, K., Sliwinski, M. K., and Baum, D. A. (2008). An exploration of LEAFY expression in independent evolutionary origins of rosette flowering in Brassicaceae. *Am. J. Bot.* **95,** 286–293.

Bowman, J. L., and Floyd, S. K. (2008). Patterning and polarity in seed plant shoots. *Ann. Rev. Plant Biol.* **59,** 67–88.

Bowman, J. L., Smyth, D. R., and Meyerowitz, E. M. (1989). Genes directing flower development in *Arabidopsis*. *Plant Cell* **1,** 37–52.

Bradley, D., Carpenter, R., Sommer, H., Hartley, N., and Coen, E. (1993). Complementary floral homeotic phenotypes result from opposite orientation of a transposon at the *plena* locus of Antirrhinum. *Cell* **72,** 85–95.

Bradshaw, H. D., Otto, K. G., Frewen, B. E., McKay, J. K., and Schemske, D. W. (1998). Quantitative trait loci affecting differences in floral morphology between two species of monkeyflower (Mimulus). *Genetics* **149,** 367–382.

Brigneti, G., Martin-Hernandez, A. M., Jin, H. L., Chen, J., Baulcombe, D. C., Baker, B., and Jones, J. D. G. (2004). Virus-induced gene silencing in Solanum species. *Plant J.* **39,** 264–272.

Broholm, S. K., Tahtiharju, S., Laitinen, R. A. E., Albert, V. A., Teeri, T. H., and Elomaa, P. (2008). A TCP domain transcription factor controls flower type specification along the radial axis of the *Gerbera* (Asteraceae) inflorescence. *Proc. Natl. Acad. Sci. USA* **105,** 9117–9122.

Burch-Smith, T. M., Anderson, J. C., Martin, G. B., and Dinesh-Kumar, S. P. (2004). Applications and advantages of virus-induced gene silencing for gene function studies in plants. *Plant J.* **39,** 734–746.

Busch, A., and Zachgo, S. (2007). Control of corolla monosymmetry in the Brassicaceae Iberis amara. *Proc. Natl. Acad. Sci. USA* **104,** 16714–16719.

Busov, V., Fladung, M., Groover, A., and Strauss, S. (2005). Insertional mutagenesis in Populus: Relevance and feasibility. *Tree Genet. Gen.* **1,** 135–142.

Candela, H., and Hake, S. (2008). The art and design of genetic screens: maize. *Nature Rev. Gen.* **9,** 192–203.

Carlson, J. E., Leebens-Mack, J. H., Wall, P. K., Zahn, L. M., Mueller, L. A., Landherr, L. L., Hu, Y., Ilut, D. C., Arrington, J. M., Choirean, S., Becker, A., Field, D., *et al.* (2006). EST database for early flower development in California poppy (Eschscholzia californica Cham., Papaveraceae) tags over 6000 genes from a basal eudicot. *Plant Mol. Biol.* **62,** 351–369.

Carpenter, R., and Coen, E. S. (1990). Floral homeotic mutations produced by transposon-mutagenesis in *Antirrhinum majus*. *Genes Dev.* **4,** 1483–1493.

Cartolano, M., Castillo, R., Efremova, N., Kuckenberg, M., Zethof, J., Gerats, T., Schwarz-Sommer, Z., and Vandenbussche, M. (2007). A conserved microRNA module exerts homeotic control over Petunia hybrida and Antirrhinum majus floral organ identity. *Nat. Genet* **39,** 901–905.

Causier, B., Castillo, R., Zhou, J. L., Ingram, R., Xue, Y. B., Schwarz-Sommer, Z., and Davies, B. (2005). Evolution in action: Following function in duplicated floral homeotic genes. *Curr. Biol.* **15,** 1508–1512.

Champagne, C. E. M., Goliber, T. E., Wojciechowski, M. F., Mei, R. W., Townsley, B. T., Wang, K., Paz, M. M., Geeta, R., and Sinhaa, N. R. (2007). Compound leaf development and evolution in the legumes. *Plant Cell* **19,** 3369–3378.

Chapman, M. A., Leebens-Mack, J. H., and Burke, J. M. (2008). Positive selection and expression divergence following gene duplication in the sunflower CYCLOIDEA gene family. *Mol. Biol. Evol.* **25,** 1260–1273.

Chitty, J. A., Allen, R. S., and Larkin, P. J. (2006). Opium poppy (Papaver somniferum). *Meth. Mol. Biol.* 383–391.

Chitwood, D. H., Guo, M. J., Nogueira, F. T. S., and Timmermans, M. C. P. (2007). Establishing leaf polarity: The role of small RNAs and positional signals in the shoot apex. *Development* **134,** 813–823.

Citerne, H. L., Luo, D., Pennington, T., Coen, E., and Cronk, Q. C. B. (2003). A phylogenomic investigation of *CYCLOIDEA*-like TCP genes in the Leguminosae. *Plant Phys.* **131,** 1042–1053.

Citerne, H. L., Moller, M., and Cronk, Q. C. B. (2000). Diversity of *cycloidea*-like genes in Gesneriaceae in relation to floral symmetry. *Ann. Bot.* **86,** 167–176.

Citerne, H. L., Pennington, R. T., and Cronk, Q. C. B. (2006). An apparent reversal in floral symmetry in the legume Cadia is a homeotic transformation. *Proc. Natl. Acad. Sci. USA* **103,** 12017–12020.

Clifford, H. T. (1987). Spiklet and floral morphology. *In* "Grass Systematics" (T. R. Soderstrom, K. W. Hilu, C. S. Campbell, and M. E. Barkworth, Eds.) pp. 21–30. Smithsonian Institution Press, Washington, D.C.

Coen, E. S., Doyle, S., Romero, J. M., Elliot, R., Magrath, R., and Carpenter, R. (1991). Homeotic genes controlling flower development in *Antirrhinum*. *Development* (Suppl. 1), 149–156.

Coen, E. S., and Meyerowitz, E. M. (1991). The war of the whorls: Genetic interactions controlling flower development. *Nature* **353,** 31–37.

Coen, E. S., and Nugent, J. M. (1994). Evolution of flowers and inflorescences. *In* "The Evolution of Developmental Mechanisms" (M. Akam, P. Holland, P. Ingham, and G. Wray, Eds.). Company of Biologists, Cambridge.

Cong, B., Barrero, L. S., and Tanksley, S. D. (2008). Regulatory change in YABBY-like transcription factor led to evolution of extreme fruit size during tomato domestication. *Nat. Gen.* **40,** 800–804.

Cove, D. (2005). The moss Physcomitrella patens. *Ann. Rev. Gen.* **39,** 339–358.

Cove, D., Bezanilla, M., Harries, P., and Quatrano, R. S. (2006). Mosses as model systems for the study of metabolism and development. *Ann. Rev. Plant Biol.* **57,** 497–520.

Cubas, P. (2002). Role of TCP genes in the evolution of morphological characters in angiosperms. *In* "Developmental Genetics and Plant Evolution" (Q. C. B. Cronk, R. M. Bateman, and J. A. Hawkins, Eds.), pp. 247–266. Taylor and Hawkins, London.

Cubas, P., Coen, E., and Martinez Zapater, J. M. (2001). Ancient asymmetries in the evolution of flowers. *Curr. Biol.* **11,** 1050–1052.

Cubas, P., Vincent, C., and Coen, E. (1999). An epigenetic mutation responsible for natural variation in floral symmetry. *Nature* **401,** 157–161.

Dahlgren, R. M. T., Clifford, H. T., and Yeo, P. F. (1984). The Families of the Monocotyledons. Springer Verlag, New York.

Davies, B., Cartolano, M., and Schwarz-Sommer, Z. (2006). Flower development: The Antirrhinum perspective. *Adv. Bot. Res.* **44,** 279–321.

Davies, B., Motte, P., Keck, E., Saedler, H., Sommer, H., and Schwarz-Sommer, Z. (1999). PLENA and FARINELLI: Redundancy and regulatory interactions between two Antirrhinum MADS-box factors controlling flower development. *EMBO.* **18,** 4023–34.

Ditta, G., Pinyopich, A., Robles, P., Pelaz, S., and Yanofsky, M. (2004). The SEP4 gene of Arabidopsis thaliana functions in floral organ and meristem identity. *Curr. Biol.* **14,** 1935–1940.

Donoghue, M. J., Ree, R. H., and Baum, D. A. (1998). Phylogeny and the evolution of flower symmetry in the Asteridae. *Tren. Plant Sci.* **3,** 311–317.

Donohue, K. (1998). Maternal determinants of seed dispersal in Cakile edentula: Fruit, plant, and site traits. *Ecology* **79,** 2771–2788.

Doust, A. N., Devos, K. M., Gadberry, M. D., Gale, M. D., and Kellogg, E. A. (2004). Genetic control of branching in foxtail millet. *Proc. Natl. Acad. Sci. USA* **101,** 9045–9050.

Doyle, J. A., and Endress, P. K. (2000). Morphological phylogenetic analysis of basal angiosperms: Comparison and combination with molecular data. *Int. J. Plant Sci.* **161,** S121–S153.

Drea, S., Hileman, L. C., de Martino, G., and Irish, V. F. (2007). Functional analyses of genetic pathways controlling petal specification in poppy. *Development* **134,** 4157–4166.

Drinnan, A. N., Crane, P. R., and Hoot, S. B. (1994). Patterns of floral evolution in the early diversification of non-magnoliid dicotyledons (eudicots). *In* "Early Evolution of Flowers" (P. K. Endress and E. M. Friis, Eds.), pp. 93–122. Springer-Verlag, New York.

Endress, P. K. (1994). Floral structure and evolution of primitive angiosperms: Recent advances. *Plant Syst. Evol.* **192,** 79–97.

Facchini, P. J., and De Luca, V. (2008). Opium poppy and Madagascar periwinkle: Model non-model systems to investigate alkaloid biosynthesis in plants. *Plant J.* **54,** 763–784.

Feng, X. Z., Zhao, Z., Tian, Z. X., Xu, S. L., Luo, Y. H., Cai, Z. G., Wang, Y. M., Yang, J., Wang, Z., Weng, L., Chen, J. H., Zheng, L. Y., *et al.* (2006). Control of petal shape and floral zygomorphy in Lotus japonicus. *Proc. Natl. Acad. Sci. USA* **103,** 4970–4975.

Ferrandiz, C., Liljegren, S. J., and Yanofsky, M. F. (2000). Negative regulation of the SHATTERPROOF genes by FRUITFULL during Arabidopsis fruit development. *Science* **289,** 436–8.

Fishman, L., Kelly, A. J., and Willis, J. H. (2002). Minor quantitative trait loci underlie floral traits associated with mating system divergence in Mimulus. *Evolution* **56,** 2138–2155.

Floyd, S. F., and Bowman, J. L. (2007). The ancestral developmental tool kit of land plants. *Int. J. Plant Sci.* **1,** 1–35.

Floyd, S. K., and Bowman, J. L. (2006). Distinct developmental mechanisms reflect the independent origins of leaves in vascular plants. *Curr. Biol.* **16,** 1911–1917.

Franklin-Tong, V. E. (2007). Inhibiting self-pollen: Self-incompatibility in Papaver involves integration of several signaling events. *J. Int. Plant Biol.* **49,** 1219–1226.

Friml, J., Benfey, P., Benkova, E., Bennett, M., Berleth, T., Geldner, N., Grebe, M., Heisler, M., Hejatko, J., Jurgens, G., Laux, T., Lindsey, K., *et al.* (2006). Apical-basal polarity: Why plant cells don't stand on their heads. *Tren. Plant Sci.* **11,** 12–14.

Fujita, T., Sakaguchi, H., Hiwatashi, Y., Wagstaff, S. J., Ito, M., Deguchi, H., Sato, T., and Hasebe, M. (2008). Convergent evolution of shoots in land plants: Lack of auxin polar transport in moss shoots. *Evol. Dev.* **10,** 176–186.

Gleissberg, S. (2004). Comparative analysis of leaf shape development in Eschscholzia californica and other Papaveraceae-Eschscholzioideae. *Am. J. Bot.* **91,** 306–312.

Gonzalez, F., and Stevenson, D. W. (2000). Perianth development and systematics of *Aristolochia*. *Flora* **195,** 370–391.

Goue, N., Lesage-Descauses, M. C., Mellerowicz, E. J., Magel, E., Label, P., and Sundberg, B. (2008). Microgenomic analysis reveals cell type-specific gene expression patterns between ray and fusiform initials within the cambial meristem of Populus. *New Phyt.* **180,** 45–56.

Gould, B., and Kramer, E. M. (2007). Virus-induced gene silencing as a tool for functional analyses in the emerging model plant Aquilegia (columbine, Ranunculaceae). *Plant Meth.* **3,** 6.

Groover, A., Fontana, J. R., Dupper, G., Ma, C. P., Martienssen, R., Strauss, S., and Meilan, R. (2004). Gene and enhancer trap tagging of vascular-expressed genes in poplar trees. *Plant Phys.* **134,** 1742–1751.

Groover, A. T. (2007). Will genomics guide a greener forest biotech? *Tren. Plant Sci.* **12,** 234–238.

Groover, A. T., Mansfield, S. D., DiFazio, S. P., Dupper, G., Fontana, J. R., Millar, R., and Wang, Y. (2006). The Populus homeobox gene ARBORKNOX1 reveals overlapping mechanisms regulating the shoot apical meristem and the vascular cambium. *Plant Mol. Biol.* **61,** 917–932.

Hall, D., Luquez, V., Garcia, V. M., St Onge, K. R., Jansson, S., and Ingvarsson, P. K. (2007). Adaptive population differentiation in phenology across a latitudinal gradient in European Aspen (Populus tremula, L.): A comparison of neutral markers, candidate genes and phenotypic traits. *Evolution* **61,** 2849–2860.

Hall, J. C., Tisdale, T. E., Donohue, K., and Kramer, E. M. (2006). Developmental basis of an anatomical novelty: Heteroarthrocarpy in Cakile lanceolata and Erucaria erucarioides (Brassicaceae). *Int. J. Plant Sci.* **167,** 771–789.

Harrison, C. J., Rezvani, M., and Langdale, J. A. (2007). Growth from two transient apical initials in the meristem of Selaginella kraussiana. *Development* **134,** 881–889.

Hay, A., and Tsiantis, M. (2006). The genetic basis for differences in leaf form between Arabidopsis thaliana and its wild relative Cardamine hirsuta. *Nat. Gen.* **38,** 942–947.

Hileman, L. C., Drea, S., de Martino, G., Litt, A., and Irish, V. F. (2005). Virus-induced gene silencing is an effective tool for assaying gene function in the basal eudicot species Papaver somniferum (opium poppy). *Plant J.* **44,** 334–341.

Hileman, L. C., Kramer, E. M., and Baum, D. A. (2003). Differential regulation of symmetry genes and the evolution of floral morphologies. *Proc. Natl. Acad. Sci. USA* **100,** 12814–12819.

Hirano, K., Nakajima, M., Asano, K., Nishiyama, T., Sakakibara, H., Kojima, M., Katoh, E., Xiang, H., Tanahashi, T., Hasebe, M., Banks, J. A., Ashikari, M., *et al.* (2007). The GID1-mediated gibberellin perception mechanism is conserved in the lycophyte Selaginella moellendorffii but not in the bryophyte Physcomitrella patens. *Plant Cell* **19,** 3058–3079.

Hodges, S. A. (1997). Floral nectar spurs and diversification. *Int'l J. Plant Sci.* **158,** S81–88.

Hodges, S. A., and Arnold, M. L. (1994). Columbines - A geographically widespread species flock. *Proc. Natl. Acad. Sci. USA* **91,** 5129–5132.

Hodges, S. A., and Kramer, E. M. (2007). Columbines. *Curr Biol.* **17,** R992–R994.

Hodges, S. A., Whittall, J. B., Fulton, M., and Yang, J. Y. (2002). Genetics of floral traits influencing reproductive isolation between *Aquilegia formosa* and *Aquilegia pubescens*. *Am. Nat.* **159**(Suppl.), S51–S60.

Hofer, J., Turner, L., Hellens, R., Ambrose, M., Matthews, P., Michael, A., and Ellis, N. (1997). *UNIFOLIATA* regulates leaf and flower morphogenesis in pea. *Curr. Biol.* **7,** 581–587.

Hoot, S., and Crane, P. R. (1995). Inter-familial relationships in the Ranunculidae based on molecular systematics. *Plant Syst. Evol. [Suppl.]* **9,** 119–131.

Hoot, S. B., Magallon, S., and Crane, P. R. (1999). Phylogeny of basal eudicots based on three molecular data sets: *atp*B, *rbc*L and 18S nuclear ribosomal DNA sequences. *Ann. MO Bot. Gard.* **86,** 1–32.

Ingvarsson, P. K., Garcia, M. V., Hall, D., Luquez, V., and Jansson, S. (2006). Clinal variation in phyB2, a candidate gene for day-length-induced growth cessation and bud set, across a latitudinal gradient in European aspen (Populus tremula). *Genetics* **172,** 1845–1853.

Janssen, B. J., Lund, L., and Sinha, N. (1998). Overexpression of a homeobox gene, LeT6, reveals indeterminate features in the tomato compound leaf. *Plant Phys.* **117,** 771–786.

Jansson, S., and Douglas, C. J. (2007). Populus: A model system for plant biology. *Ann. Rev. Plant Biol.* **58,** 435–458.

Jaramillo, M. A., and Kramer, E. M. (2004). APETALA3 and PISTILLATA homologs exhibit novel expression patterns in the unique perianth in Aristolochia (Aristolochiaceae). *Evol. Dev.* **6,** 449–458.
Jaramillo, M. A., and Kramer, E. M. (2007). The role of developmental genetics in understanding homology and morphological evolution in plants. *Int. J. Plant Sci.* **168,** 61–72.
Juarez, M. T., Twigg, R. W., and Timmermans, M. (2004). Specification of adaxial cell fate during maize leaf development. *Development* **131,** 4533–4544.
Kang, H.-G., Jeon, J.-S., Lee, S., and An, G. (1998). Identification of class B and class C floral organ identity genes from rice plants. *Plant Mol. Biol.* **38,** 1021–1029.
Kardailsky, I., Shukla, V. K., Ahn, J. H., Dagenais, N., Christensen, S. K., Nguyen, J. T., Chory, J., Harrison, M. J., and Weigel, D. (1999). Activation tagging of the floral inducer FT. *Science* **286,** 1962–5.
Kellogg, E. A. (2000). The grasses: A case study in macroevolution. *Ann. Rev. Ecol. Syst.* **31,** 217–238.
Kim, M., Cui, M. L., Cubas, P., Gillies, A., Lee, K., Chapman, M. A., Abbott, R. J., and Coen, E. (2008). Regulatory Genes Control a Key Morphological and Ecological Trait Transferred Between Species. *Science* **322,** 1116–1119.
Kim, M., McCormick, S., Timmermans, M., and Sinha, N. (2003a). The expression domain of PHANTASTICA determines leaflet placement in compound leaves. *Nature* **424,** 438–443.
Kim, M., Pham, T., Hamidi, A., McCormick, S., Kuzoff, R. K., and Sinha, N. (2003b). Reduced leaf complexity in tomato wiry mutants suggests a role for PHAN and KNOX genes in generating compound leaves. *Development* **130,** 4405–4415.
Kimura, S., Koenig, D., Kang, J., Yoong, F. Y., and Sinha, N. (2008). Natural variation in leaf morphology results from mutation of a novel KNOX gene. *Curr. Biol.* **18,** 672–677.
Knapp, S. (2002). Tobacco to tomatoes: A phylogenetic perspective on fruit diversity in the Solanaceae. *J. Exp. Bot.* **53,** 2001–2022.
Knapp, S., Bohs, L., Nee, M., and Spooner, D. M. (2004). Solanaceae - a model for linking genomics with biodiversity. *Comp. Func. Gen.* **5,** 285–291.
Koch, M., Bishop, J., and Mitchell-Olds, T. (1999). Molecular systematics and evolution of Arabidopsis and Arabis. *Plant Biol.* **1,** 529–537.
Kosuge, K. (1994). Petal evolution in Ranunculaceae. *Plant Syst. Evol. (Suppl.)* **8,** 185–191.
Kramer, E. M. (2006). Wood grain pattern formation: A brief review. *J. Plant Grow. Reg.* **25,** 290–301.
Kramer, E. M. (2009). Aquilegia: A new model for plant development, ecology, and evolution. *Ann. Rev. Plant Biol.* **60,** in press.
Kramer, E. M., Di Stilio, V. S., and Schluter, P. (2003). Complex patterns of gene duplication in the APETALA3 and PISTILLATA lineages of the Ranunculaceae. *Int. J. Plant Sci.* **164,** 1–11.
Kramer, E. M., Dorit, R. L., and Irish, V. F. (1998). Molecular evolution of genes controlling petal and stamen development: Duplication and divergence within the *APETALA3* and *PISTILLATA* MADS-box gene lineages. *Genetics* **149,** 765–783.
Kramer, E. M., Holappa, L., Gould, B., Jaramillo, M. A., Setnikov, D., and Santiago, P. (2007). Elaboration of B gene function to include the identity of novel floral organs in the lower eudicot Aquilegia (Ranunculaceae). *Plant Cell* **19,** 750–766.
Kramer, E. M., and Irish, V. F. (1999). Evolution of genetic mechanisms controlling petal development. *Nature* **399,** 144–148.
Kramer, E. M., and Irish, V. F. (2000). Evolution of the petal and stamen developmental programs: Evidence from comparative studies of the lower eudicots and basal angiosperms. *Int. J. Plant Sci.* **161,** S29–S40.
Kramer, E. M., and Jaramillo, M. A. (2005). The genetic basis for innovations in floral organ identity. *J. Exp. Zool. (Mol. Dev. Evol.)* **304B,** 526–535.

Kramer, E. M., Jaramillo, M. A., and Di Stilio, V. S. (2004). Patterns of gene duplication and functional evolution during the diversification of the AGAMOUS subfamily of MADS-box genes in angiosperms. *Genetics* **166,** 1011–1023.

Kramer, E. M., Su, H.-J., Wu, J. M., and Hu, J. M. (2006). A simplified explanation for the frameshift mutation that created a novel C-terminal motif in the APETALA3 gene lineage. *BMC Evol. Biol.* **6,** 30.

Kress, W. J., Prince, L. M., and Williams, K. J. (2002). The phylogeny and a new classification of the gingers (Zingiberaceae): Evidence from molecular data. *Am. J. Bot.* **89,** 1682–1696.

Krizek, B. A., and Meyerowitz, E. M. (1996). The Arabidopsis homeotic genes APETALA3 and PISTILLATA are sufficient to provide the B class organ identity function. *Development* **122,** 11–22.

Laitinen, R. A. E., Broholm, S., Albert, V. A., Teeri, T. H., and Elomaa, P. (2006). Patterns of MADS-box gene expression mark flower-type development in Gerbera hybrida (Asteraceae). *BMC Plant Biol.* **6,** (09 June 2006).

Langlade, N. B., Feng, X. Z., Dransfield, T., Copsey, L., Hanna, A. I., Thebaud, C., Bangham, A., Hudson, A., and Coen, E. (2005). Evolution through genetically controlled allometry space. *Proc. Natl. Acad. Sci. USA* **102,** 10221–10226.

Lee, M. M., and Schiefelbein, J. (2001). Developmentally distinct MYB genes encode functionally equivalent proteins in Arabidopsis. *Development* **128,** 1539–1546.

Lee, S., Jeon, J.-S., An, K., Moon, Y.-H., Lee, S., Chung, Y.-Y., and An, G. (2003). Alteration of floral organ identity in rice through ectopic expression of OsMADS16. *Planta* **217,** 904–911.

Liang, C. Z., Jaiswal, P., Hebbard, C., Avraham, S., Buckler, E. S., Casstevens, T., Hurwitz, B., McCouch, S., Ni, J. J., Pujar, A., Ravenscroft, D., Ren, L., *et al.* (2008). Gramene: A growing plant comparative genomics resource. *Nuc. Acids Res.* **36,** D947–D953.

Liljegren, S. J., Ditta, G. S., Eshed, Y., Savidge, B., Bowman, J. L., and Yanofsky, M. F. (2000). *SHATTERPROOF* MADS-box genes control seed dispersal in *Arabidopsis*. *Nature* **404,** 766–770.

Litt, A. (2007). An evaluation of A-function: Evidence from the APETALA1 and APETALA2 gene lineages. *Int'l J. Plant Sci.* **168,** 73–91.

Liu, J. P., Van Eck, J., Cong, B., and Tanksley, S. D. (2002). A new class of regulatory genes underlying the cause of pear-shaped tomato fruit. *Proc. Natl. Acad. Sci. USA* **99,** 13302–13306.

Loconte, H., Campbell, L. M., and Stevenson, D. W. (1995). Ordinal and familial relationships of Ranunculid genera. *Plant Syst. Evol. [Suppl.]* **9,** 99–118.

Luo, D., Carpenter, R., Copsey, L., Vincent, C., Clark, J., and Coen, E. (1999). Control of organ asymmetry in flowers of Antirrhinum. *Cell* **99,** 367–376.

Luo, D., Carpenter, R., Vincent, C., Copsey, L., and Coen, E. (1996). Origin of floral asymmetry in *Antirrhinum*. *Nature* **383,** 794–799.

MacLeod, B. P., and Facchini, P. J. (2006). Methods for regeneration and transformation in Eschscholzia californica - A model plant to investigate alkaloid biosynthesis. *Meth. Mol. Biol.* 357–368.

Magallon, S., Crane, P. R., and Herendeen, P. S. (1999). Phylogenetic pattern, diversity, and diversification of eudicots. *Ann. MO Bot. Gard.* **86,** 297–372.

Magnani, E., and Hake, S. (2008). KNOX lost the OX: The Arabidopsis KNATM gene defines a novel class of KNOX transcriptional regulators missing the homeodomain. *Plant Cell* **20,** 875–887.

Maizel, A., Busch, M. A., Tanahashi, T., Perkovic, J., Kato, M., Hasebe, M., and Weigel, D. (2005). The floral regulator LEAFY evolves by substitutions in the DNA binding domain. *Science* **308,** 260–263.

Malcomber, S. T., and Kellogg, E. A. (2004). Heterogeneous expression patterns and separate roles of the SEPALLATA gene LEAFY HULL STERILE1 in Grasses. *Plant Cell* **16,** 1692–1706.

Malcomber, S. T., and Kellogg, E. A. (2005). SEPALLATA gene diversification: Brave new whorls. *Tren. Plant Sci.* **10,** 427–435.

Malcomber, S. T., Preston, J. C., Reinheimer, R., Kossuth, J., and Kellogg, E. A. (2006). Developmental gene evolution and the origin of grass inflorescence diversity. *Adv. Bot. Res.* **44,** 425–481.

Marella, H. H., Sakata, Y., and Quatrano, R. S. (2006). Characterization and functional analysis of ABSCISIC ACID INSENSITIVE3-like genes from Physcomitrella patens. *Plant J.* **46,** 1032–1044.

McSteen, P. (2006). Branching out: The ramosa pathway and the evolution of grass inflorescence morphology. *Plant Cell* **18,** 518–522.

Minami, A., Nagao, M., Arakawa, K., Fujikawa, S., and Takezawa, D. (2003). Abscisic acid-induced freezing tolerance in the moss Physcomitrella patens is accompanied by increased expression of stress-related genes. *J. Plant Phys.* **160,** 475–483.

Ming, R., Hou, S. B., Feng, Y., Yu, Q. Y., Dionne-Laporte, A., Saw, J. H., Senin, P., Wang, W., Ly, B. V., Lewis, K. L. T., Salzberg, S. L., Feng, L., *et al.* (2008). The draft genome of the transgenic tropical fruit tree papaya (Carica papaya Linnaeus). *Nature* **452,** 991–996.

Mitchell-Olds, T., and Schmitt, J. (2006). Genetic mechanisms and evolutionary significance of natural variation in Arabidopsis. *Nature* **441,** 947–952.

Mondragon-Palomino, M., and Theissen, G. (2008). MADS about the evolution of orchid flowers. *Tren. Plant Sci.* **13,** 51–59.

Moore, M. J., Bell, C. D., Soltis, P. S., and Soltis, D. E. (2007). Using plastid genome-scale data to resolve enigmatic relationships among basal angiosperms. *Proc. Natl. Acad. Sci. USA* **104,** 19363–19368.

Mouchel, C. F., Osmont, K. S., and Hardtke, C. S. (2006). BRX mediates feedback between brassinosteroid levels and auxin signalling in root growth. *Nature* **443,** 458–461.

Nagasawa, N., Miyoshi, M., Sano, Y., Satoh, H., Hirano, H., Sakai, H., and Nagato, Y. (2003). *SUPERWOMAN1* and *DROOPING LEAF* genes control floral organ identity in rice. *Development* **130,** 705–718.

Nakazato, T., Jung, M. K., Housworth, E. A., Rieseberg, L. H., and Gastony, G. J. (2006). Genetic map-based analysis of genome structure in the homosporous fern Ceratopteris richardii. *Genetics* **173,** 1585–1597.

Nilsson, J., Karlberg, A., Antti, H., Lopez-Vernaza, M., Mellerowicz, E., Perrot-Rechenmann, C., Sandberg, G., and Bhalerao, R. P. (2008). Dissecting the molecular basis of the regulation of wood formation by auxin in hybrid aspen. *Plant Cell* **20,** 843–855.

Oyama, R. K., and Baum, D. A. (2004). Phylogenetic relationships of north American Antirrhinum (Veronicaceae). *Am. J. Bot.* **91,** 918–925.

Pavy, N., Johnson, J. J., Crow, J. A., Paule, C., Kunau, T., MacKay, J., and Retzel, E. F. (2007). ForestTreeDB: A database dedicated to the mining of tree transcriptomes. *Nuc. Acids Res.* **35,** D888–D894.

Pelaz, S., Ditta, G. S., Baumann, E., Wisman, E., and Yanofsky, M. (2000). B and C floral organ identity functions require SEPALLATA MADS-box genes. *Nature* **405,** 200–203.

Peterson, P. A. (2005). The plant genetics discovery of the century: Transposable elements in maize. Early beginnings to 1990. *Maydica* **50,** 321–337.

Pinyopich, A., Ditta, G. S., Savidge, B., Liljegren, S. J., Baumann, E., Wisman, E., and Yanofsky, M. F. (2003). Assessing the redundancy of MADS-box genes during carpel and ovule development. *Nature* **424,** 85–88.

Prazmo, W. (1965). Cytogenetic studies on the genus *Aquilegia*. IV. Fertility relationships among the *Aquilegia* species. *Acta Soc. Bot. Pol.* **34,** 667–685.

Qiu, Y.-L., Li, L. B., Chen, Z. D., Dombrovska, O., Lee, J., Kent, L., Li, R. Q., Jobson, R. W., Hendry, T. A., Taylor, D. W., Testa, C. M., and Ambros, M. (2007). A nonflowering land plant phylogeny inferred from nucleotide sequences of seven chloroplast, mitochondrial, and nuclear genes. *Int'l J. Plant Sci.* **168,** 691–708.

Quatrano, R. S., McDaniel, S. F., Khandelwal, A., Perroud, P. F., and Cove, D. J. (2007). Physcomitrella patens: Mosses enter the genomic age. *Curr. Opin. Plant Biol.* **10,** 182–189.

Rasmussen, D. E., Kramer, E. M., and Zimmer, E. A. (2009). One size fits all? Molecular evidence for a commonly inherited petal identity program in the Ranunculales. *Am. J. Bot.* **96,** 1–14.

Rensing, S. A., Lang, D., Zimmer, A. D., Terry, A., Salamov, A., Shapiro, H., Nishiyama, T., Perroud, P. F., Lindquist, E. A., Kamisugi, Y., Tanahashi, T., Sakakibara, K., *et al.* (2008). The Physcomitrella genome reveals evolutionary insights into the conquest of land by plants. *Science* **319,** 64–69.

Riechmann, J. L., Krizek, B. A., and Meyerowitz, E. M. (1996). Dimerization specificity of Arabidopsis MADS domain homeotic proteins APETALA1, APETALA3, PISTILLATA, and AGAMOUS. *Proc. Natl. Acad. Sci. USA* **93,** 4793–4798.

Rieseberg, L. H., Raymond, O., Rosenthal, D. M., Lai, Z., Livingstone, K., Nakazato, T., Durphy, J. L., Schwarzbach, A. E., Donovan, L. A., and Lexer, C. (2003). Major ecological transitions in wild sunflowers facilitated by hybridization. *Science* **301,** 1211–1216.

Rijpkema, A. S., Royaert, S., Zethof, J., van der Weerden, G., Gerats, T., and Vandenbussche, M. (2006). Functional divergence within the DEF/AP3 lineage: An analysis of PhTM6 in Petunia hybrida. *Plant Cell* **18,** 1819–1832.

Robertson, D. (2004). VIGS vectors for gene silencing: Many targets, many tools. *Ann. Rev. Plant Biol.* **55,** 495–519.

Rutherford, G., Tanurdzic, M., Hasebe, M., and Banks, J. A. (2004). A systemic gene silencing method suitable for high throughput, reverse genetic analyses of gene function in fern gametophytes. *BMC Plant Biol.* **4,** 6.

Sablowski, R. (2007). Flowering and determinacy in Arabidopsis. *J. Exp. Bot.* **58,** 899–907.

Schrader, J., Baba, K., May, S. T., Palme, K., Bennett, M., Bhalerao, R. P., and Sandberg, G. (2003). Polar auxin transport in the wood-forming tissues of hybrid aspen is under simultaneous control of developmental and environmental signals. *Proc. Natl. Acad. Sci. USA* **100,** 10096–10101.

Schrader, J., Nilsson, J., Mellerowicz, E., Berglund, A., Nilsson, P., Hertzberg, M., and Sandberg, G. (2004). A high-resolution transcript profile across the wood-forming meristem of poplar identifies potential regulators of cambial stem cell identity. *Plant Cell* **16,** 2278–2292.

Schranz, M. E., and Mitchell-Olds, T. (2006). Independent ancient polyploidy events in the sister families Brassicaceae and Cleomaceae. *Plant Cell* **18,** 1152–1165.

Schranz, M. E., Song, B. H., Windsor, A. J., and Mitchell-Olds, T. (2007). Comparative genomics in the Brassicaceae: A family-wide perspective. *Curr. Opin. Plant Biol.* **10,** 168–175.

Schwarz-Sommer, Z., Davies, B., and Hudson, A. (2003). An everlasting pioneer: The story of Antirrhinum research. *Nature Rev. Gen.* **4,** 657–666.

Schwinn, K., Venail, J., Shang, Y. J., Mackay, S., Alm, V., Butelli, E., Oyama, R., Bailey, P., Davies, K., and Martin, C. (2006). A small family of MYB-regulatory genes controls floral pigmentation intensity and patterning in the genus Antirrhinum. *Plant Cell* **18,** 831–851.

Shu, G., Amaral, W., Hileman, L. C., and Baum, D. A. (2000). LEAFY and the evolution of rosette flowering in violet cress (Jonopsidium acaule, Brassicaceae). *Am. J. Bot.* **87,** 634–641.

Siegfried, K. R., Eshed, Y., Baum, S. F., Otsuga, D., Drews, G. N., and Bowman, J. L. (1999). Members of the YABBY gene family specify abaxial cell fate in Arabidopsis. *Development* **126,** 4117–4128.

Simon, M., Lee, M. M., Lin, Y., Gish, L., and Schiefelbein, J. (2007). Distinct and overlapping roles of single-repeat MYB genes in root epidermal patterning. *Dev. Biol.* **311,** 566–578.

Simpson, G. G. (2003). Evolution of flowering in response to day length: Flipping the CONSTANS switch. *BioEssays* **25,** 829–832.

Sinha, N. (1997). Simple and compound leaves: Reduction or multiplication? *Tren. Plant Sci.* **2,** 396–402.

Soltis, D. E., Soltis, P. S., Bennett, M. D., and Leitch, I. J. (2003). Evolution of genome size in angiosperms. *Am. J. Bot.* **90,** 1596–1603.

Stewart, W. N., and Rothwell, G. W. (1993). Paleobotany and the evolution of plants, 2nd ed. Cambridge University Press, Cambridge.

Stout, S. C., Clark, G. B., Archer-Evans, S., and Roux, S. J. (2003). Rapid and Efficient Suppression of Gene Expression in a Single-Cell Model System, Ceratopteris richardii. *Plant Phys.* **131,** 1165–1168.

Sylvester, A. W., Cande, W. Z., and Freeling, M. (1990). Division and differentiation during normal and liguleless-1 maize leaf development. *Development* **110,** 985–1000.

Tamura, M. (1965). Morphology, ecology, and phylogeny of the Ranunculaceae IV. *Sci. Rep. Osaka Univ.* **14,** 53–71.

Tanahashi, T., Sumikawa, N., Kato, M., and Hasebe, M. (2005). Diversification of genie function: Homologs of the floral regulator FLO/LFY control the first zygotic cell division in the moss Physcomitrella patens. *Development* **132,** 1727–1736.

Tang, C. L., Toomajian, C., Sherman-Broyles, S., Plagnol, V., Guo, Y. L., Hu, T. T., Clark, R. M., Nasrallah, J. B., Weigel, D., and Nordborg, M. (2007a). The evolution of selfing in Arabidopsis thaliana. *Science* **317,** 1070–1072.

Tang, W., Newton, R. J., and Weidner, D. A. (2007b). Genetic transformation and gene silencing mediated by multiple copies of a transgene in eastern white pine. *J. Exp. Bot.* **58,** 545–554.

Tanurdzic, M., and Banks, J. A. (2004). Sex-determining mechanisms in land plants. *Plant Cell* **16,** S61–S71.

Teeri, T. H., Kotilainen, M., Uimari, A., Ruokolainen, S., Ng, Y. P., Malm, U., Pollanen, E., Broholm, S., Laitinen, R., Elomaa, P., and Albert, V. A. (2006a). Floral developmental genetics of Gerbera (Asteraceae). *Adv. Bot. Res.* **44,** 323–351.

Teeri, T. H., Uimari, A., Kotilainen, M., Laitinen, R., Help, H., Elomaa, P., and Albert, V. A. (2006b). Reproductive meristem fates in Gerbera. *J. Exp. Bot.* **57,** 3445–3455.

True, J. R., and Haag, E. S. (2001). Developmental system drift and flexibility in evolutionary trajectories. *Evol. Dev.* **3,** 109–119.

Tsai, W. C., Hsiao, Y. Y., Pan, Z. J., Hsu, C. C., Yang, Y. P., Chen, W. H., and Chen, H. H. (2008). Molecular biology of orchid flowers: With emphasis on Phalaenopsis. *Adv. Bot. Res.* **47,** 99–145.

Tsai, W. C., Kuoh, C. S., Chuang, M. H., Chen, W. H., and Chen, H. H. (2004). Four DEF-Like MADS box genes displayed distinct floral morphogenetic roles in Phalaenopsis orchid. *Plant Cell Phys.* **45,** 831–844.

Tsiantis, M., Schneeberger, R., Golz, J. F., Freeling, M., and Langdale, J. A. (1999). The maize rough sheath2 gene and leaf development programs in monocot and dicot plants. *Science* **284,** 154–156.

Uimari, A., Kotilainen, M., Elomaa, P., Yu, D., Albert, V. A., and Teeri, T. H. (2004). Integration of reproductive meristem fates by a SEPALLATA-like MADS-box gene. *Proc. Natl. Acad. Sci. USA* **101,** 15817–15822.

van der Knaap, E., and Tanksley, S. D. (2003). The making of a bell pepper-shaped tomato fruit: Identification of loci controlling fruit morphology in Yellow Stuffer tomato. *Theor. Appl. Gen.* **107,** 139–147.

van der Krol, A. R., and Chua, N.-H. (1993). Flower development in petunia. *Plant Cell* **5,** 1195–1203.

Vandenbussche, M., Theissen, G., Van de Peer, Y., and Gerats, T. (2003a). Structural diversification and neo-functionalization during floral MADS-box gene evolution by C-terminal frameshift mutations. *Nuc. Acids Res.* **31,** 4401–4409.

Vandenbussche, M., Zethof, J., Royaert, S., Weterings, K., and Gerats, T. (2004). The duplicated B-class heterodimer model: Whorl-specific effects and complex genetic interactions in Petunia hybrida flower development. *Plant Cell* **16,** 741–754.

Vandenbussche, M., Zethof, J., Souer, E., Koes, R., Torinelli, G. B., Pezzotti, M., Ferrario, S., Angenent, G. C., and Gerats, T. (2003b). Toward the analysis of the Petunia MADS box gene family by reverse and forward transposon insertion mutagenesis approaches: B, C, and D function organ identity functions require SEPALLATA-like MADS box genes in Petunia. *Plant Cell* **15,** 2680–2693.

Veluthambi, K., Gupta, A. K., and Sharma, A. (2003). The current status of plant transformation technologies. *Curr. Sci.* **84,** 368–380.

Vollbrecht, E., Springer, P. S., Goh, L., Buckler, E. S., and Martienssen, R. (2005). Architecture of floral branch systems in maize and related grasses. *Nature* **436,** 1119–1126.

Waites, R., and Hudson, A. (1995). PHANTASTICA - A gene required for dorsoventrality of leaves in Antirrhinum majus. *Development* **121,** 2143–2154.

Waites, R., Selvadurai, H. R. N., Oliver, I. R., and Hudson, A. (1998). The PHANTASTICA gene encodes a MYB transcription factor involved in growth and dorsoventrality of lateral organs in Antirrhinum. *Cell* **93,** 779–789.

Wang, W. M., Tanurdzic, M., Luo, M. Z., Sisneros, N., Kim, H. R., Weng, J. K., Kudrna, D., Mueller, C., Arumuganathan, K., Carlson, J., Chapple, C., de Pamphilis, C., *et al.* (2005). Construction of a bacterial artificial chromosome library from the spikemoss Selaginella moellendorffii: A new resource for plant comparative genomics. *BMC Plant Biol.* **5,** 10.

Wang, Z., Luo, Y. H., Li, X., Wang, L. P., Xu, S. L., Yang, J., Weng, L., Sato, S. S., Tabata, S., Ambrose, M., Rameau, C., Feng, X. Z., *et al.* (2008). Genetic control of floral zygomorphy in pea (Pisum sativum L.). *Proc. Natl. Acad. Sci. USA* **105,** 10414–10419.

Warren, R., Nagy, L., Selegue, J., Gates, J., and Carroll, S. (1994). Evolution of homeotic gene regulation and function in flies and butterflies. *Nature* **372,** 458–461.

Weatherbee, S. D., Halder, G., Kim, J., Hudson, A., and Carroll, S. (1998). Ultrabithorax regulates genes at several levels of the wing-patterning hierarchy to shape the development of the Drosophila haltere. *Genes Dev.* **12,** 1474–1482.

Wege, S., Scholz, A., Gleissberg, S., and Becker, A. (2007). Highly efficient virus-induced gene silencing (VIGS) in california poppy (Eschscholzia californica): An evaluation of VIGS as a strategy to obtain functional data from non-model plants. *Ann. Bot.* **100,** 641–649.

Whibley, A. C., Langlade, N. B., Andalo, C., Hanna, A. I., Bangham, A., Thebaud, C., and Coen, E. (2006). Evolutionary paths underlying flower color variation in Antirrhinum. *Science* **313,** 963–966.

Whipple, C. J., Zanis, M. J., Kellogg, E. A., and Schmidt, R. J. (2007). Conservation of B class gene expression in the second whorl of a basal grass and outgroups links the origin of lodicules and petals. *Proc. Natl. Acad. Sci. USA* **104,** 1081–1086.

Whittall, J. B., and Hodges, S. A. (2007). Pollinator shifts drive increasingly long nectar spurs in columbine flowers. *Nature* **447,** 706–710.

Whittall, J. B., Voelckel, C., Kliebenstein, D. J., and Hodges, S. A. (2006). Convergence, constraint and the role of gene expression during adaptive radiation: floral anthocyanins in Aquilegia. *Mol. Ecol.* **15,** 4645–4657.

Worsdell, W. C. (1903). The origin of the perianth of flowers, with special reference to the Ranunculaceae. *New Phyt.* **2,** 42–48.

Wu, C. A., Lowry, D. B., Cooley, A. M., Wright, K. M., Lee, Y. W., and Willis, J. H. (2008). Mimulus is an emerging model system for the integration of ecological and genomic studies. *Heredity* **100,** 220–230.

Xiao, H., Jiang, N., Schaffner, E., Stockinger, E. J., and van der Knaap, E. (2008). A retrotransposon-mediated gene duplication underlies morphological variation of tomato fruit. *Science* **319,** 1527–1530.

Yanofsky, M. F., Ma, H., Bowman, J. L., Drews, G. N., Feldmann, K. A., and Meyerowitz, E. M. (1990). The protein encoded by the *Arabidopsis* homeotic gene *agamous* resembles transcription factors. *Nature* **346,** 35–39.

Yoon, H.-S., and Baum, D. A. (2004). Transgenic study of parallelism in plant morphological evolution. *Proc. Natl. Acad Sci. USA* **101,** 6524–6529.

Yu, D., Kotilainen, M., Pollanen, E., Mehto, M., Elomaa, P., Helariutta, Y., Albert, A., and Teeri, T. (1999). Organ identity genes and modified patterns of flower development in *Gerbera hybrida* (Asteraceae). *Plant J.* **17,** 51–62.

Yu, J., Wang, J., Lin, W., Li, S. G., Li, H., Zhou, J., Ni, P. X., Dong, W., Hu, S. N., Zeng, C. Q., Zhang, J. G., Zhang, Y., *et al.* (2005). The Genomes of Oryza sativa: A history of duplications. *PLOS Biol.* **3,** 266–281.

Zahn, L. M., Leebens-Mack, J., dePamphilis, C. W., Ma, H., and Theissen, G. (2005). To B or not to B a flower: The role of DEFICIENS and GLOBOSA orthologs in the evolution of the angiosperms. *J. Heredity* **96,** 225–240.

CHAPTER FIVE

Patterning the Spiralian Embryo: Insights from *Ilyanassa*

J. David Lambert

Contents

Abstract

The spiralian developmental program is a highly conserved mode of early development that is characterized by regularities in cleavage pattern, fate map, and larval morphology. It is found in a number of animal phyla, and was

Department of Biology, University of Rochester, Rochester, New York, USA

Current Topics in Developmental Biology, Volume 86
ISSN 0070-2153, DOI: 10.1016/S0070-2153(09)01005-9

likely present in the last common ancestor of the large superphylum Lophotrochozoa. Despite this key position for understanding the evolution of development in animals, and the intrinsic advantages for using spiralian embryos to study embryonic development and asymmetric cell division, very little is known about the molecular mechanisms of spiralian embryogenesis. The snail *Ilyanassa* has typical spiralian development, as well as a number of practical and experimental advantages that have made it a useful model for understanding spiralian embryogenesis and basic processes in metazoan development. Here, I describe the key embryological experiments that inform our understanding of spiralian development in *Ilyanassa*, and I review recent progress in understanding the molecular basis of patterning this embryo.

1. Introduction

The spiralian developmental program is shared by a number of protostome animal phyla. It is characterized by a set of striking similarities in developmental traits. Spiralian embryos share regularities in the proportion and angles of early cell divisions, known collectively as *spiral cleavage* (e.g., Fig. 5.1). In these embryos, the large cells present at the four-cell stage (called macromeres) divide synchronously to generate smaller cells (called micromeres) in sets of four called quartets. The angle of the divisions is also characteristic; all the cleavages that produce the micromeres in a given quartet are oriented at the same slight angle away from the animal pole, and this angle alternates in successive cleavage cycles. Spiralian development is also characterized by strong similarities in the fate map of the blastula, even among distantly related spiralian taxa (some examples are discussed below). Finally, spiralians also share similarities in larval morphology. The most striking of these is a band of one or more rows of ciliated cells that encircle the larva, at the level of the mouth.

Recent phylogenetic reconstructions of the evolutionary history of the animal kingdom indicate that the bilateral animals are comprised of three clades. These are the Deuterostomes (the vertebrates, ascidians, urchins, and hemichordates), the Ecdysozoans (including arthropods and nematodes and several other phyla), and the Lophotrochozoans. The latter clade is the largest, and includes molluscs, annelids, platyhelminth flatworms, nemerteans, entoprocts, brachiopods, bryozoans, and phoronids. Spiralian development is the dominant mode of early development in the Lophotrochozoa, and may be ancestral for this group (Dunn *et al.*, 2008). Despite the large fraction of animal diversity that displays spiralian development, and the key position of this character for understanding the evolution of early development in the metazoans, the mechanisms of spiralian development remain poorly understood.

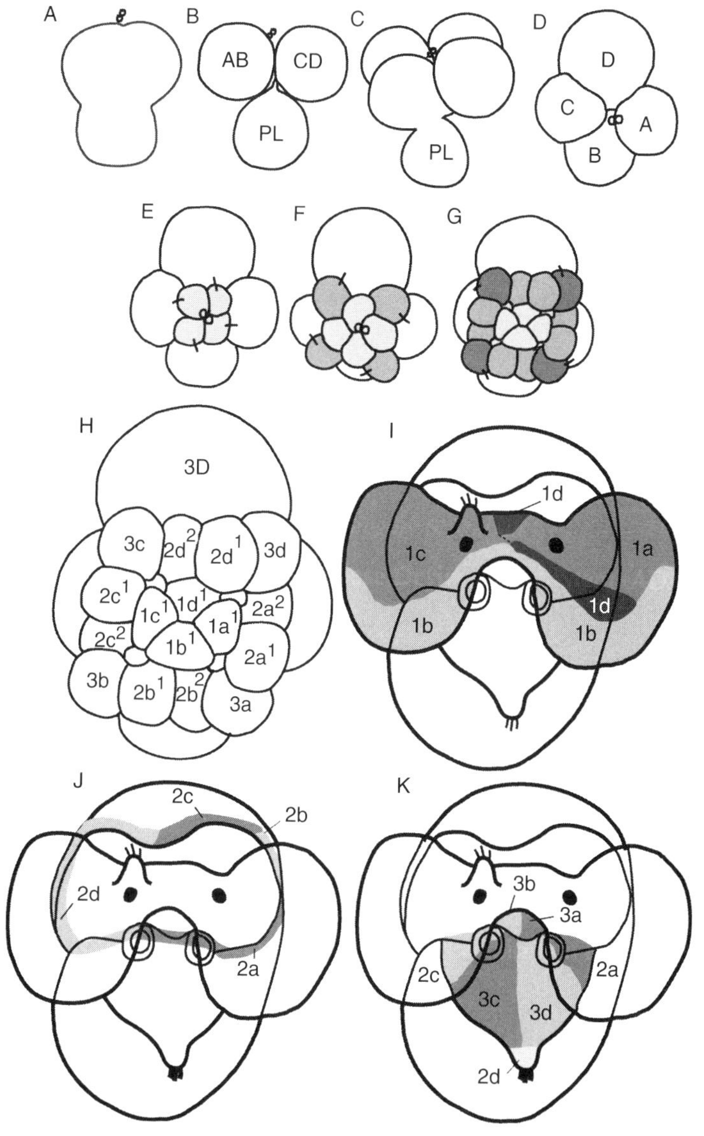

Figure 5.1 Early cleavage of *Ilyanassa*, and the fates of the micromeres. Polar lobes (PLs) are produced in the meiotic divisions (not shown) and during the first two cleavages (A–C). (D) At the four-cell stage, the cells are named A–D and known as macromeres. Successive cleavage cycles of the macromeres produce quartets of smaller cells called micromeres. (E) The eight-cell stage, after the birth of the first quartet (1a, 1b, 1c, and 1d, or collectively "1q"). Hatch marks indicate sister cell relationships from the preceding division. (F) The 12-cell stage, after the birth of the second quartet ("2q"). The first quartet cells will divide next to produce the 16-cell stage (not shown). (G and H) The 24-cell stage follows the birth of the third quartet ("3q"). At this stage, the

Ilyanassa obsoleta is a marine snail that displays typical spiralian development, including the conserved cleavage pattern, fate map, and larval characteristics. It has several key experimental and practical advantages that make it one of the most powerful models available for studying spiralian development, and a history of investigations over the last century has made this one of the best characterized spiralian embryos. The larval fates of the cells in the early embryo have been determined by lineage tracing with fluorescent markers (Render, 1991, 1997; Fig. 5.1 and Table 5.1). The developmental effects of deleting these cells has also been assayed (Clement, 1967, 1976, 1986a,b; Table 5.1). *Ilyanassa* is abundant and easy to collect in North America and elsewhere, or can be obtained from commercial suppliers. While the natural spawning season is short, high-quality embryos can now be obtained nearly year round, which is unusual for a marine invertebrate. In the last decade, a number of key molecular methods and resources have been developed for the *Ilyanassa* embryo (Gharbiah *et al.*, 2009). Perhaps most importantly, a robust microinjection protocol has been established which allows several different approaches for testing the function of specific genes in development (Rabinowitz *et al.*, 2008). Here, I review recent progress understanding the molecular basis of embryonic patterning in *Ilyanassa*, and discuss implications for understanding spiralian development.

2. Early Development in *Ilyanassa*: The Embryological Perspective

The major focus of classical embryological studies on *Ilyanassa* has been to understand how the cells of the early embryo are specified. Fate mapping studies have been performed for the cells born in the first five cleavage cycles (up to the 24-cell stage). They show that these cells have

first and second quartets have each divided into animal and vegetal tiers of cells: 1q produces $1q^1$ and $1q^2$, and 2q produces $2q^1$ and $2q^2$ (the $1q^2$ tier is comprised of the four small unlabeled cells adjacent to each $1q^1$ cell). The key characters that define spiral cleavage are the quadrilateral symmetry of the embryo, the macromere divisions to produce quartets of micromeres, and the alternation of the angle of the micromere divisions with respect to the animal pole in successive cleavage cycles. (I–K) Approximate contributions of micromere cells to selected organs in the veliger larva, which hatches about 7 days after fertilization. Clonal contributions are based on Render (1991, 1997). (I) Fates of first quartet cells in the head. Diagram based on Goulding (2003). (J) Contributions of second quartet cells to the mantle edge. 3c also makes a contribution to the mantle edge (not shown). (K) Contributions of the second and third quartet cells to the foot and esophagus. Clonal boundaries are approximate, because they are not based on simultaneous labeling of neighboring clones and cellular resolution was limited. Many aspects of the fate map are omitted for clarity; see Table 5.1 and references above for details. Also see Hejnol *et al.* (2007) for micromere fate maps in the closely related snail *Crepidula*. Views are from the side in (A) and (B), from the animal pole with the D quadrant up in (C–H), and from the anterior in (I–K).

Table 5.1 Fate map contributions and cell ablation phenotypes for micromere cells in *Ilyanassa*

Cell	Larval structures labeled after lineage tracing[a]	Effects of deletion[b]
1a	Left eye and velar lobe	Lacking left eye (8/8) Smaller left velar lobe (7/7)
1b	Ventral medial region of the head, including ventral edge of the velum	Synophthalmia (eyes too close) (14/14) Other variable defects (11/14)
1c	Right eye and velar lobe	Right eye and tentacle lacking (4/4), right velar lobe smaller (3/3)
1d	Medial and left velar ectoderm	No consistent morphological defect
2a	Ventral inner most edge of left side of velum	Left eye missing (12/17), left velum often smaller
	Left side of stomodeum	Left statolith smaller (8/11) or absent (1/11)
	Left upper edge of foot and upper half of statocyst capsule Ventral mantle edge	Anomalous internal birefringent masses in left anterior region (10/17)
2b	Posterior edge of velum around the dorsal and left and right sides Dorsal stomodeum Left mantle edge Larval retractor muscle, near site of attachment	No consistent morphological defect ($n = 15$)
2c	Ventral innermost edge of the right half of the velum	Stomodeum everted (16/16)
	Right half of the stomodeum	Shell small (16/16), often poorly formed
	Dorsal edge of the mantle	No beating heart observed
	Heart	Right statolith smaller (12/16) or absent (2/16)
	Right upper edge of the foot and upper half of statocyst capsule	Anomalous interior birefringent bodies in the anterior right region (8/16)
2d	Right edge of the mantle Tip of the foot	Shell is small, rudimentary or absent (21/21)

(*continued*)

Table 5.1 *(continued)*

Cell	Larval structures labeled after lineage tracing[a]	Effects of deletion[b]
3a	Left posterior ventral edge of the velum Left side of esophagus	Smaller left velar lobe (8/11)
3b	Right posterior ventral edge of the velum Right side of esophagus	Smaller right velar lobe
3c	Right posterior velum Right half of the foot and right statocyst Dorsal mantle edge	Right half of the foot including statocyst absent or defective (23/23) Stomodeum everted (10/22)
3d	Left posterior velum Left half of the foot and left statocyst Heart	Left half of the foot including statocyst absent or defective (26/26) Left velar lobe smaller (6/23)

[a] Lineage tracing data are from Render (1991, 1997).
[b] Deletion studies are from Clement (1967, 1976, 1986a,b).

defined contributions in the larva, indicating that their fates are specified at these early stages (Fig. 5.2 and Table 5.1). Our working model for the specification of these fates involves three steps, described below.

2.1. Polar lobes are required to specify the D macromere, founder of the D quadrant lineage

During the first few divisions, a region of vegetal cytoplasm is sequestered during cytokinesis in structures called the polar lobes (PLs; Fig. 5.1A–C). The PL material is inherited by one cell at the four-cell stage, which is called the D macromere and will be the founder of the D quadrant lineage (Fig. 5.1B). The other cells are called the A, B, and C macromeres, and each will found its own lineage or quadrant. The other three cells are equivalent, but the D cell has a slightly different cleavage program and exerts a special role in development (see below). Removal of the polar lobe during the first division blocks the specification of the D macromere, and has severe consequences for embryonic patterning (Atkinson, 1986; Clement, 1952; Crampton, 1896).

2.2. The animal–vegetal axis is subdivided into tiers of equivalent cells

After the second division, the four macromere cells each divide to produce smaller daughter cells called micromeres (Fig. 5.1). The macromeres divide synchronously, and the four micromeres produced simultaneously in a given

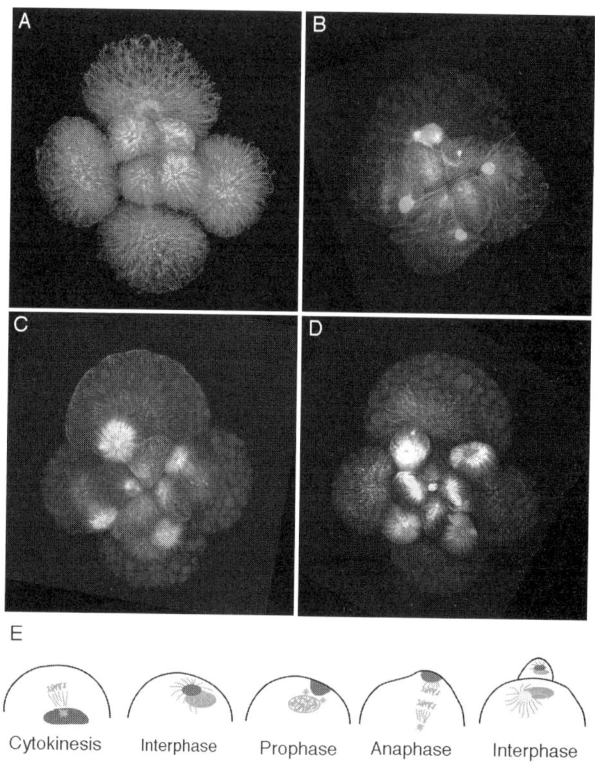

Figure 5.2 Centrosomal localization and asymmetric segregation of RNAs in early cleavage. The IoUbiquitin ligase RNA is localized by *in situ* hybridization detected with fluorescent tyramide precipitation (red). This RNA was formerly known as IoLR2 (Kingsley *et al.*, 2007) but significant homology to ubiquitin ligases was identified in additional flanking sequence recovered in an EST sequencing project (unpublished data). DNA is stained with DAPI (blue). Microtubules are stained with an antibody against β-tubulin (green). Images are projections of confocal *Z*-stacks. Yellow indicates that RNA and microtubules are colocalized, or superimposed in different sections. (A) During cytokinesis of the third cleavage cycle, the IoLR2 RNA surrounds the spindle poles in the macromeres. (B) At the interphase–prophase transition of the eight-cell stage, two macromeres are in interphase (1A and 1B, lower and right) and show IoLR2 RNA localization to the large spherical centrosomes. The other two are in prophase: at this stage, the RNA is moving from the centrosomes to the cortex and the prophase asters are visible as two small foci of microtubules under the RNA in 1D (upper macromere). (C) In metaphase of the fourth division, the IoLR2 RNA is on the cortex in all four macromeres, and the spindles are aligned toward the RNA, which changes from a disk-shaped patch to a ring, with the spindle pole at its center. (D) During cytokinesis of the fourth division, the IoLR2 RNA is on the cortex of the second quartet micromeres, which will inherit all of this RNA. (E) Diagram of RNA localization and segregation events during early cleavage cycles, showing the RNA, microtubules and DNA. Data are from Kingsley *et al.* (2007). (See Color Insert.)

cleavage cycle are called a quartet (Fig. 5.1E–G). The birth of the quartets, and the division of cells in the quartets, organizes the spiralian embryo into tiers of equivalent cells; tiers include the quartets of micromeres, and the two sets of four equivalent progeny cells generated when a tier divides. For example, the first quartet (1q) is a tier, and when the 1q cells divide, the set of four equivalent daughter cells that form closer to the animal pole ($1q^1$) is a tier, as are the four vegetal daughters ($1q^2$; Figs. 5.1 and 5.2A).

The micromeres within each tier have similar developmental properties. In general, tier mates divide with the same tempo and geometry, which differ from other tiers. Cells in a tier also tend to generate a distinct set of larval fates. In the first quartet, the $1q^2$ tier cells are fated to become part of the ciliated band of the larva, and the $1q^1$ tier will generate head structures like eyes. Similarly, the third quartet tier generates esophagus and foot structures. These tier-specific properties appear to be dependent on birth order (i.e., quartet membership, see below and Sweet, 1998). One class of exceptions to the general pattern of similar cleavage patterns and cell fates within a tier are the D quadrant members of the first and second quartets, which differ slightly from their quartet mates in cleavage pattern, in a polar lobe-dependent manner (Clement, 1952). The other class of exceptions are differences within a quartet that are caused by signaling from the embryonic organizer.

2.3. Signals from the D macromere organize micromere fates along the secondary embryonic axis

After the third quartet of micromeres is produced, the D macromere signals to multiple cells in the embryo to establish the normal pattern of cell fates among the micromeres. Removal of the D macromere during the 4-, 8-, 16-, or early 24-cell stages prevents the development of many ectodermal larval structures such as eyes, external shell, and foot (Clement, 1962). Subsequent fate mapping studies have shown that this effect is a consequence of blocking inductive interactions, since these structures are not derived from the 3D macromere but deletions of this cell early in its life can prevent their development. If this cell is deleted progressively later during its life, the resulting larvae show increasingly improved development of the affected structures (Labordus and van der Wal, 1986; Sweet, 1996). This indicates that induction of the proper micromere fates is occurring, at least in part, in this interval.

In a series of experiments that were both technically impressive and extremely informative, Sweet (1998) showed that the effect of organizer signaling on micromeres depends on their position and quartet identity. By transplanting micromeres into various positions in the embryo, she was able to demonstrate that specification of organizer-dependent fates depends on proximity to the organizer cell, and that the response to the signal depended on a cell's quartet identity. For instance, the 1b cell does not form an eye but the 1a cell, which is nearer the organizer, does generate an eye. If 1b is transplanted into the position of 1a, then it too will generate an eye. Importantly, if shell-forming second quartet cells were transplanted into the position of 1a, they did not make eyes, but often made masses of shell material. Finally, Sweet was able to show that 1d, the closest first quartet cell to the organizer, is prevented from making an eye because it is born from the macromere that inherits the polar lobe. This was extended by Goulding (2003), who showed that the inability of 1d to form an eye is related to its smaller size, which is a consequence of cell contacts during the period immediately preceding its birth. These contacts may allow extracellular signals to take place, or they may influence the cleavage geometry and the size of 1d.

The secondary axis of the early embryo is determined by the specification of the D quadrant, and runs through the D and B quadrants, though it is slightly offset from the axis that runs through the exact center of the D and B macromeres. In the micromeres, it is patterned in part by signaling from the 3D macromere. The secondary axis is usually referred to as the dorsal–ventral axis, but this is not precisely true for all lineages in the embryo. For first quartet derivatives, the generalization holds, because the D quadrant is dorsal; A and C generate the left and right sides of the head, respectively; and B derivatives are ventral (Fig. 5.1H and I). However, the relative position of the second and third quartets along the dorsal–ventral axis is shifted dramatically by gastrulation movements, especially the movement of the blastopore to the anterior–ventral side of the embryo. The D quadrant second quartet cells cover much of the posterior of the embryo after gastrulation, on both the dorsal and ventral sides. Second quartet domains are further shifted by the rotation of shell and visceral mass in relation to the head and foot during the process of torsion. In the third quartet, 3a and 3b are adjacent to the B macromere and these cells largely contribute to the esophagus. The other third quartet cells 3c and 3d flank the D macromere and generate foot structures. Despite the fact that 3a and 3b are on the opposite side of the blastula's secondary axis from 3c and 3d, both pairs of cells generate structures that are basically ventral in the larva, with the foot lying just posterior to the mouth and esophagus on the ventral surface. Available evidence indicates that fates which are normally established in cells in the B quadrant are default, and organizer signaling overrides these fates. For example, in the case of the third quartet, blocking organizer activity by either ablation or by inhibiting the MAPK pathway results in abundant esophageal tissue, but foot structures are not found (Clement, 1952, 1962; Lambert and Nagy, 2001).

2.4. Proliferation and interaction of micromere lineages during gastrulation and organogenesis

The three steps described above are thought to specify most micromere fates during early cleavage stages. After these events, the micromeres continue to divide and form clones with predictable behaviors and contributions to larval organs. We assume that intrinsic properties of micromere lineages drive much of the morphogenesis of the embryo, but this has not yet been shown for any case. The clones seem to have regular cleavage patterns, at least for the first few divisions (Clement, 1952; Goulding, 2001; Rabinowitz *et al.*, 2008). Deletion studies indicate that these lineages are largely autonomous (Table 5.1), though some results hint at interactions between clones. For instance, the second quartet cells 2a and 2c contribute to the statocysts and the proximal portion of the foot, but deletion of these cells has relatively mild effects on foot morphology and reduces statocyst size somewhat (Clement, 1986b; Table 5.1). In contrast, 3c and 3d contribute to the right and left sides, respectively, of the more distal portion of the foot, including parts of the statocyst, and deletion of these cells has a much larger developmental effect than 2a and 2c deletions, with larvae lacking any recognizable foot structures on the expected side. This suggests that 3c and 3d derivatives may be required to organize foot-forming cells from the 2a and 2c micromeres in normal development.

3. The Molecular Basis of Spiralian Development in *Ilyanassa*: Progress and New Problems

The above steps in our model assign fates to the ectodermal micromere cells, by superimposing three patterning mechanisms—intrinsic differences in developmental potential between tiers of cells, polar lobe-dependent differences in D quadrant cells compared to quartet mates, and position-dependent induction of particular fates by the organizer cell 3D. The interaction of these three patterning systems could in principle generate all of the different fates in the blastula, but real understanding of these processes will clearly require identification of genes that are involved studies of their function in the embryo. In this section, I review some recent advances in understanding the molecular basis of early development in *Ilyanassa*.

3.1. Specification of quartet-specific properties: A role for RNAs on the centrosome?

A mechanism has been described that may localize determinants which specify the predicted quartet-specific developmental properties (Kingsley *et al.*, 2007; Lambert and Nagy, 2002). Numerous RNAs are specifically localized to centrosomes during interphase periods of early macromere cleavage cell cycles, and then are asymmetrically inherited during division (Fig. 5.2). Before localization, they are diffuse and ubiquitous in the cytoplasm, but during late cytokinesis, these RNAs abruptly localize to the centrosome, where they remain during interphase. In prophase, the RNA-containing centrosomal material moves to the region of cortex that directly overlies the centrosome, where it forms a disk-shaped plaque. After nuclear envelope breakdown, the spindle forms and division proceeds with the spindle oriented toward the patch of centrosomal material on the cortex. In the ensuing division, the centrosomal material is inherited entirely by the micromere daughter cell.

RNA localization and segregation is widespread in the *Ilyanassa* embryo. The three RNAs that were first shown to be segregated in this fashion were IoDpp, IoTld, and IoEve—all conserved developmental regulatory proteins (Lambert and Nagy, 2002). To determine the frequency of centrosomal localization among RNAs with a wider range of predicted functions, and to find more examples of localized RNAs, an *in situ* screen was performed (Kingsley *et al.*, 2007). In this experiment, probes were generated from randomly picked cDNA clones and used for *in situ* hybridization on early cleavage stages to look for localized RNAs. In an initial screen of 103 unique sequences, four RNAs were very specifically localized to centrosomes in at least some cleavages cycles. Of the remainder, about half were not enriched on centrosomes (though many had other patterns of subcellular localization), and about half were nonspecifically localized to centrosomes. The specific centrosomal RNAs were generally segregated asymmetrically in the next division, and other RNAs were not. These results indicate that a large fraction of mRNAs in this embryo are specifically localized to centrosomes, and asymmetrically segregated. The overall level of RNA subcellular localization, including centrosomal localization and other modes, was around 65% which is similar to the levels of RNA localization reported in the fly embryo (Lecuyer *et al.*, 2007).

RNA segregation in the *Ilyanassa* embryo generates a large amount of potential patterning information. As described above, the spiralian embryo is organized into tiers of equivalent cells, including the quartets, and the sets of cells generated when the cells of a quartet divide. The patterns of RNA localization parallel this organization, since most localized RNAs are specific to one or more particular tiers of four cells in the early embryo. Examination of the 19 patterns characterized so far (Kingsley *et al.*, 2007; Lambert and Nagy, 2002) reveals that all cells in the early embryo up to the 27-cell stage have specific centrosomal localization, and all divisions after the second have specific segregation of centrosomal RNA. While two RNAs may have identical localization patterns at one stage (i.e., four-cell macromeres), all of the known specific localization patterns are unique when considered across early cleavage stages. This indicates that the mechanisms of RNA localization in the embryo are remarkably intricate. The large fraction of RNA localization that is tier-specific indicates that cells within a tier inherit the same set of molecules, and this set is different from the set inherited by other tiers. It seems likely that inheritance of centrosomally localized RNA determinants is involved with the specification of tier-specific developmental potentials, but this has not yet been demonstrated. There are also some quadrant-specific aspects of RNA localization in the embryo, for instance the cortical localization of several RNAs that occurs in 1d but not 1abc (i.e., Fig. 5.3A). This kind of RNA segregation may be involved in specifying quadrant-specific developmental potential, but this also remains to be tested.

These results indicate that RNA segregation in *Ilyanassa* is more extensive than in any other embryo known, even other well-characterized systems with largely invariant cell lineages. In *Caenorhabditis elegans*, most cell divisions are not associated with RNA segregation, and few RNAs are known to be asymmetrically segregated, especially outside of the germline (Schisa *et al.*, 2001; Seydoux and Fire, 1994). In the ascidian *Halocynthia*, the frequency of cytoplasmic localization is similar to what we have observed in *Ilyanassa*, but RNA segregation occurs in a smaller subset of embryonic divisions (Makabe *et al.*, 2001). It is important to point out that we still lack a system-level appreciation for the extent of asymmetric *protein* segregation in *Ilyanassa* and other animal embryos, so the relative contribution of this mechanism to embryonic patterning cannot yet be compared among animal embryos.

Localization of RNA to the centrosome before cortical localization and segregation is a conserved mechanism in animal oogenesis. In vertebrates, insects, and many other animals, a structure called the Balbiani body serves as a site of accumulation for molecules which are subsequently localized in the oocyte (see Kloc *et al.*, 2004b for a recent review, also see Guraya, 1979; Raven, 1961). We propose that localization to the centrosome during *Ilyanassa* cleavage is homologous to the formation of the Balbiani body. This structure (also called the yolk nucleus) forms as an accumulation of material on the oocyte centrosome (Kloc *et al.*, 2004a). Structurally, it is rich in germline-destined mitochondria, Golgi vesicles, and RNA granules, and contains centrioles at its core. The ultrastructure of the *Ilyanassa* interphase centrosomes is very similar (though they lack the germline mitochondria). Indeed, when the centrosomes of *Ilyanassa* early cleavage stages were first described at the TEM level, they were called yolk nuclei (i.e., Balbiani bodies), rather than centrosomes (Fioroni, 1974). One role of the Balbiani bodies in *Xenopus* and *Drosophila* is to assemble RNAs that are destined for localization, which then move to the cortex of the oocyte (Cox and Spradling, 2003; Kloc and Etkin, 1995). Remarkably, this directly parallels the role of the *Ilyanassa* centrosome, to which various RNAs localize in advance of movement to the cellular cortex. In *Ilyanassa* oocytes, some RNAs are localized to the oocyte centrosome (i.e., Balbiani body), further supporting the connection between the events observed in the cleavage stages and Balbiani body function in oocytes (Fig. 5.3E and F). Insights gained into the mechanisms and function of centrosomal localization in

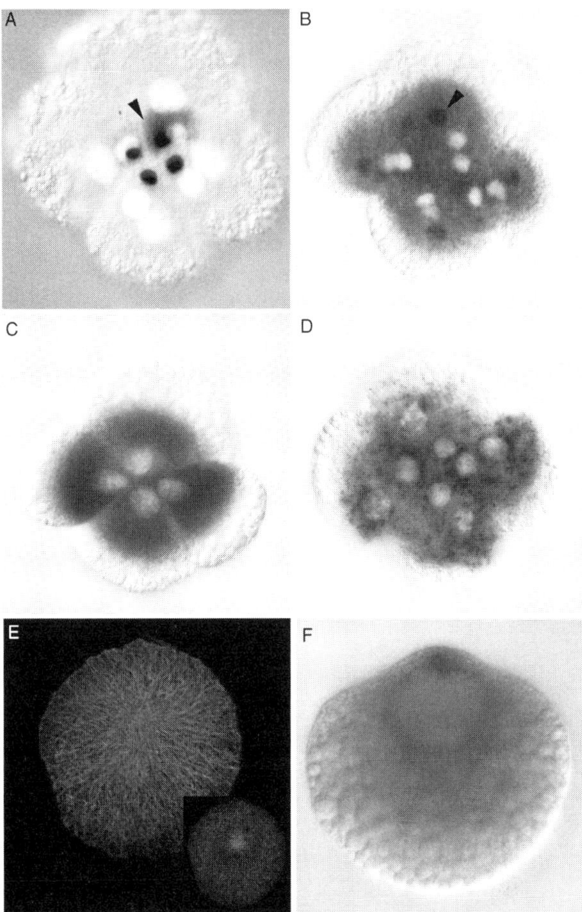

Figure 5.3 RNA localization in early embryos and oocytes. (A) Specific localization of IoLR1 RNA to the centrosomes of the first quartet micromeres at the eight-cell stage, visualized with *in situ* hybridization and detection with chromogenic substrate (see Kingsley *et al.*, 2007 for details). Nuclei are stained with DAPI and appear white, and the stained centrosomes are black. RNA is also localized to the cortex in 1d (arrowhead). (B) Nonspecific localization of IoEST0056 (no significant homology) RNA to macromere centrosomes at the eight-cell stage. The 1D centrosome is indicated with an arrowhead. (C) Unlocalized α-tubulin mRNA at the four-cell stage. (D) The IoGCN1-like RNA is subcellularly localized in a granular pattern, but not localized to centrosomes and not segregated in the ensuing division. (E) Single confocal section of the microtubules of the mature oocyte visualized with anti-β-tubulin staining shows that the center of the microtubule array is at the animal pole, above the germinal vesicle. The inset shows the distribution of the IoEST00134 RNA (similar to autoantigen La), showing that the RNA is localized to the center of the microtubule array. (F) IoEST00134 RNA detected with a chromogenic stain in an oocyte viewed from the side. (A)–(D) are from Kingsley *et al.* (2008).

Ilyanassa should thus shed light on oocyte patterning in other animals. Moreover, the broad conservation of Balbiani body formation increases the likelihood that in other groups, centrosomal localization is used later in development as a strategy for asymmetric segregation during cell division.

Recently, it has been reported that several RNAs specifically localize to centrosomes in oocytes of the clam *Spisula* (Alliegro and Alliegro, 2008; Alliegro *et al.*, 2006). This was interpreted as evidence for the existence of a population of centrosome-specific RNAs that may be related to centrosome biogenesis or function. Recognition of the Balbiani body as a centrosome-derived structure highlights the conserved role of the oocyte centrosome in RNA localization, and thus favors the view that the RNA localization reported in *Spisula* is relevant for oocyte function or embryonic patterning rather than for some intrinsically centrosomal process. It will be interesting to learn more about the eventual fate and function of these RNAs in the future.

3.2. The role of ERK1/2 MAPK signaling from the D quadrant macromere

As described above, ablation of the D macromere at the start of the 24-cell stage prevents the development of many larval structures that are derived from micromeres that overlie this cell. This indicates that this cell is inducing these fates in micromere cells, and this signaling center is called the D quadrant organizer. Analysis of the ERK1/2 MAPK pathway has corroborated the existence of the spiralian organizer, and provided the first insight into the molecular basis of its activity (Lambert and Nagy, 2001; Fig. 5.4). The diphosphorylated, activated MAPK is initially detected in 3D at the start of the 24-cell stage, shortly before the induction of micromeres can first be detected based on deletion studies. Activation is then detected in overlying micromeres, and in the next 3 h, the activation spreads into a dorsal and lateral arc of micromeres that includes all of the cells that are predicted to require organizer signaling, based on the fates that are missing after organizer ablation. When 3D divides, activation is observed in its daughter 4d. Polar lobe deletion blocks activation in 3D, showing that activation is associated with organizer specification. 3D deletion blocks MAPK activation in the micromeres, indicating that activation in those cells is downstream of signaling from the organizer. Blocking the pathway with inhibitors precisely copies the results of organizer deletions, showing that organizer signaling requires activation of the pathway. Addition of inhibitors at sequential time points during and after the 24-cell stage results in increasingly complete differentiation of organizer-dependent structures, indicating that MAPK signaling is required for a progressive specification of target cells in this interval. This is similar to the effects of deleting 3D at different points in this period (Clement, 1962; Labordus and van der Wal, 1986; Sweet, 1996).

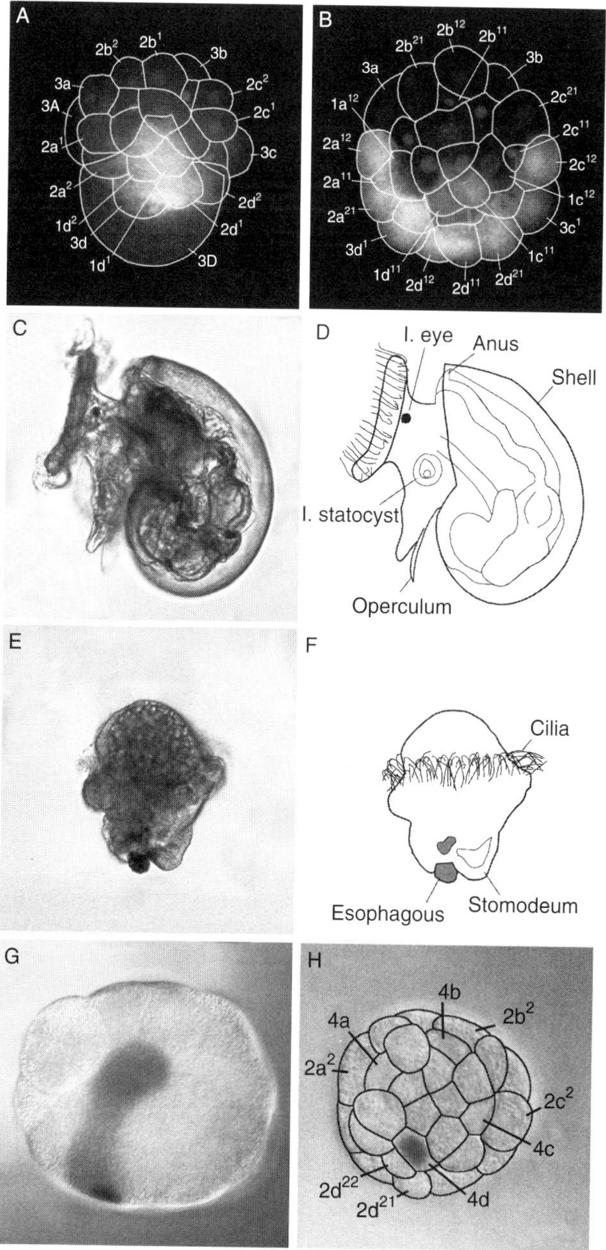

Figure 5.4 The ERK1/2 MAPK pathway in the spiralian organizer. (A) Antidiphosphorylated ERK1/2 MAPK staining in the 3D macromere of a 24-cell stage embryo. (B) A 39-cell embryo (about 2 h after the stage shown in (A)), with staining in an arc of dorsal and lateral micromeres. Activation is also observed in this stage in 4d

What might be activating MAPK in 3D? The polar lobe is required for D macromere specification, so in principle MAPK might be activated in an autonomous fashion by some polar lobe-dependent event. However, it is not known if inheritance of the polar lobe is sufficient for specification, or if other cues are required. In molluscs embryos where the first two cleavages are equal ("equally cleaving" embryos), the quadrants are initially equipotent, but an early cell-signaling event specifies the D macromere (reviewed in Freeman and Lundelius, 1992). In such embryos, preventing the contact of first quartet daughter micromeres with the macromeres blocks 3D specification (Henry *et al.*, 2006; Martindale *et al.*, 1985; van den Biggelaar and Guerrier, 1979). In contrast, it has been reported that ablation of the first quartet cells in *Ilyanassa* affects only the structures derived from the ablated cells—structures that require organizer signaling like external shell and foot are still present (Sweet, 1998). Intriguingly, ablation of the first quartet micromeres along with one second quartet cell blocks MAPK activation in 3D and organizer-dependent specification of micromeres (J. Wandelt, A. Nakamoto, and L. Nagy, personal communication). These experiments suggest that 3D specification in *Ilyanassa* may be more similar to equally cleaving molluscs than previously appreciated. Similarly, it has recently been shown that in *Crepidula*—a caenogastropod-like *Ilyanassa*—the polar lobe is not required for D quadrant specification, but contact with the first quartet micromere derivatives is necessary (Henry *et al.*, 2006). These results show that even mollusc embryos with a polar lobe may specify the D quadrant inductively, in the manner of equally cleaving spiralians, with the polar lobe only serving to bias this process in some way.

In *Ilyanassa*, inhibitor treatments block MAPK in 3D or its daughter cell 4d, at the same time that the treatments block activation in overlying micromere cells of the first three quartets. This complicates the use of these reagents to test which aspects of the MAPK activation pattern are functionally required. We found that inhibition of MAPK activation after 3D divides to produce 4d still caused defects in the micromere cells with activated MAPK, and available evidence suggested that 4d was not required for organizer signaling (Clement, 1962, 1986b; Lambert and Nagy, 2001). Based on these observations, we argued that the activation in the

(not shown). (C, D) Wild-type (solvent control) 8-day-old veliger larva viewed from the left side. (E, F) Eight-day-old larva after inhibition of MAPK activation starting during early 24-cell stage with the compound U0126 at 10 uM. (G) Lateral through-focus view of a 36-cell embryo of the chiton *Chaetopleura apiculata*, stained brown for activated MAPK. The staining is in the 3D macromere, which is extended through the blastocoel cavity to make contact with the micromeres of the first quartet. (H) Vegetal view of an embryo of the polychaete annelid *Hydroides hexagonus* about 40 min after the birth of 4d, with staining in the nucleus of 4d. No MAPK activation was observed in *Hydroides* during several time points during the life of 3D. (A)–(F) are from Lambert and Nagy (2001), and (G) and (H) are from Lambert and Nagy (2003).

micromeres was functionally important. Recently, we revisited the question of whether 4d was involved in organizer signaling, and found that deletion of this cell often disrupted organizer-dependent fates, suggesting that it was continuing the signaling of its mother cell (Rabinowitz *et al.*, 2008). In fact, the phenotypic effects of inhibiting the MAPK pathway after the birth of 4d closely resemble the effects of 4d deletion (Lambert and Nagy, 2001; Rabinowitz *et al.*, 2008). Thus, it now seems that all of the effects of MAPK inhibition could be explained by blocking organizer signaling in 3D and 4d, and there is no clear evidence for a functional requirement for MAPK activation in the micromeres in *Ilyanassa*.

3.3. Control of the cleavage pattern and cell fate specification in the 4d micromere lineage by the IoNanos protein

The 4d micromere lineage is particularly important for establishing the body plan of the embryo, and it is one of the conserved aspects of the spiralian fate map. It contributes the major larval muscle, and multiple other mesodermal organs including the heart. It also generates the intestine, making this lineage a classic case of mesendoderm. The 4d cell divides along the embryo's axis of bilateral symmetry to form two cells that undergo a regular series of highly asymmetric cell divisions to generate paired bands of cells (Fig. 5.5). This is an example of a *blast* cell lineage, where a large mother cell makes repeated asymmetric cell divisions to produce a stereotyped series of daughter cells. Examples of these cells in other systems include insect neuroblasts and leech teloblasts.

A factor that is required for normal development of the 4d lineage was recently identified (Rabinowitz *et al.*, 2008). IoNanos is an *Ilyanassa* ortholog of the highly conserved Nanos protein. Nanos proteins are thought to act as translational regulators and have been implicated in germline development in diverse animals (Forbes and Lehmann, 1998; Subramaniam and Seydoux, 1999; Tsuda *et al.*, 2003; Wang *et al.*, 2007). Nanos also has roles in posterior patterning of somatic tissues, but evidence for this has been limited to insects (Lall *et al.*, 2003; Lehmann and Nusslein-Volhard, 1991). IoNanos was recovered in a screen for RNAs with subcellular localization (Kingsley *et al.*, 2007). The IoNanos mRNA is initially ubiquitous, then becomes specifically localized to centrosomes in the 3A, 3B, and 3C macromeres, and abundant and unlocalized in the 3D macromere. When this cell divides, the RNA is largely restricted to one of its daughters, the 4d micromere. The IoNanos protein also becomes restricted to 4d. It is largely specific to the teloblast cells during the next two rounds of division, and then becomes undetectable.

Knockdown of IoNanos using either translation-blocking morpholino oligos or long antisense RNA reproducibly prevented the normal development of the 4d-derived structures that were scored: the larval retractor muscle, the heart, and the intestine (Fig. 5.5). These results show that

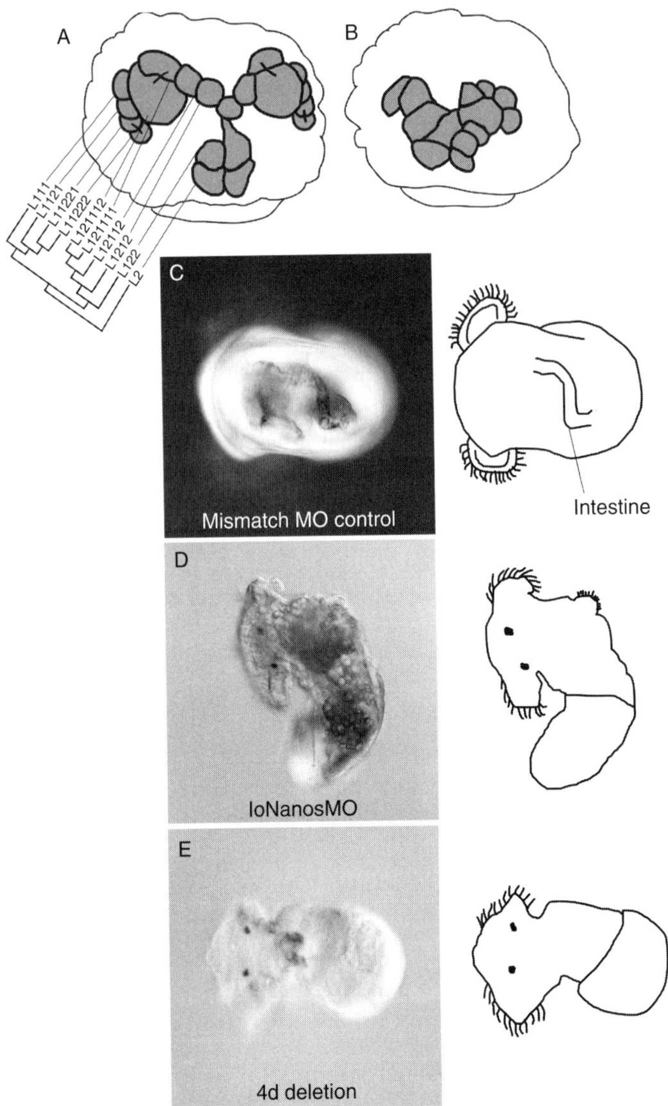

Figure 5.5 The effect of IoNanos knockdown in the *Ilyanassa* embryo. (A) Drawing of the 4d clone at 4d + 24 h, when 20 cells are present, based on confocal imaging of the labeled lineage. (B) Typical 4d clone at 4d + 24 h after injection of 0.1 mM IoNanosMO, with disorganized cleavage pattern, fewer cells, and reduced size asymmetry among cells compared to control. (C) Dorsal view of a larva with wild-type morphology after mismatch control morpholino injection into the zygote. (D) Typical larva after injection of IoNanosMO. Ectodermal structures like head and shell are present but smaller than control. Structures derived from 4d are absent, like intestine, normal retractor muscle, and heart (the latter two are not shown here). (E) Larva after deletion of 4d, showing some ectodermal defects, and absence of 4d-derived structures. From Rabinowitz *et al.* (2008).

somatic tissues require IoNanos for normal development in *Ilyanassa*. The 4d lineage is the likely source of the germline cells in *Ilyanassa*, but the germline has not been identified in this embryo (Swartz *et al.*, 2008), so the effects of IoNanos knockdown on germline specification could not be scored. The role for Nanos in the soma may be shared by other spiralian embryos, since a Nanos ortholog in leech is also required for normal embryonic development (Agee *et al.*, 2006).

While more sampling within the Lophotrochozoa is obviously needed, these results, together with observations in insects, suggest that a role for Nanos in somatic patterning may have been in place at the base of the protostome clade. From this perspective, it is notable that blast cells of the 4d lineage, where IoNanos is mainly expressed, roughly define the posterior side of the embryo during gastrulation. It will be interesting to learn the role of Nanos in basal arthropods that also have somatic blast cell lineages in the posterior, like crustaceans (Anderson, 1973). The role of Nanos in somatic development may be even more ancient than the origin of the protostomes, since Nanos in the cnidarian *Nematostella* is expressed in somatic lineages in early embryogenesis (Extavour *et al.*, 2005).

IoNanos knockdown also caused less-penetrant effects on structures which are not derived from 4d but do require signaling from the 3D organizer. Since 3D is the mother cell of 4d, it would not be surprising if 4d carried on some organizer signaling, especially since MAPK remains activated in this cell. Evidence from the literature was conflicting on this point. Two reports suggested that late ablation of 3D, or ablation of 4d did not seem to interfere with organizer-dependent fates (Clement, 1962, 1986b). Another set of experiments suggested that at least in some cases, the requirement for organizer signaling was not met when 3D divides (Sweet, 1996). To address this discrepancy, a series of 4d ablations were performed and scored along with the IoNanos knockdown larvae. Deletion of 4d always prevented development of 4d-derived organs, as expected. But there were also less-penetrant defects in organizer-dependent structures derived from other micromeres. The most striking example was that only 17% of 4d deletion larvae had a normal foot, and half lacked one or both statocysts. Cell deletions and inhibitor studies indicate that foot structures are among the last to be specified by organizer signaling (Clement, 1962; Lambert and Nagy, 2001), so this is consistent with interfering with late organizer activity. The finding that 4d is signaling to the micromeres is interesting in a comparative context. In the closely related gastropod *Crepidula*, none of organizer signaling has been completed at the point 3D divides, suggesting that the organizer activity is in the 4d cell in this embryo (Henry *et al.*, 2006). Similarly, in the polychaete annelid *Hydroides*, MAPK activation is first observed in 4d, suggesting that this cell may be the organizer in this embryo (Lambert, 2008; Lambert and Nagy, 2003).

IoNanos protein seems to be required for some key blast cell characteristics in the 4d lineage. The behavior of the clones in morpholino knockdown embryos was different in several respects from their wild-type controls (Fig. 5.5; Rabinowitz *et al.*, 2008). The highly regular cleavage pattern was completely disorganized, so that bilateral symmetry was lost, and a typical cleavage pattern could not be identified. The proliferation rate of the lineage was lower after knockdown, with controls having an average of around 12 cells compared to 20 cells in the controls at 24 h after the birth of 4d. One striking aspect of the 4d lineage is the size asymmetry between the teloblast cells and their progeny cells. Since the clones were too variable to identify the cleavage pattern of the lineage after knockdown, the overall level of size asymmetry was measured by reconstructing cell volumes using confocal microscopy. In control embryos, the two largest cells in the lineage at 4d + 24 were always the paired teloblast cells, and these were always considerably bigger than the remaining cells in the lineage. In knockdown lineages, the largest cells were significantly smaller than the teloblast cells in control embryos, and smaller cells were significantly larger than the progeny cells in the control embryo, showing that the normal size asymmetry of the 4d lineage requires IoNanos activity. Blast cell lineages are a kind of stem cell lineage, and display many basic stem cell characteristics. They have high proliferative capacity, they divide asymmetrically, and they often maintain pluripotency to generate daughter cells with different fates. IoNanos seems to be required for all of these behaviors in the 4d blast cell lineages, since knockdown lowers proliferation, asymmetry of cell division, and blocks all of the fates generated by the progeny cells in the lineage. Recently, Nanos has been shown to be required for stem cell properties in the female germline of *Drosophila* (Wang and Lin, 2004; Gilboa and Lehmann, 2004), so the characterization of Nanos in *Ilyanassa* suggests that this protein may have a conserved role in maintaining stem cell properties in animals.

Many exciting questions remain about the 4d lineage, and the role of IoNanos. It will be important to determine the fates of the 4d derivatives, and find molecular markers for various 4d sublineages. This will allow direct tests of whether the blast cells are maintaining pluripotency, and help explain how IoNanos effects daughter cell specification. Further study of the 4d lineage in *Ilyanassa* will also likely lead to interesting comparative studies, since this lineage is one of the most conserved aspects of spiralian development (Lambert, 2008; Wilson, 1899).

4. Perspectives on Evolution and Development from the Spiralia

The Spiralia includes a large fraction of the extant diversity of animal body plans, and several extremely large and diverse phyla. Descriptive studies performed over the last 125 years have generated a preliminary characterization of diversity in spiralian cell lineages and fate maps. As the understanding of the molecular basis of spiralian development progresses, this group promises to be an exceptionally rich area for studies of the evolution of development.

4.1. Evolution and development at the level of lineage

One of the most remarkable characteristics of the spiralian embryo is the extreme conservation of cell lineage. Even when comparing embryos of distantly related organisms, the very regular cleavage pattern allows corresponding cells to be identified based on quartet membership and quadrant lineage. These cells are obviously homologous at the level of cell lineage—they correspond to a cell in the same position in the cell lineage of the common ancestor, but they are also often homologous in their fate or developmental potential. In general, the first, second, and third quartets are considered to be homologous across the Spiralia, and they have similar positions in postgastrulation embryos and contributions to larval structures in divergent spiralian taxa (e.g., Wierzejski, 1905; Wilson, 1892). Other examples of highly conserved cell lineages include: the 4d micromere lineage (described above), the cells of the vegetal tier of the first quartet ($1q^2$), which generate much of the ciliated band across the Spiralia; and the 2d cell, which generates much of the trunk ectoderm in spiralians. This high degree of homology has several implications for the study of the evolution of developmental mechanisms. Perhaps most importantly, the strong conservation of lineages and fates in the spiralian embryo means that we can identify evolutionary changes in the developmental program at the resolution of single cells.

Cell lineages are obviously an important unit of organization in animal embryos, but very little is known about the patterns of evolutionary change at the level of lineage, or what mechanisms may influence these patterns. Our recent results, combined with the body of classical spiralian embryology, suggest that the spiralian embryo may rely to an unusual degree on lineage-dependent mechanisms. As the depth and breadth of our understanding of the spiralian embryo increases, we will be able to test specific hypotheses about how the reliance on lineage-dependent patterning influences patterns of evolutionary change. For instance, as we learn more details about the mechanisms that specify various cell types in different spiralians, we will be able to test whether autonomous specification leads to stronger conservation of the cleavage pattern. It will also be interesting to address other questions about the evolution of cell lineages, including whether the sets of cues that sequentially determine a lineage during a series of cell divisions can be co-opted for a novel use; or whether we can detect constraint on the evolution of lineages whose member cells interact via cell signaling later in development.

4.2. Sampling spiralian development: The ERK1/2 MAPK pathway as test case?

The ERK1/2 MAPK cascade was one of the first molecular mechanisms to be implicated in spiralian development. Since then, comparative studies of the role of this pathway in several different animals are providing some of the first glimpses of the patterns of evolutionary change in the spiralian embryo (Henry and Perry, 2008; Koop *et al.*, 2007; Lambert and Nagy, 2003; Lartillot *et al.*, 2002). Overall, these results not only highlight the homology of cell lineages in spiralians, but also demonstrate a surprising degree of variation in patterning mechanism underlying the conserved early cleavage pattern.

One striking finding about the evolution of the ERK MAPK pathway is that the pattern of activation that was first described in *Ilyanassa*—in the 3D macromere and overlying micromeres—is not the ancestral state for molluscs. In other gastropods from more basally branching groups, like *Lymnaea, Patella, Lottia* (a.k.a. *Tectura*), and *Haliotis*, the pathway is activated in 3D, but not in the overlying micromeres (Koop *et al.*, 2007; Lambert and Nagy, 2003; Lartillot *et al.*, 2002). This is also true in the chiton *Chaetopleura*, a member of the basally branching molluscan class Polyplacophora, clearly indicating that the activation pattern observed in the micromeres of *Ilyanassa* is derived (Lambert and Nagy, 2003). The pathway has also been examined in *Crepidula fornicata*, which is a caenogastropod-like *Ilyanassa* (Henry and Perry, 2008). In this embryo, MAPK is activated in the 3D macromere and briefly in its daughter 4d, similar to *Ilyanassa*. However, in *Crepidula* the activation observed in the micromeres is significantly different. It first appears before 3D is born, at the late 16-cell stage, then disappears from the micromeres, remaining only in 3D during the midlate 24 cells stage. Weak activation reappears in micromeres after the birth of 4d, but in a restricted set of micromeres compared to what is observed in *Ilyanassa*. Surprisingly, inhibiting the pathway at various time points shows that MAPK activation is not required after the early 24-cell stage, unlike *Ilyanassa*, where treatments throughout the life of 3D and later reproducibly impair organizer-dependent micromere specification (Lambert and Nagy, 2001). This result indicates that in *Crepidula*, the MAPK pathway is required for specification of 3D, but not organizer activity, which is mediated by the 3D daughter cell 4d (described above; Henry and Perry, 2008; Henry *et al.*, 2006).

There is only one nonmollusc spiralian where we have data on the MAPK pathway in the early embryo, but it does suggest another notable evolutionary change. In the polychaete annelid *Hydroides*, the pathway was not detected in the micromeres, similar to the predicted ancestral state in molluscs. However, it was also not detected in 3D, but was activated in 4d. This cell has a similar fate in polychaetes as in molluscs, but it may also act as an embryonic organizer, a function that is thought to reside in 3D in basal gastropods (Martindale, 1986; van den Biggelaar and Guerrier, 1979).

The deletion experiments to determine when organizer signaling is occurring have not been done in a polychaete. However, the shift from quadrilateral (quadrant) symmetry to bilateral symmetry in the cleavage pattern can be shown to reflect organizer signaling (Martindale *et al.*, 1985; van den Biggelaar, 1976). This shift happens shortly after 3D is born in equally cleaving molluscs, but shortly after 4d is born in equally cleaving polychaetes. Together with the change in MAPK staining, this signals a heterochronic shift in organizer activity between molluscs and polychaete annelids.

The Spiralia is a vast group of animals, and sampling of the molecular mechanisms of spiralian development has just begun. We expect that in the near future, a number of spiralian model systems will be developed to the point where molecular studies are possible, and this is exciting from the perspective of comparative biology and development. The spiralian community will be able to test whether newly discovered molecular mechanisms are conserved across the Spiralia, and how they might be modified. In this way, we should be able to arrive at a rigorous understanding of the basic mechanisms of the spiralian embryo more efficiently, and simultaneously learn about the diversification of patterning mechanisms in this clade. The broad survey of spiralian development will be particularly powerful because individual homologous cells can be compared at multiple taxonomic levels, from genera to phyla—unlike any known animal model system.

There are several protostome phyla that are closely related to spiralians, but seem to have lost much or all of the spiralian developmental program, based on recent phylogenetic studies (Dunn *et al.*, 2008). These include phoronids, bryozoans, and brachiopods. These will obviously provide interesting material for studying the modification of spiralian developmental mechanisms. There are also several important spiralian groups where we still lack models for mechanistic studies, including entoprocts, sipunculans, and various classes of molluscs and annelids. One particularly important group is the polyclad flatworms. While most polyclad flatworms have typical spiralian development, indicating that this is ancestral for the phylum, the planarian flatworms that are most commonly studied have highly modified embryogenesis that cannot be recognized as spiralian. Since phylogenetic studies often place the platyhelminth flatworms at the base of the Spiralia (Dunn *et al.*, 2008; Helmkampf *et al.*, 2008), establishing systems for molecular studies in a spiralian flatworm seems particularly important for comparative purposes.

5. Conclusions

This is an exciting time for students of the spiralian embryo. A new level of interest in these systems has been kindled by large amounts of new genomic and transcriptome sequence, the development of molecular and functional approaches, and a new appreciation for the phylogenetic significance of this group. While some progress has been made recently in understanding the mechanisms of spiralian development, these advances are only toeholds that might help us approach the beautiful but hard problems that remain. We still do not know how the spiralian organizer is specified, or the nature of its signaling activity. The inherited factors (and/or signals) that distinguish the micromere quartets from each other are not known. And, we still do not know how any of the micromere fates are specified, or how the lineages that descend from them are involved in morphogenesis and development of the body plan. It is fortunate that some of us find spiralian embryos so beautiful to study, because we certainly have a lot of hours at the microscope and the lab bench ahead of us.

ACKNOWLEDGMENTS

I thank Lisa Nagy for sharing unpublished results, and Morgan Q. Goulding and Jon Q. Henry for insightful comments on the manuscript.

REFERENCES

Agee, S. J., Lyons, D. C., and Weisblat, D. A. (2006). Maternal expression of a NANOS homolog is required for early development of the leech *Helobdella robusta*. *Dev. Biol.* **298,** 1–11.

Alliegro, M. C., and Alliegro, M. A. (2008). Centrosomal RNA correlates with intron-poor nuclear genes in *Spisula* oocytes. *Proc. Natl. Acad. Sci. USA* **105,** 6993–6997.

Alliegro, M. C., Alliegro, M. A., and Palazzo, R. E. (2006). Centrosome-associated RNA in surf clam oocytes. *Proc. Natl. Acad. Sci. USA* **103,** 9034–9038.

Anderson, D. T. (1973). "Embryology and Phylogeny in Annelids and Arthropods." Pergamon Press, Oxford.

Atkinson, J. W. (1986). An atlas of light micrographs of normal and lobeless larvae of the marine gastropod *Ilyanassa obsoleta*. *Int. J. Invert. Reprod. Dev.* **9,** 169–178.

Clement, A. C. (1952). Experimental studies on germinal localization in *Ilyanassa*. I. The role of the polar lobe in determination of the cleavage pattern and its influence in later development. *J. Exp. Zool.* **132,** 427–446.

Clement, A. C. (1962). Development of *Ilyanassa* following the removal of the D macromere at successive cleavage stages. *J. Exp. Zool.* **149,** 193–216.

Clement, A. C. (1967). The embryonic value of micromeres in *Ilyanassa obsoleta*, as determined by deletion experiment. I. The first quartet cells. *J. Exp. Zool.* **166,** 77–88.

Clement, A. C. (1976). Cell determination and organogenesis in molluscan development—Reappraisal based on deletion experiments in *Ilyanassa*. *Am. Zool.* **16,** 447–453.

Clement, A. C. (1986a). The embryonic value of the micromeres in *Ilyanassa obsoleta*, as determined by deletion experiments. II. The second quartet cells. *Int. J. Invert. Reprod. Dev.* **9,** 139–153.

Clement, A. C. (1986b). The embryonic value of the micromeres in *Ilyanassa obsoleta*, as determined by deletion experiments. III. The third quartet cells and the mesentoblast cell, 4d. *Int. J. Invert. Reprod. Dev.* **9,** 155–168.

Cox, R. T., and Spradling, A. C. (2003). A Balbiani body and the fusome mediate mitochondrial inheritance during *Drosophila* oogenesis. *Development* **130,** 1579–1590.

Crampton, H. E. (1896). Experimental studies on gastropod development. *Roux Arch. EntwMech.* **3,** 1–19.

Dunn, C. W., Hejnol, A., Matus, D. Q., Pang, K., Browne, W. E., Smith, S. A., Seaver, E., Rouse, G. W., Obst, M., Edgecombe, G. D., Sorensen, M. V., Haddock, S. H., *et al.* (2008). Broad phylogenomic sampling improves resolution of the animal tree of life. *Nature* **452,** 745–749.

Extavour, C. G., Pang, K., Matus, D. Q., and Martindale, M. Q. (2005). *vasa* and *nanos* expression patterns in a sea anemone and the evolution of bilaterian germ cell specification mechanisms. *Evol. Dev.* **7,** 201–215.

Fioroni, L. S. A. P. (1974). The ultrastructure of the yolk nucleus during early cleavage of *Nassarius reticulatus* L. (Gastropoda, Prosobranchia). *Cell Tissue Res.* **153,** 79–88.

Forbes, A., and Lehmann, R. (1998). Nanos and Pumilio have critical roles in the development and function of *Drosophila* germline stem cells. *Development* **125,** 679–690.

Freeman, G., and Lundelius, J. W. (1992). Evolutionary implications of the mode of D quadrant specification in coelomates with spiral cleavage. *J. Evol. Biol.* **5,** 205–247.

Gharbiah, M., Cooley, J., Leise, E. M., Nakamoto, A., Rabinowitz, J. S., Lambert, J. D., and Nagy, L. M. (2009). The snail *Ilyanassa*: A reemerging model for studies in development. "Emerging Model Organisms" Vol. 1, pp. 592. Cold Spring Harbor Press, Cold Spring Harbor, NY.

Gilboa, L., and Lehmann, R. (2004). Repression of primordial germ cell differentiation parallels germ line stem cell maintenance. *Curr. Biol.* **14,** 981–986.

Goulding, M. (2001). "Comparative and Experimental Analysis of Precocious Cell-Lineage Diversification in the Embryonic Dorsoventral Axis of the Gastropod *Ilyanassa*." Ph.D. Thesis. Zoology Department, University of Texas, Austin.

Goulding, M. (2003). Cell contact-dependent positioning of the D cleavage plane restricts eye development in the Ilyanassa embryo. *Development* **130,** 1181–1191.

Guraya, S. S. (1979). Recent advances in the morphology, cytochemistry, and function of Balbiani's vitelline body in animal oocytes. *Int. Rev. Cytol.* **59,** 249–321.

Hejnol, A., Martindale, M. Q., and Henry, J. Q. (2007). High-resolution fate map of the snail *Crepidula fornicata*: The origins of ciliary bands, nervous system, and muscular elements. *Dev. Biol.* **305,** 63–76.

Helmkampf, M., Bruchhaus, I., and Hausdorf, B. (2008). Phylogenomic analyses of lophophorates (brachiopods, phoronids and bryozoans) confirm the Lophotrochozoa concept. *Proc. Biol. Sci.* **275,** 1927–1933.

Henry, J. J., and Perry, K. J. (2008). MAPK activation and the specification of the D quadrant in the gastropod mollusc, *Crepidula fornicata*. *Dev. Biol.* **313,** 181–195.

Henry, J. Q., Perry, K. J., and Martindale, M. Q. (2006). Cell specification and the role of the polar lobe in the gastropod mollusc *Crepidula fornicata*. *Dev. Biol.* **297,** 295–307.

Kingsley, E. P., Chan, X. Y., Duan, Y., and Lambert, J. D. (2007). Widespread RNA segregation in a spiralian embryo. *Evol. Dev.* **9,** 527–539.

Kloc, M., and Etkin, L. D. (1995). Two distinct pathways for the localization of RNAs at the vegetal cortex in *Xenopus* oocytes. *Development* **121,** 287–297.

Kloc, M., Bilinski, S., Dougherty, M. T., Brey, E. M., and Etkin, L. D. (2004a). Formation, architecture and polarity of female germline cyst in *Xenopus*. *Dev. Biol.* **266,** 43–61.

Kloc, M., Bilinski, S., and Etkin, L. D. (2004b). The Balbiani body and germ cell determinants: 150 years later. *Curr. Top. Dev. Biol.* **59,** 1–36.

Koop, D., Richards, G. S., Wanninger, A., Gunter, H. M., and Degnan, B. M. (2007). The role of MAPK signaling in patterning and establishing axial symmetry in the gastropod *Haliotis asinina*. *Dev. Biol.* **311,** 200–212.

Labordus, V., and van der Wal, U. P. (1986). The determination of the shell field cells during the first hour in the sixth cleavage cycle of eggs of *Ilyanassa obsoleta*. *J. Exp. Zool.* **239,** 65–75.

Lall, S., Ludwig, M. Z., and Patel, N. H. (2003). Nanos plays a conserved role in axial patterning outside of the Diptera. *Curr. Biol.* **13,** 224–229.

Lambert, J. D. (2008). Mesoderm in spiralians: The organizer and the 4d cell. *J. Exp. Zool. B Mol. Dev. Evol.* **310,** 15–23.

Lambert, J. D., and Nagy, L. M. (2001). MAPK signaling by the D quadrant embryonic organizer of the mollusc *Ilyanassa obsoleta*. *Development* **128,** 45–56.

Lambert, J. D., and Nagy, L. M. (2002). Asymmetric inheritance of centrosomally localized mRNAs during embryonic cleavages. *Nature* **420,** 682–686.

Lambert, J. D., and Nagy, L. M. (2003). The MAPK cascade in equally cleaving spiralian embryos. *Dev. Biol.* **263,** 231–241.

Lartillot, N., Lespinet, O., Vervoort, M., and Adoutte, A. (2002). Expression pattern of *Brachyury* in the mollusc *Patella vulgata* suggests a conserved role in the establishment of the AP axis in Bilateria. *Development* **129,** 1411–1421.

Lecuyer, E., Yoshida, H., Parthasarathy, N., Alm, C., Babak, T., Cerovina, T., Hughes, T. R., Tomancak, P., and Krause, H. M. (2007). Global analysis of mRNA localization reveals a prominent role in organizing cellular architecture and function. *Cell* **131,** 174–187.

Lehmann, R., and Nusslein-Volhard, C. (1991). The maternal gene nanos has a central role in posterior pattern formation of the *Drosophila* embryo. *Development* **112,** 679–691.

Makabe, K. W., Kawashima, T., Kawashima, S., Minokawa, T., Adachi, A., Kawamura, H., Ishikawa, H., Yasuda, R., Yamamoto, H., Kondoh, K., Arioka, S., Sasakura, Y., *et al.* (2001). Large-scale cDNA analysis of the maternal genetic information in the egg of *Halocynthia roretzi* for a gene expression catalog of ascidian development. *Development* **128,** 2555–2567.

Martindale, M. Q. (1986). The organizing role of the D quadrant in an equal-cleaving spiralian, *Lymnaea stagnalis* as studied by UV laser deletion of macromeres at intervals between third and fourth quartet formation. *Int. J. Invert. Reprod. Dev.* **9,** 229–242.

Martindale, M. Q., Doe, C. Q., and Morrill, J. B. (1985). The role of animal–vegetal interaction with respect to the determination of dorsoventral polarity in the equal-cleaving spiralian, *Lymnaea palustris*. *Roux Arch. Dev. Biol.* **194,** 281–295.

Rabinowitz, J. S., Chan, X. Y., Kingsley, E. P., Duan, Y., and Lambert, J. D. (2008). Nanos is required in somatic blast cell lineages in the posterior of a mollusk embryo. *Curr. Biol.* **18,** 331–336.

Raven, C. P. (1961). "Oogenesis: The Storage of Developmental Information." Pergamon Press, New York.

Render, J. (1991). Fate maps of the first quartet micromeres in the gastropod *Ilyanassa obsoleta*. *Development* **113,** 495–501.

Render, J. (1997). Cell fate maps in the *Ilyanassa obsoleta* embryo beyond the third division. *Dev. Biol.* **189,** 301–310.

Schisa, J. A., Pitt, J. N., and Priess, J. R. (2001). Analysis of RNA associated with P granules in germ cells of *C. elegans* adults. *Development* **128,** 1287–1298.

Seydoux, G., and Fire, A. (1994). Soma–germline asymmetry in the distributions of embryonic RNAs in *Caenorhabditis elegans*. *Development* **120,** 2823–2834.

Subramaniam, K., and Seydoux, G. (1999). nos-1 and nos-2, two genes related to *Drosophila* nanos, regulate primordial germ cell development and survival in *Caenorhabditis elegans*. *Development* **126,** 4861–4871.

Swartz, S. Z., Chan, X. Y., and Lambert, J. D. (2008). Localization of *Vasa* mRNA during early cleavage of the snail *Ilyanassa*. *Dev. Genes Evol.* **218,** 107–113.

Sweet, H. C. (1996). Regional specification of the first quartet micromeres in embryos of the gastropod *I. obsoleta*. *In* "Department of Zoology." University of Texas, Austin, TX.

Sweet, H. C. (1998). Specification of first quartet micromeres in *Ilyanassa* involves inherited factors and position with respect to the inducing D macromere. *Development* **125,** 4033–4044.

Tsuda, M., Sasaoka, Y., Kiso, M., Abe, K., Haraguchi, S., Kobayashi, S., and Saga, Y. (2003). Conserved role of nanos proteins in germ cell development. *Science* **301,** 1239–1241.

van den Biggelaar, J. A. M. (1976). The development of dorsoventral polarity preceding the formation of the mesentoblast in *Lymnaea stagnalis*. *Proc. Kon. Ned. Akad. Wet. C* **79,** 112–126.

van den Biggelaar, J. A. M., and Guerrier, P. (1979). Dorsoventral polarity and mesentoblast determination as concomitant results of cellular interactions in the mollusk *Patella vulgata*. *Dev. Biol.* **68,** 462–471.

Wang, Z., and Lin, H. (2004). Nanos maintains germline stem cell self-renewal by preventing differentiation. *Science* **303,** 2016–2019.

Wang, Y., Zayas, R. M., Guo, T., and Newmark, P. A. (2007). nanos function is essential for development and regeneration of planarian germ cells. *Proc. Natl. Acad. Sci. USA* **104,** 5901–5906.

Wierzejski, A. (1905). Embryologie von *Physa fontinalis* L. *Z. Wiss. Zool.* **83,** 502–706. plates 18–27.

Wilson, E. B. (1892). The cell lineage of *Nereis*. *J. Morphol.* **6,** 361–480.

Wilson, E. B. (1899). Cell-lineage and ancestral reminiscence. *In* "Biological Lectures 1898; The Marine Biological Laboratory, Wood's Holl, Mass." The Athenaeum Press, Boston, MA.

CHAPTER SIX

The Origin and Diversification of Complex Traits Through Micro- and Macroevolution of Development: Insights from Horned Beetles

Armin P. Moczek

Contents

Abstract

Understanding how development and ecology shape organismal evolution is a central goal of evolutionary developmental biology. This chapter highlights a class of traits and organisms that are emerging as new models in evo-devo and eco-devo research: beetle horns and horned beetles. Horned beetles are morphologically diverse, ecologically rich, and developmentally and genetically increasingly accessible. Recent studies have begun to take advantage of these attributes and are starting to link the microevolution of horned beetle development to the macroevolution of novel features, and to identify the genetic, developmental, and ecological mechanisms, and the interactions between them, that mediate organismal innovation and diversification in

Department of Biology, Indiana University, Bloomington, Indiana, USA

Current Topics in Developmental Biology, Volume 86
ISSN 0070-2153, DOI: 10.1016/S0070-2153(09)01006-0

natural populations. Here, I review the most significant recent findings and their contributions to current frontiers in evolutionary developmental biology.

1. Introduction

Organismal form and function emerge during ontogeny through complex interactions between genotype, environmental conditions, and ontogenetic processes (Raff, 1996; West-Eberhard, 2003). These interactions are central themes in many biological and medical disciplines, and occupy a particularly prominent position in evolutionary biology: ultimately, evolutionary diversification of organismal form and function is possible only through changes in the nature of at least some of these interactions. This poses a particular challenge in the origin and diversification of novel, complex traits. Evolutionary novelties not only beg the question as to how they are made during ontogeny, but also how whatever it takes to make them was able to arise from whatever genetic and developmental tool box existed in the ancestor prior to their first origin. As outlined below, beetle horns and horned beetles offer an unusual opportunity to integrate genetic, developmental, physiological, and environmental mechanisms into a holistic understanding of how complex traits are generated, integrated, and modified during both development and evolution. In this chapter, I highlight and synthesize recent advances in our understanding of the genetic, developmental, and ecological origins of horns and horn diversity, as well as their consequences for diversification and radiation of horned beetles. Before doing so, however, I will briefly review what it is about beetles and their horns that makes them a promising window into the mechanics of innovation and diversification in nature.

2. Uniqueness and Diversity of Horns

Beetle horns combine several characteristics that make them outstanding models for exploring the origin, integration, and diversification of novel traits. First, beetle horns are massive, solid, three-dimensional outgrowths that often severely transform the shape of whoever bears them (Fig. 6.1; e.g., Mizunuma, 1999). Horns are often as long or longer than other appendages such as legs, can double the length of an individual, and can make up more than 30% of body mass. Not surprisingly, beetle horns often dominate the morphological and behavioral phenotype of their bearers. Second, beetle horns are *unique* structures lacking clear homology to existing traits in insects. They are not modified mouthparts or legs; instead, they exist *alongside* these structures in body regions in which insects normally do

Figure 6.1 Examples of horned beetles illustrating diversity and magnitude of horn expression in adult beetles. Clockwise from top: *Trypoxylus (Allomyrina) dichotoma, Onthophagus watanabei, Golofa claviger*, and *Phanaeus imperator*. (See Color Insert.)

not produce any outgrowths (Moczek, 2005). Hence, horns can be looked at as an evolutionary innovation that occurred at some point during the history of beetles and which fueled one of the most impressive radiations of secondary sexual traits known in the animal kingdom. It is the resulting diversity of horn phenotypes that adds a third major rationale for horned beetles' utility as a model system for understanding the origins of organismal diversity (Arrow, 1951). Horn expression is restricted to relatively few beetle families such as the Tenebrionidae, Staphilinidae, Passalidae, Curculionidae, Chrysomelidae, and Scarabaeidae (reviewed in Moczek, 2005). However, within these families, and especially within the family

Scarabaeidae, horn expression is frequent and highly diverse. Moreover, much diversity can be found over remarkably short phylogenetic distances. For example, in the scarab genus *Onthophagus* species differ in the body regions that participate in horn growth (e.g., head or thorax), differ in horn number (single, paired, or combinations thereof), or differ in how horns scale with body size (e.g., isometric or sigmoidal; Balthasar, 1963). Amazingly, much of this diversity is also found within species where it is manifest in the expression of dramatic sexual dimorphisms as well as alternative male phenotypes (male dimorphism), suggesting possibly important links between the origins of diversity that exist within species to those existing between. This diversity in horn expression among conspecifics and congeners thus provides a remarkable opportunity to identify genetic and developmental mechanisms that generate variation in horn growth between individuals, as well as the ecological and behavioral causes that ultimately underlie this variation. To appreciate these causes, however, we must first learn more about the ecology of horned beetles and understand what, if anything, they use their horns for (Fig. 6.2).

3. Form and Function

Several hypotheses have been proposed to explain the adaptive significance of beetle horns (reviewed in Arrow, 1951). Horns have been thought to allow beetles to defend themselves against predators, indicate male quality to choosy females, or facilitate digging through soil. Alternatively, Arrow (1951) suggested that horns may actually have no function and may simply be the product of selection toward larger body size. However, little evidence exists in support of any of these hypotheses. In contrast, much evidence has now accumulated across a range of beetle families that suggests that horns are used as weapons in male–male combat over access to females (Cook, 1990; Eberhard, 1978; Emlen, 1997; Moczek and Emlen, 2000; Palmer, 1978; Siva-Jothy, 1987). Specifically, depending on horn size, shape, and fighting context, beetles use their horns to push, prod, lift, grab, stab or otherwise reduce their rivals' ability to access nearby females. For example, males of many species, including all members of the genus *Onthophagus* studied so far, fight in subterranean tunnels to gain or maintain access to breeding chambers and females. Here, fights take place within a confined space and horns are predominantly used as blocking and positioning devices. Fights then consist primarily of shoving contests, which can take a long time and appear energetically expensive injuries are rare to absent (Emlen, 1997; Moczek and Emlen, 2000; Palmer, 1978). In contrast, many species in the subfamily Dynastinae, such as the famous *Chalcosoma* species, fight arboreally. Here, horns are used to dislodge, lift, and throw

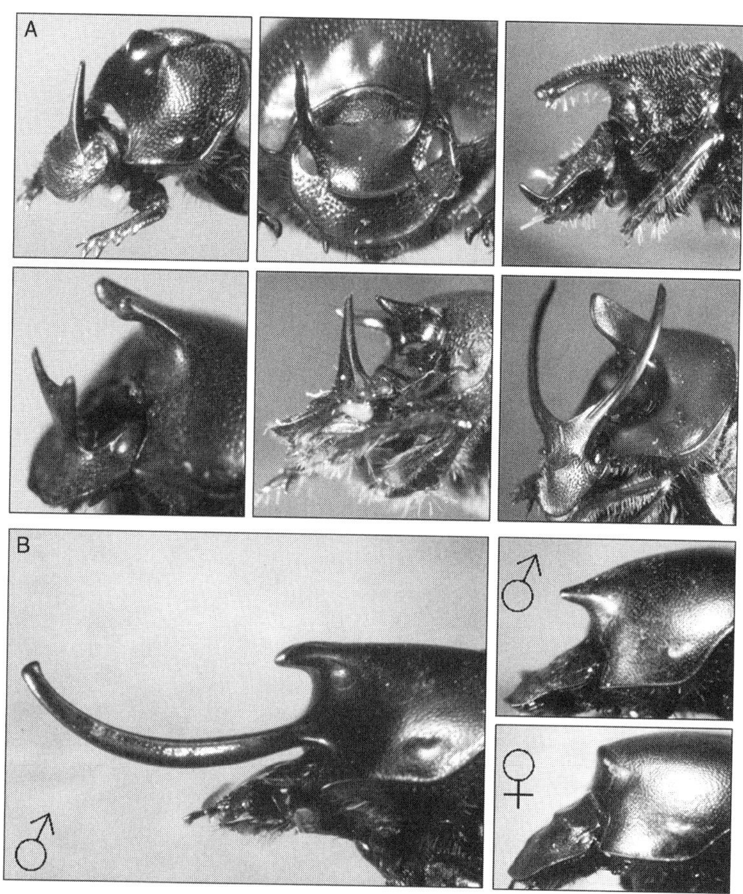

Figure 6.2 Diversity between and within *Onthophagus* species. (A) Six *Onthophagus* species illustrating the diversity of horn types that exist within the genus. (B) Sexual and male dimorphism in *Onthophagus nigriventris*. (See Color Insert.)

rivals off of branches. Fights can be brief but have the potential to inflict severe injury when males crack their exoskeleton upon hitting the ground (Beebe, 1944; Siva-Jothy, 1987). Both direct behavioral observations on several species (Emlen, 1997; Moczek and Emlen, 2000) and fitness estimates on at least one species (Hunt and Simmons, 2001) confirm that horn possession is indeed adaptive in these aggressive contests and improves a given male's chances of succeeding in fights.

Not all males within a species, however, express a full set of horns. In fact, horn dimorphisms are common in natural populations, resulting in the occurrence of two relatively discrete horned (also called major) and hornless (minor) morphs. Importantly, these alternative male morphs do *not* reflect allelic variants but instead are the product of environmental differences—predominantly larval feeding conditions (Emlen, 1994; Moczek and Emlen, 1999). Larvae with access to optimal feeding conditions eclose to adult larger than a certain size threshold and thus express a full set of horns, whereas larvae limited to suboptimal conditions eclose at smaller adult sizes and remain largely hornless. This horn polyphenism is not restricted to morphological differences, but also results in discrete behavioral and physiological differences between morphs. For example, in contrast to the aggressive fighting behavior employed by horned males, small hornless males employ nonaggressive sneaking behaviors to access females (Moczek and Emlen, 2000). Similarly, hornless males produce disproportionately larger ejaculate volumes during copulation (Simmons *et al.*, 1999), and recent work also showed that the presence or absence of horns has profound consequences for individual thermoregulatory properties (Shepherd *et al.*, 2008).

In summary, the horns of beetles represent an evolutionary novelty of extraordinary diversity within and between species, and play a significant role in the behavioral ecology of individuals as well as populations. We are now in a good position to begin exploring the developmental and genetic basis of horns, an effort that began only a few years ago (Moczek and Nagy, 2005), but that has already yielded important insights into the origins of novel features.

4. The Ontogeny of Horns

Beetles are holometabolous insects, and as such the larval stage constitutes their main feeding stage. In general, the larvae of horned beetles use relatively low-nutrition food sources such as dung (e.g., *Onthophagus*), decaying plant matter (e.g., *Chalcosoma, Trypoxylus*), or carrion (e.g., *Coprophanaeus*), and depending on the final adult size, larval development may take anywhere from weeks (*Onthophagus*) to several years (*Chalcosoma*). With respect to horn development, however, little happens during this period. Instead, most if not all horn patterning and growth takes place very late in larval development when the animal is nearing the transition to the pupal stage. Here, two brief and temporally dissociated stages are primarily responsible for generating and differentiating horn primordia during development. The transient *prepupal* stage at the very end of larval development marks the first of these two. At this point, all larval epidermis detaches from the larval cuticle—a phenomenon known as apolysis—and selected regions in the head and/or thorax undergo dramatic cell proliferation to generate the pupal precursors of adult horns (Moczek and Nagy, 2005). The *pupal* stage then marks the onset of the second developmental phase important for adult horn expression. During this stage, the pupal epidermis apolyses once more, but instead of the rapid growth marking earlier stages, apolysis is followed by sculpting and remodeling of the pupal epidermis into the final adult shape. Remodeling can be dramatic and is capable of removing large amounts of pupal horn tissue over a period of just a few days. In many species, pupal remodeling allows fully horned pupae to molt into entirely hornless adults (Moczek, 2006b). Degree of horn expression among adult beetles is thus the consequence of both prepupal growth late in larval development and the pupal remodeling phase just prior to the final, adult molt. Importantly, even congeneric species can differ widely in the degree to which they rely on one or the other mechanisms in generating intra-and interspecific diversity.

More generally, however, beetle horns originate and differentiate in a manner surprisingly similar to the primordia of adult legs, mouthparts, wings, or antennae of most insect orders (Svácha, 1992). Like horns, traditional appendages such as legs and mouthparts are epidermal outgrowths that form during late larval development and are remodeled during the pupal stage. The only dramatic deviation from this pattern occurs in all appendages produced by higher flies as well as in the wings of Hymenoptera, Lepidoptera, and some Coleoptera (Svácha, 1992). In these cases, appendages develop from imaginal disks, which represent a highly derived mode of appendage formation absent in the majority of insect orders (Kojima, 2004). Imaginal disks are epidermal invaginations specified during embryonic development which grow throughout larval development. Moreover, many important patterning steps take place while the disk is essentially a two-dimensional sheet of tissue, and all disk growth occurs while the disk is *in*vaginated into the body interior (Fig. 6.3A). Beetle horns differ in that they (a) appear not to be specified during embryonic development, (b) grow from the start as three-dimensional epidermal outbuddings, (c) have their growth confined to the relatively brief prepupal stage (~48 h), and (d) as they grow, evaginate into the space between epidermis and larval cuticle (Moczek, 2006a; Fig. 6.3B). Consequently, the *Drosophila* model of limb development has likely limited applicability for beetle horns. Instead, given their growth as epidermal outbuddings, beetle horns develop more like the appendages of most other insect orders (Svácha, 1992). Unfortunately, most of our understanding of insect appendage formation comes from studies of imaginal disk development in *Drosophila* (Kojima, 2004). Consequently, even though faced with serious limitations when applied to beetle horns, the *Drosophila* model of limb development represents our best starting point to begin exploring the regulation of horn growth and differentiation (Fig. 6.4).

5. The Regulation of Prepupal Horn Growth

As introduced above, one way to think of beetle horns is as simplified appendages. Unlike traditional appendages, beetle horns lack muscles, nerves, or joints, but like traditional appendages, beetle horns are three-dimensional

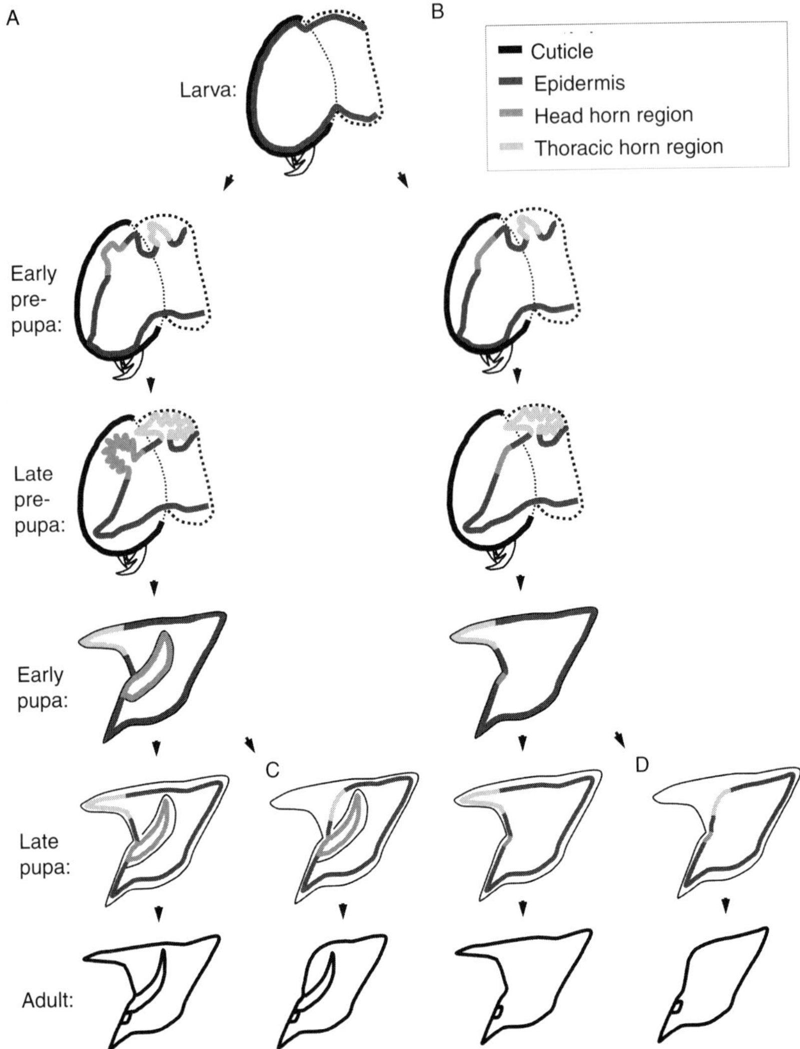

Figure 6.3 Development of horns and horn dimorphisms in *Onthophagus* beetles. (A) Apolysis is followed by rapid cell proliferation of selected epidermal tissue regions (shown here for a head horn and thoracic horn). Horn primordia expand during the pupal molt and become externally visible. During the pupal stage epidermal cells apolyse once more, followed by remodeling of the pupal epidermis into the final adult shape. The pupa then undergoes one last molt to the final adult stage. (B) Development of horn dimorphisms through differential proliferation of prepupal horn tissue (illustrated here for head horns). During the prepupal stage presumptive horn tissue proliferates little, resulting in the absence of external horns in pupae and resulting adults. (C, D) Development of horn dimorphisms through differential remodeling of pupal horn tissue (illustrated here for thoracic horns). Pupal horn epidermis is resorbed prior to the secretion of the adult cuticle. This mechanism generates sexual dimorphisms for thoracic horns in many species, and can occur in the presence or absence of (differential) head horn development (modified after Moczek, 2005). (See Color Insert.)

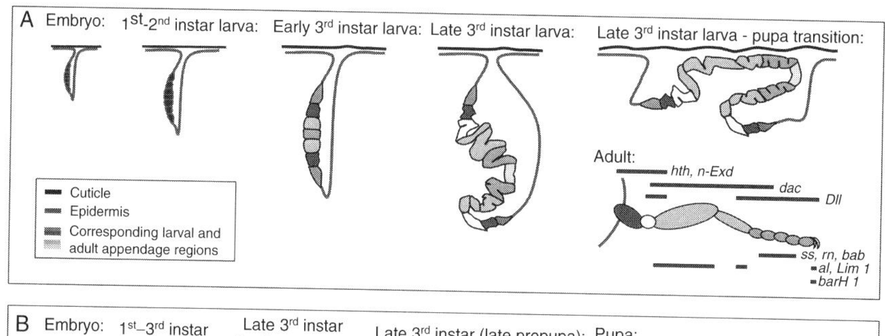

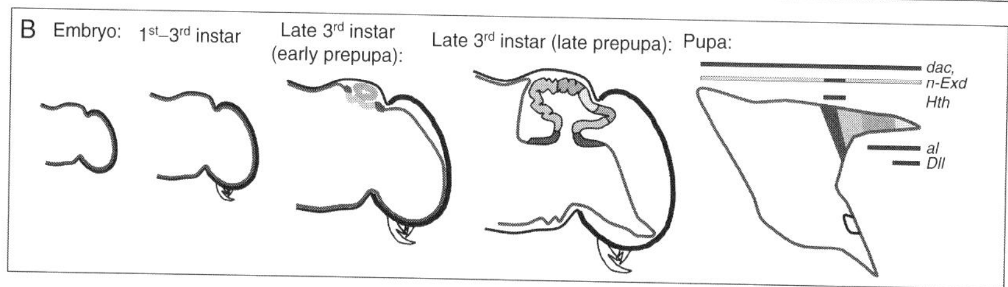

Figure 6.4 Differences and similarities in the development of the (A) *Drosophila* leg and (B) thoracic horns in beetles (see text for details). Colors indicate tissue types and regional relationships between immature and mature appendage. Also indicated is the approximate relationship between expression domains of common p/d patterning genes during development and the corresponding adult appendage region (modified after Moczek, 2006a,b). (See Color Insert.)

outgrowths of epidermal origin with clearly defined proximodistal, mediolateral, and anterior–posterior polarities. This raises the possibility that the regulation of beetle horns may rely at least in part on the same genetic and developmental mechanisms that regulate the expression of more traditional appendages. Recent data, focusing on the establishment of the proximodistal (p/d) axis, strongly support this hypothesis.

In *Drosophila*, establishment of the proximodistal axis begins with the concentration-dependent and combined action of two diffusible morphogens: *wingless* (*wg*) and *decapentaplegic* (*Dpp*). These subdivide imaginal disks into roughly concentric, nested domains of expression of several transcription factors including *Distal-less* (*Dll*), *dachshund* (*dac*), and *homothorax* (*hth*). The center of the leg disk, characterized by *Dll* expression, eventually gives rise to the distal region of the adult appendage, while progressively more peripheral disk regions, characterized by *dac* and *hth* expression, form progressively more proximal appendage regions once the imaginal disk telescopes outwards to form the adult appendage (reviewed in Kojima, 2004). As emphasized above, in most other insects, adult appendages develop not from imaginal disks but via the outbudding of selected epidermal regions during larval development (e.g., Fristrom and Fristrom, 1993; Nagy and Williams, 2001). Despite these fundamental differences in the morphogenesis of appendages, there remain many similarities in the underlying patterning mechanisms. For example, *Dll* expression in the distal region and *hth* expression in the proximal region occurs during the development of appendages in a wide range of insects and noninsect arthropods (Abzhanov and Kaufman, 2000; Inoue *et al.*, 2002; Jockusch *et al.*, 2000; Mittmann and Scholtz, 2001; Prpic and Tautz, 2003; Suzuki and Palopoli, 2001), and *Dll* activity is functionally required for distal leg formation in beetles and spiders (Beermann *et al.*, 2001; Schoppmeier and Damen, 2001). *Dll, dac,* and *hth* therefore represented legitimate candidate genes for the regulation of p/d axis formation and growth during beetle horn development (Fig. 6.5).

Expression studies lend first support to an involvement of at least two, and possibly all three of these transcription factors during horn development (Moczek and Nagy, 2005; Moczek *et al.*, 2006). In several *Onthophagus* species, *Dll* expression was found in the regions of prepupal horn primordia that later would form the part of the adult horn, while *hth* expression was confined to incipient proximal horn regions. In contrast, *dachshund*, a transcription factor normally involved in pattering medial appendage identity clearly violated the *Drosophila* model and was expressed well outside its predicted medial domain. These results suggested that partial redeployment of p/d patterning genes may have played a role in the origin of beetle horns, however, in the absence of functional assays any extrapolation from gene expression to gene function had to remain tentative at best. Recently, larval RNA interference (RNAi)-mediated gene function analyses have been used to further examine possible roles of *dac, hth,* and *Dll* in beetle horn development, with many interesting results (Moczek and Rose, unpublished data).

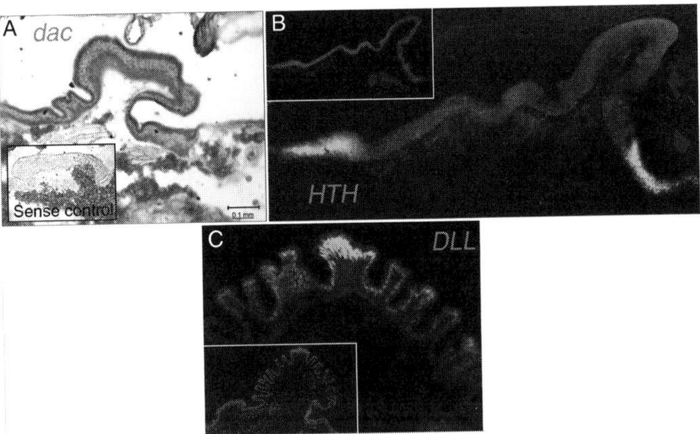

Figure 6.5 Examples of p/d genes expressed during horn development. (A) *Dachshund in situ* hybridization of the transient thoracic horn primordium in *O. taurus*. (B) Anti-HTH immunostaining of the persisting thoracic horn primordium of *O. binodis*. (C) Anti-DLL immunostaining of one of two head horn primordia in of *O. taurus*. (See Color Insert.)

First, irrespective of any involvement in horn development, larval RNAi-mediated transcript depletion of all three patterning genes generated phenotypic effects very similar to those documented by previous studies. For instance, *Dll* and *dac*RNAi resulted in loss or fusion of distal and medial regions, respectively, in the leg and antenna (Angelini and Kaufman, 2004; Kojima, 2004; Prpic *et al.*, 2001). Similarly, *hth*RNAi accelerated eye differentiation (Bessa *et al.*, 2002) and induced ectopic wing tissue on the first thoracic segment (Ryoo *et al.*, 1999; Yao *et al.*, 1999). These results documented for the first time both the feasibility and power of RNAi-mediated gene knockdown in horned beetles. In addition, this study yielded the first functional insights into the regulation of horn development. Specifically, the study showed that despite being widely expressed throughout prepupal horn primordia in *Onthophagus* (Moczek *et al.*, 2006) *dac* does not appear to play any obvious role in the regulation of size, shape, or identity of horns. Instead, *Otdac*RNAi individuals expressed thoracic and head horns of precisely the same size and overall shape as control animals despite severe *dac* knockdown phenotypes elsewhere in their body. In contrast, *hth* transcript depletion had a dramatic effect on horn expression, but only in one horn type: thoracic horns. *hth* transcript depletion resulted in drastically shortened thoracic horns over the entire range of body sizes, but had no effect on head horn expression. Instead, *Othth*RNAi individuals expressed head horns indistinguishable from control individuals despite severe effects on other head appendages. The results of *Dll*RNAi complicated things even further. Unlike *hth, Dll* transcript depletion affected the expression of both head and thoracic horns, but not in the same individuals or even species. In *Onthophagus taurus*, head horn expression was only affected in large males otherwise fated to express a full set of head horns, whereas horn expression in small- and medium-sized males was unaffected, as was the expression of pupal thoracic horns in both males and females regardless of body size. In the congener *O. binodis*, however, *Dll*RNAi affected thoracic horn expression and did so in both males and females, though the effect was strongest in large individuals. Combined, these results suggest that *Onthophagus Dll* and *hth*, but not *dac*, alter horn expression in a sex-, body region-, and body size-specific manner, and that even closely related species can diverge rather substantially in aspects of this regulation (Fig. 6.6).

These results are the first to suggest that horn development evolved via differential co-option of at least some p/d patterning genes normally involved in traditional appendage formation. On one side, these results are not surprising and confirm a general phenomenon in the evolution of novel traits: new morphologies do not require new genes or developmental pathways and instead may arise by recruiting existing developmental mechanisms into new contexts. On the other, these results also highlighted an unexpected degree of evolutionary lability, ranging from the absence of

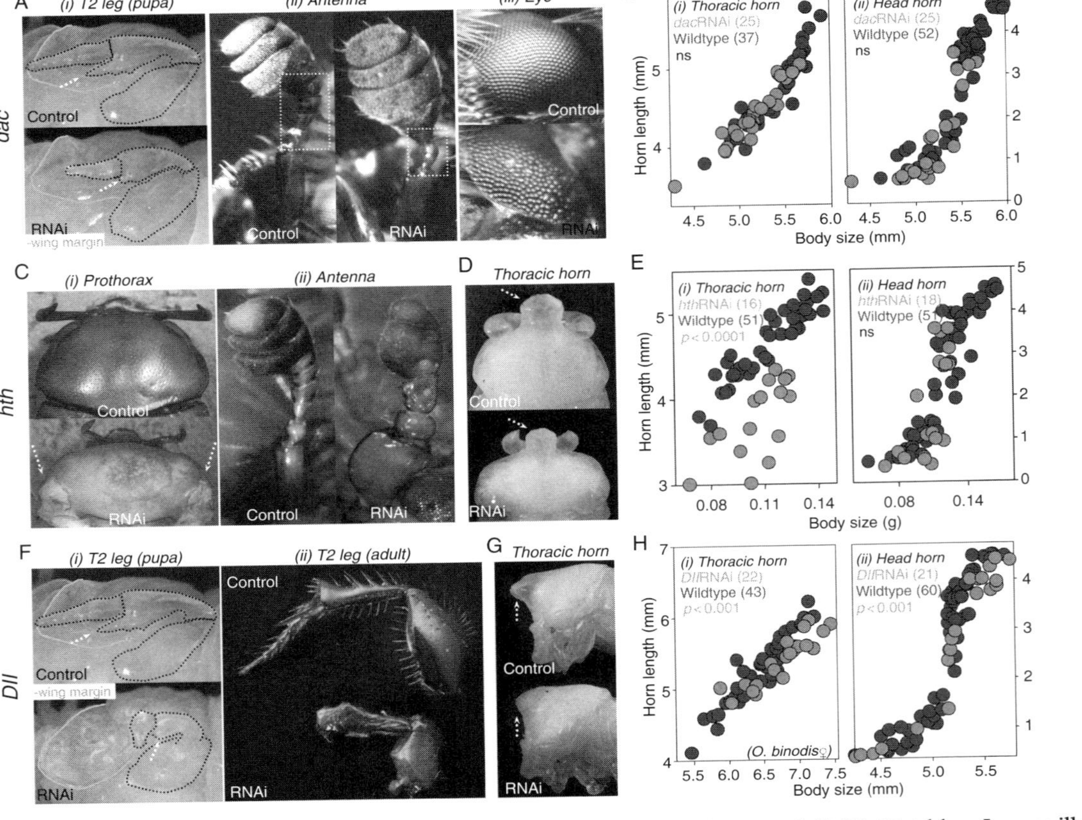

Figure 6.6 Larval RNAi-mediated transcript depletion of (A, B) *dachshund*, (C–E) *homothorax*, and (F–H) *Distal-less*. Images illustrate typical phenotypes observed in each experiment compared to wild-type phenotypes. Graphs depict scaling relationships between pupal body size and horn length for thoracic horns (i) and head horns (ii). Wild-type is shown in blue and RNAi-treated individuals are shown in red. All data are from male *O. taurus* except H(i) which were collected from female *O. binodis*. Sample sizes are given in parentheses (modified after Moczek and Rose, unpublished data). (See Color Insert.)

patterning function (*dac*) to patterning function in selected horn types only (*hth, Dll*) to function in one size class, sex, or species but not another (*Dll*). Most specifically, this suggests that different horn types, and even the same horn type in different species, may be regulated at least in part by different pathways. Different horn types may therefore have experienced distinct, and possibly independent, evolutionary histories.

It is important to realize that this is clearly just the beginning of a more detailed analysis of the developmental regulation and diversification of beetle horns. The recent development of *Onthophagus* EST libraries now provides access to members of many regulatory pathways known to be involved in insect development, ranging from genes involved in axis specification, patterning, and morphogenesis, to genes involved in many prominent signaling pathways, to genes involved in endocrine regulation of development. Furthermore, *Onthophagus* microarrays developed from these libraries have added a critical tool for rapid comparative transcriptional profiling across species, sexes, morphs, and even tissue regions within individuals. Clearly, much work lies ahead before we will have achieved a solid understanding of beetle horn development and its similarities and differences to other developmental processes. However, the most critical resources are now in place that promise that this goal will be attained within the near future. It is due to these same resources that we are already beginning to gain a much better insight into the regulation of the second developmental stage crucial for adult horn expression: the pupal remodeling stage.

6. The Regulation of Pupal Remodeling

As introduced above, the pupal stage marks the second developmental period critical to defining the final size and shape of adult horns. During this stage, animals undergo the same basic developmental steps as during the previous molts such as apolysis of the epidermis, secretion of a new cuticle, and eclosion to the next developmental stage. However, unlike in previous molts, there is no proliferation stage, and horns, just like other body parts, do not grow significantly during the pupal stage. Secondly, in at least one horn type, those extending from the thorax, there is frequent differential loss, or resorption, of presumptive horn tissue. In such cases, fully horned pupae molt into thorax horn-less adults lacking any signs of the previous existence of a thoracic horn primordium. Of 19 *Onthophagus* species studied thus far, four species utilized differential, sex-specific resorption of thoracic horn tissue to generate sexual dimorphism. The remaining 15 species use the same process to remove thoracic horn primordia in *both* sexes. In at least one of those, *O. taurus*, pupal thoracic horn resorption *eliminates* a pronounced sexual dimorphism in thoracic horns evident in pupae, but not in the resulting adults (Moczek *et al.*, 2006). Recent work now strongly implicates programmed cell death (PCD) in the destruction and removal of horn primordial tissue (Fig. 6.7).

PCD involves the coordinated destruction of cytoplasmic contents including organelles and their membranes as well as nuclear DNA degradation (Potten and Wilson, 2004). As such, PCD relies on a complex cascade of developmental and cellular processes. Despite this apparent complexity, PCD is an ancient physiological process employed by all metazoan organisms to dispose of cells during development. A recent study showed that primordial epidermis of horns programmed to be resorbed undergoes premature PCD during the first 24 h of the pupal stage (Moczek, 2006b). Relying on two different biochemical assays, the same study then showed that PCD is considerably more frequent among horn primordial cells of transient horns compared to individuals whose pupal horns are being converted into an adult structure, supporting the hypothesis that PCD is the most likely mechanisms by which horn resorption and remodeling are achieved. At the same time, comparisons across species suggested that the exact position and timing of PCD-mediated horn remodeling can differ remarkably from one species to the next. Combined, the regulation of pupal remodeling reveals many of the same features highlighted above for the regulation of horn growth. On one side, pupal remodeling and resorption of horns appears to rely on an ancient developmental mechanism, PCD, which has been recruited into a new developmental context. On the other, results suggest the existence of considerable variation within and between species regarding when, where, and how much remodeling of horns occurs. By extension, this variation suggests the existence of modifier mechanisms that regulate species-, sex-, and body region-specific resorption of horns. The identity and nature of these modifier mechanisms are currently being investigated, and many interesting preliminary data have already been collected (Fig. 6.8).

For instance, previous work on *Drosophila* has shown that the Hox genes *Deformed* (*Dfd*) and *Abdominal-B* (*Abd-B*) regulate segment boundaries through the regional activation of PCD (Lohmann *et al.*, 2002), suggesting regional Hox genes as possible gene candidates for the regulation of PCD-mediated resorption of pupal horns. Indeed, preliminary data on *Onthophagus* now suggest that the Hox gene *Sex combs reduced* (*Scr*), traditionally responsible for patterning the first thoracic segment in insects, has acquired the function to regulate PCD-mediated pupal horn remodeling, and that it exerts this function in a sex- and species-dependent manner (B. Wasik, D. Rose, and A. P. Moczek, unpublished data). Similarly, research on a variety of insects has shown PCD to be regulated by endocrine factors, in particular ecdysteroids and juvenile hormone (JH; e.g., Lobbia *et al.*, 2007; Oliver *et al.*, 2007). In *Onthophagus*, at least JH appears to play an important

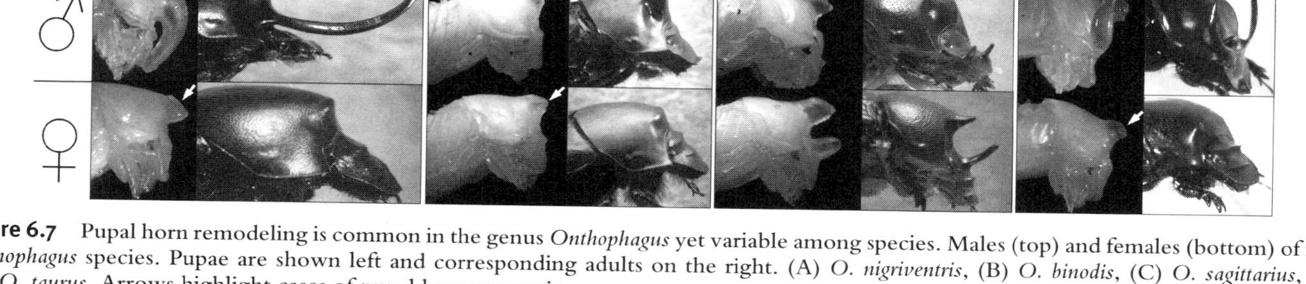

Figure 6.7 Pupal horn remodeling is common in the genus *Onthophagus* yet variable among species. Males (top) and females (bottom) of four *Onthophagus* species. Pupae are shown left and corresponding adults on the right. (A) *O. nigriventris*, (B) *O. binodis*, (C) *O. sagittarius*, and (D) *O. taurus*. Arrows highlight cases of pupal horn resorption.

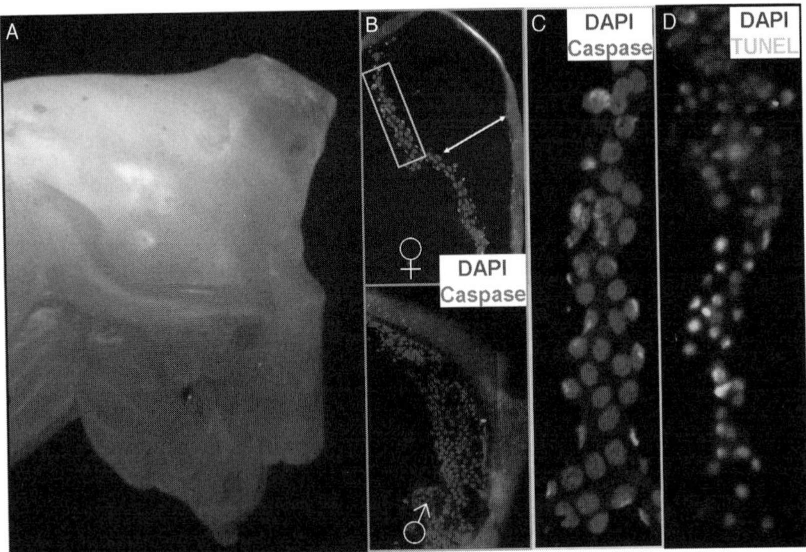

Figure 6.8 Programmed cell death appears to mediate sex-specific pupal remodeling in *O. binodis*. (A) Pupa indicating distal thoracic horn. (B) Anti-DRICE (activated caspase-3) staining in thoracic horn epidermis on pupal day 1 in (top) and (bottom). (C) Detail of Bè. (D) Corresponding region stained with TUNEL assay to detect PCD-specific DNA fragmentation. (See Color Insert.)

role in the regulation of horn expression, and both sexes and species differ in degree and nature of changes in horn expression that are induced by the same JH manipulation (Emlen and Nijhout, 1999; Shelby *et al.*, 2007). In summary, while existing data clearly provide only a very superficial understanding of the developmental regulation and diversification of pupal remodeling of beetle horns, promising avenues for future research exist that should soon make this an exciting area of study.

7. The Regulation of Plasticity

The horns of male beetles are as much famous for their extravagance and splendor as they are for the incredible variation in horn expression that exists between males of the same species. In fact, diversity between conspecific males often parallels differences between species, especially in cases in which discrete large, horned (major) and small, hornless (minor) morphs co-occur in the same population. In such instances, the existence of alternative morphs has occasionally resulted in them being described as different species (Paulian, 1935). As explained earlier, differences in body size and horn expression between conspecific males, including the expression of discrete morphs, occur in response to differences in environmental conditions, especially larval feeding conditions (Emlen, 1994; Moczek and Emlen, 1999). Larvae with access to optimal feeding conditions eclose to larger body sizes and express larger, and often disproportionately larger, horns, whereas larvae with access to poorer conditions eclose to smaller adult sizes and express no or greatly reduced horns. This *plasticity* in body size and horn expression gives rise to particular allometric scaling relationships between body size and horn length. Such allometries can differ dramatically between species and sexes, ranging from isometric allometries (in species in which large males are proportionally enlarged versions of smaller males) to positive allometry (in species in which large male have disproportionally enlarged horns compared to small males) to sigmoidal allometries (in species in which alternative morphs are separated by a threshold size). Importantly, closely related species and even populations of the same species can diverge in aspects of these scaling relationships such as the allometric slope or body size threshold separating alternative morphs (Moczek and Nijhout, 2003; Moczek *et al.*, 2002). This suggests that even though the individual differences in body size and horn length are brought about by environmental differences, heritable variation exists between genotypes causing them to respond differently to the same nutritional variation. Selection or drift can then act on this variation and shape species-specific allometries in nature. For instance, a particularly illuminating case of allometric divergence has been documented in *O. taurus*, in which three exotic populations established less than 50 years ago from the native range of the species have diverged heritably in body size thresholds to a degree normally only observed between species (Moczek and Nijhout, 2003). Ecological studies in the field suggest that these divergences were driven by differences in the intensity of intra- and interspecific competition for breeding opportunity, and thus changes in the social context within which horned (fighting) and hornless (sneaking) male morphs function (Moczek, 2003). More generally, these and other findings highlight that besides the origin and diversification of horns *per se*, the evolution and diversification of *plasticity* in horn expression has contributed massively to extant patterns of morphological diversity. This implies the existence of independent regulatory mechanisms which, rather than controlling the expression of horns, regulate the degree and nature of *plasticity* in horn expression in response to environmental gradients. *Onthophagus* beetles again provide outstanding opportunities to explore the developmental and genetic regulation of plasticity given the enormous variation in *plasticity* of horn expression that exists between species, ranging from absence of environmental sensitivity to complete determination by nutritional conditions. In some cases, both extremes of sensitivity can even be found in different horn types expressed *by the same individual*, such as the nutrition-insensitive pupal thoracic horns of male *O. taurus* and the highly nutrition-sensitive head horns in the same individuals. Identifying the nature and mechanics of developmental pathways underlying plasticity in horn expression, and evolutionary changes in horn expression, are therefore major foci of current research, and several important insights have already been gained.

For instance, comparing the two most divergent *O. taurus* populations mentioned above, Moczek and Nijhout (2002) found that allometric divergences correlated with evolved differences in degree and timing of sensitivity to JH. In both populations, artificial applications of a JH analog induced horns in male larvae otherwise fated to eclose into hornless adults. However, populations in which males already expressed horns at relatively small body sizes were more sensitive to JH manipulations, and were sensitive *earlier* in development compared to populations that confined horn expression to only but the largest males. These findings supported the hypothesis that a JH titer-mediated threshold response underlies the expression of alternative male morphs. Moreover, it suggested that aspects of this threshold response, such as degree and timing of sensitivity to JH, are capable of undergoing remarkably rapid evolution in natural populations. A recent study by Shelby *et al.* (2007) extended this perspective to sexual dimorphisms and interspecific differences in horn size and shape.

These and other observations therefore suggest that endocrine factors such as JH mediate between nutritional variation experienced by larvae and morphological, behavioral, and physiological variation that exists among the resulting adults. The mechanisms by which endocrine factors adjust development to environmental conditions are presently not understood, but many critical resources exist that will allow researchers to make headway in this direction over the next few years. For instance, the *Onthophagus* cDNA libraries and microarrays introduced earlier contain *Onthophagus* orthologs of many genes involved in a plethora of developmental processes likely to be crucial for horn formation as well as many genes likely involved in endocrine regulation via ecdysteroid-, JH-, and Insulin-signaling. Studies are now under way to use these and other resources to identify genes, pathways, and gene networks whose expression change in response to nutritional changes, to characterize the level of conservation of this induction across body regions, morphs, sexes, and species, and ultimately to identify the functions of the most promising gene candidates.

8. The Origins of Novelty and Diversity

The preceding sections highlighted several areas in which we are beginning to get a better understanding of the regulation of developmental processes relevant to growth and differentiation of horns, such as the function of p/d axis patterning genes during prepupal growth, the activation of PCD during pupal remodeling, or the endocrine underpinning of plasticity in horn expression. Each of these cases illustrates a by-now-familiar pattern in the evolution of development, including the evolution of novel features: novel traits do not require new genes or developmental pathways to come into being, but instead may arise from co-option of pre-existing developmental machinery into new contexts. P/d patterning genes and PCD still carry their ancestral function of instructing axis polarity or removing superfluous cells, but what is new is the location and timing of their action. Further research into the regulation of beetle horn development will undoubtedly add additional examples. In addition, we are also beginning to see examples of possibly truly novel functions, acquired by old regulators during the evolution of beetle horns. For example, if current research further confirms that the Hox gene *Scr* regulates PCD during the pupal remodeling phase of development, this may well emerge as a regulatory function of *Scr* that is unique to horned beetles and which has no parallels to its ancestral functions during insect development. If correct, this would suggest that the evolution of beetle horns involved the recruitment of conserved developmental mechanisms into new contexts enriched by novel regulatory interactions acquired by pre-existing regulatory genes. Lastly, we do not yet know of any genes or pathways whose expression and functions are entirely unique to beetle horns, but we should not lose sight of this possibility. Current *Onthophagus* arrays contain several hundred ESTs with large open reading frames yet lacking obvious orthology to existing databases, and it is conceivable that some of those may represent horn-specific genes and regulators that evolved solely in the context of *Onthophagus* horn development. In addition to identifying conserved or putatively novel regulatory properties of interesting genes and pathways, the studies on beetle horn development summarized above have also unearthed tremendous *variation* in these properties between morphs, sexes, populations, and species. Combined, these findings have three major implications. First, they contradict the notion that highly upstream regulators, such as p/d patterning genes or Hox genes, should be evolutionarily entrenched given their importance in the regulation of basic aspects of animal architecture and thus resistant to the acquisition of novel functions (Davidson and Erwin, 2006). Instead, they illustrate that regulatory genes whose functions are otherwise highly conserved nevertheless retain the capacity to acquire additional functions. Second, results to date suggest that little phylogenetic distance is necessary for the evolution of sex- and species-specific differences in these functions. If confirmed, this would argue that even master-regulator genes and their interactions can diversify on the level of populations and species with unexpected ease. Third, many of the developmental differences seen between species, such as the presence or absence of horns or horn expression in different body regions, have striking parallels in sexual dimorphisms or male dimorphisms. This raises the possibility that the developmental capacity to generate macroevolutionary differences may originate well within species, between sexes, and—fueled by developmental plasticity—across alternative morphs.

Understanding the developmental, behavioral, and ecological basis of horns and horn diversity, however incomplete, now puts us in a position to address the questions posed at the beginning of this chapter. What are the genetic, developmental, and ecological mechanisms, and the interactions between them that brought about the first transition from a hornless ancestor to a horned descendant, and that since have shaped the subsequent diversification of beetle horns? Recent work has begun to provide some surprising answers to both of these questions, and I will end this chapter by highlighting where we have made the most progress. I will begin, appropriately, with the origin of horns. Specifically, I will focus on one particular horn type, those protruding from the thorax, where recent work has made the greatest headway toward understanding the possible origin of these structures.

9. Thoracic Horns as an Exaptation

As introduced above, PCD-mediated resorption of pupal thoracic horn primordia appears common, if not ubiquitous, among *Onthophagus* species, raising the question as to the adaptive significance, if any, of transient horn expression. Experimental approaches have now revealed that pupal horns, irrespective of whether they give rise to a corresponding adult structure or not, actually play a crucial role during the larval-to-pupal molt and the shedding of the larval head capsule (Moczek *et al.*, 2006). Unlike in larval–larval and pupal–adult molts, larvae that molt into pupae have little muscle tissue left that could aid in the shedding of the larval cuticle, as most larval muscles have already undergone histolysis. Instead, the animal uses peristaltic contractions to pump hemolymph to inflate selected body regions and to force old cuticle to rupture. This is sufficient to remove the highly membranous thoracic and abdominal cuticle of larval scarab beetles. However, the larval head capsule may pose additional challenges as it is composed of extremely thick cuticle used between molts to anchor powerful jaw muscles. Such muscles, and the corresponding head capsule, may be particularly strong in fiber-feeding scarab larvae such as *Onthophagus*, and this is where the thoracic horn primordia of *Onthophagus* beetles unexpectedly enter the stage. Carefully staged sections showed that during *Onthophagus*' prepupal stage, thoracic horn primordia insert themselves into the space vacated between the larval head capsule and corresponding epidermis and subsequently fill with hemolymph and expand. Eventually, this expansion forces the larval head capsule to fracture along prepatterned suture lines. As a consequence, as the larval head molts into a pupal head, the first pupal structure visible from the outside is not a part of the head, but instead the thoracic horn primordium as it breaks through the head capsule. Experimental elimination of thoracic horn primordia prior to the prepupal stage resulted in pupae that (a) lack a thoracic horn and (b) failed to shed their larval head capsule (Moczek *et al.*, 2006). Replicating this approach in and outside the genus *Onthophagus* showed that this putative dual function of thoracic horn primordia appears unique to onthophagine beetles. Further phylogenetic analyses suggested that the pupal molting function of horns preceded the horns-as-a-weapon function of the adult counterparts, and that ancestrally, pupal horns were always resorbed prior to the adult molt (Moczek *et al.*, 2006). If correct, this would explain why so many *Onthophagus* species grow thoracic horns even though those outgrowths are not used to form a functional structure in the adult.

These results also raise the possibility that the origin of *adult* horns could have been the result of a simple failure to remove otherwise pupal-specific projections through PCD. A survey of the available literature suggests that such events actually occur in natural populations frequently enough to be detected by entomologists (e.g., Ballerio, 1999; Paulian, 1945; Ziani, 1994). Even though such an outgrowth would initially have been rather small, behavioral studies have shown that if used in the context of a fight, even very small increases in horn length bring about significant increases in fighting success and fitness (Emlen, 1997; Moczek and Emlen, 2000). Behavioral studies have also shown that aggressive fighting behavior is widespread among beetles and occurs well outside horned taxa. Possession of adult horns is therefore not a prerequisite for fighting, instead male beetles most likely fought each other well before the first adult horn ever surfaced, creating a selective environment in which the first pupal horn that failed to be removed before the adult molt could have provided an immediate fitness advantage. Thoracic beetle horns may thus be a good example of a novelty that arose as an exaptation from traits originally selected for providing a completely different function during a completely different stage of development. It is equally important to realize, however, that none of these arguments appear to hold for other horn types such as head horns. Head horns, at least in *Onthophagus*, only undergo mild remodeling if any, and morphological differences among adults are already largely established in the preceding pupal stage (Moczek, 2007). These basic differences underscore the likely evolutionary and developmental independence that characterizes different types of horns, and most likely different lineages of horned beetles. More generally, the possible origin of adult thoracic horns from ancestral molting devices provides a vivid example of the crooked routes that developmental evolution is capable of taking as it generates what we in the end perceive as an evolutionary novelty. The same complexity in the interactions between development, morphology, and ecology emerges when we examine the diversification of already existing horns, as the next example hopes to illustrate.

10. Developmental Tradeoffs and the Diversification of Horns and Horned Beetles

Holometabolous insects such as beetles provide an exceptional opportunity to study a phenomenon believed to have shaped phenotypic diversity well beyond the insects and that is likely important for all metazoan organisms: resource allocation tradeoffs during development. Resource allocation tradeoffs arise during development when two or more structures compete for a shared and limited resource to sustain their growth. As such, resource allocation tradeoffs not only have the potential to alter ontogenetic outcomes, as developmental enlargements of one structure may only be feasible at the expense of another, but also evolutionary trajectories, as development may only be able to accommodate evolutionary enlargements of one structure through compensatory reduction of another. Resource allocation tradeoffs are likely ubiquitous during metazoan development, but are possibly particularly important in the development of holometabolous insects. Here, all growth of adult structures is confined to a time period during which larvae no longer take in nutrients and so represents essentially a closed system with a finite pool of resources to fuel all of metamorphosis. While the exact nature of resource allocation tradeoffs remains obscure, growing evidence exists that they have real potential to bias developmental outcomes and long-term evolutionary trajectories (Nijhout and Emlen, 1998). Recent work on horned beetles has begun to implicate resource allocation tradeoffs in the diversification of horns and other body parts, with intriguing implications for the diversification of horned beetle species (Emlen, 2001; Kawano, 2002; Moczek and Nijhout, 2004; Parzer and Moczek, 2008; Simmons and Emlen, 2006; Simmons *et al.*, 2007).

In 2002, Kazuo Kawano showed that two species of giant rhinoceros beetles (genus *Chalcosoma*) had diverged in both relative horn sizes and copulatory organ sizes, and that this divergence was more pronounced between sympatric (overlapping) than allopatric (separated) populations. His findings were perfectly consistent with reproductive character displacement reinforced in sympatry but not allopatry. What was intriguing, however, was the observation that the species which had evolved relatively longer horns had also evolved relatively shorter copulatory organs, and vice versa. In other words, male horn sizes and copulatory organ sizes had coevolved *antagonistically*. Subsequent experimental work on *O. taurus*

(Moczek and Nijhout, 2004) suggested that this antagonistic coevolution may not have been a coincidence. In this study, surgical ablation of the genital precursor tissue during development resulted in males with disproportionately longer horns. The magnitude of the effect depended on timing of ablation, which contradicted an earlier study that emphasized physical proximity as the main determinant of tradeoff intensity (Emlen, 2001). Rather than growing close to each other, it seemed that growing at the same time was more important in determining whether tradeoffs would occur or not. These arguments aside, the available data suggested that there may be a connection between how horns and copulatory organs developed, and therefore how they evolved. This was particularly intriguing because changes in male copulatory organs are thought to play a major role in the evolution of reproductive isolation, and thus, speciation (Eberhard, 1985). In fact, copulatory organ morphology is often the only way to distinguish cryptic and recent species, suggesting that whatever mechanism is able to influence copulatory organ expression in a population may have immediate repercussions for that population's ability to interbreed with others.

The strongest evidence to date that suggests exactly that kind of interaction between horn evolution and copulatory organ evolution now comes from a very recent study examining both within- and between-species covariation in horn investment versus copulatory organ investment (Parzer and Moczek, 2008). Specifically, this study focused first on the rapidly diverging exotic *O. taurus* populations introduced earlier. Recall that these populations were introduced from their native Mediterranean range to the Eastern US as well as to Eastern and Western Australia, and that these introductions occurred less than 50 years ago. Also, recall that all of these populations have evolved significant differences in male horn investment due to diversifying selection acting on the horn-length switch point, with Western Australian males growing the relatively shortest horns whereas Eastern US males grow the relatively longest, with the other two populations intermediate (Moczek, 2003; Moczek and Nijhout, 2003). Add to this the realization that there is no sympatry between any of these populations, and you have the perfect test situation to answer whether evolutionary changes in horn investment may cause correlated changes in copulatory organ size independent of possible reproductive character displacement in sympatry. And the answer is: they do! Among the four populations examined, there was a perfect negative correlation between relative investment into horns and relative investment into copulatory organ size. As a second step, the study applied the same approach to nine different *Onthophagus* species, and the same highly significant negative correlation between relative investment into horns and copulatory organ size emerged. Intriguingly, the greatest differences observed between *O. taurus* populations were similar in nature and magnitude to some of the differences detected between *Onthophagus* species. These results had three major

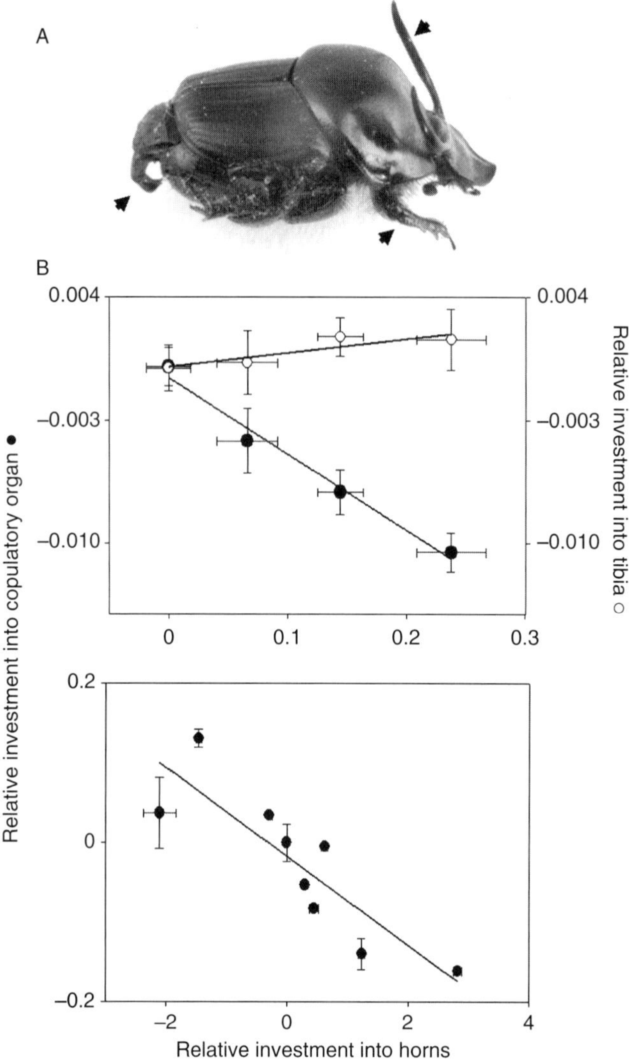

Figure 6.9 Tradeoffs between primary and secondary sexual characters in populations and species of *Onthophagus* beetles. (A) Horned male *Onthophagus taurus*. Arrows highlight horns, copulatory organ, and fore tibia. (B) Relative investment into copulatory organ size (left, g) and fore tibia size (right, o) as a function of relative investment into horn size in four different populations of *O. taurus*. Error bars represent one standard error. (C) Relative investment into copulatory organ size as a function of relative investment into horn size in nine different *Onthophagus* species. Data are corrected for differences in body size (modified after Parzer and Moczek, 2008).

implications. First, they suggest that copulatory organ size, a primary sexual trait, may diverge as a byproduct of evolutionary changes occurring in horns, a secondary sexual trait. Second, these findings illustrate that the resulting signatures of antagonistic coevolution are detectable both during microevolutionary divergences between populations operating on a time-scale of years, as well as macroevolutionary divergence between species operating on a timescale of tens of millions of years. Third, and most remarkable, given the extreme importance of copulatory organ morphology for reproductive isolation, these findings begin to raise the possibility that secondary sexual trait evolution may promote speciation as a byproduct. If tradeoffs between horns and male copulatory organs are indeed driving speciation in *Onthophagus* this might help explain how this genus, famous for its dramatic diversity in secondary sexual traits, was able to radiate into over 2400 extant species, making it the most speciose genus in the animal kingdom (Arrow, 1951) (Fig. 6.9).

11. Conclusions

In this chapter, I hope to have shown that horned beetles in general—and the genus *Onthophagus* in particular—offer a rich microcosm in which to explore the mechanisms of evolutionary innovation and diversification. Combining extreme morphological diversity with a rich ecology and natural history as well as developmental and genetic accessibility, research on *Onthophagus* beetles is now beginning to permit an increased integration across levels of biological organization as well as timescales, allowing us to integrate genetic, endocrine and ecological contributions to phenotypic diversity, and to bridge micro- and macroevolutionary perspectives on development. Given the diversity of questions that can be addressed with these organisms and the experimental tools available to researchers, I hope that *Onthophagus* beetles will attract the attention of the next generation of students in evolution and development. It will be up to them to fully realize what we have barely begun to imagine.

ACKNOWLEDGMENTS

Matthew Stansbury provided constructive comments on earlier drafts of this chapter. Research presented here was supported by National Science Foundation grants IOS 0445661 and IOS 0718522.

REFERENCES

Abzhanov, A., and Kaufman, T. C. (2000). Homologs of *Drosophila* appendage genes in the patterning of arthropod limbs. *Dev. Biol.* **227,** 673–689.

Angelini, D. R., and Kaufman, T. C. (2004). Functional analyses in the hemipteran *Oncopeltus fasciatus* reveal conserved and derived aspects of appendage patterning in insects. *Dev. Biol.* **271,** 306–321.

Arrow, G. H. (1951). "Horned Beetles." W. Junk, The Hague, Netherlands.

Ballerio, A. (1999). Revision of the genus *Pterorthochaetes* first contribution (Coleoptera: Scarabaeoidea: Ceratocanthidae). *Folia Heyrovskyana* **7,** 221–228.

Balthasar, V. (1963). "Monographie der Scarabaeidae und Aphodiidae der palaearktischen und orientalischen Region (Coleoptera: Lamellicornia). Band 2, Coprinae." Verlag der tschechoslowakischen Akademie der Wissenschaften, Prag.

Beebe, W. (1944). The function of secondary sexual characters in two species of Dynastidae (Coleoptera). *Zoologica.* **29,** 53–58.

Beermann, A., Jay, D. G., Beerman, R. W., Huelskamp, M., Tautz, D., and Juergens, G. (2001). The *Short antenna* gene of *Tribolium* is required for limb development and encodes the orthologue of the *Drosophila Distal-less* protein. *Development* **128,** 287–297.

Bessa, J., Gebelein, B., Pichaud, F., Casares, F., and Mann, R. S. (2002). Combinatorial control of *Drosophila* eye development by *eyeless, homothorax*, and *teashirt*. *Genes Dev.* **16,** 2415–2427.

Cook, D. (1990). Differences in courtship, mating and postcopulatory behavior between male morphs of the dung beetle *Onthophagus binodis* Thunberg (Coleoptera: Scarabaeidae). *Anim. Behav.* **40,** 428–436.

Davidson, E. H., and Erwin, D. H. (2006). Gene regulatory networks and the evolution of animal body plans. *Science* **311,** 796–800.

Eberhard, W. G. (1978). Fighting behavior of male *Golofa porteri* beetles (Scarabaeidae: Dynastinae). *Psyche* **83,** 292–298.

Eberhard, W. G. (1985). "Sexual Selection and Animal Genitalia." Harvard University Press, Cambridge, MA.

Emlen, D. J. (1994). Environmental control of horn length dimorphism in the beetle *Onthophagus acuminatus* (Coleoptera, Scarabaeidae). *Proc. R. Soc. Lond. B* **256,** 131–136.

Emlen, D. J. (1997). Alternative reproductive tactics and male dimorphism in the horned beetle *Onthophagus acuminatus*. *Behav. Ecol. Sociobiol.* **41,** 335–341.

Emlen, D. J. (2001). Costs and the diversification of exaggerated animal structures. *Science* **291,** 1534–1536.

Emlen, D. J., and Nijhout, H. F. (1999). Hormonal control of male horn length dimorphism in the dung beetle *Onthophagus taurus* (Coleoptera: Scarabaeidae). *J. Insect Physiol.* **45,** 45–53.

Fristrom, D., and Fristrom, J. W. (1993). The metamorphic development of the adult epidermis. *In* "The Development of *Drosophila melanogaster*" (M. Bate and A. M. Arias, Eds.), pp. 843–897. Cold Spring Harbor Laboratory Press, New York.

Hunt, J., and Simmons, L. W. (2001). Status-dependent selection in the dimorphic beetle *Onthophagus taurus*. *Proc. R. Soc. Lond. B* **268,** 2409–2414.

Inoue, Y., Mito, T., Miyawaki, K., Terasawa, T., Matsushima, K., Shinmyo, Y., Niwa, N., Mito, T., Ohuchi, H., and Noji, S. (2002). Correlation of expression patterns of *homothorax, dachshund*, and *Distal-less* with the proximodistal segmentation of the cricket leg bud. *Mech. Dev.* **113,** 141–148.

Jockusch, E., Nulsen, C., and Nagy, L. M. (2000). Leg development in flies vs. grasshoppers: Differences in *dpp* expression do not lead to differences in the expression of downstream components of the leg patterning pathway. *Development* **127,** 1617–1626.

Kawano, K. (2002). Character displacement in giant rhinoceros beetles. *Am. Nat.* **159,** 255–271.

Kojima, T. (2004). The mechanism of *Drosophila* leg development along the proximodistal axis. *Dev. Growth Differ.* **46,** 115–129.

Lobbia, S., Futahashi, R., and Fujiwara, H. (2007). Modulation of the ecdysteroid-induced cell death by juvenile hormone during pupal wing development of Lepidoptera. *Arch. Insect Biochem. Physiol.* **65,** 152–163.

Lohmann, I., McGinnis, N., Bodmer, M., and McGinnis, W. (2002). The *Drosophila* Hox gene *deformed* sculpts head morphology via direct regulation of the apoptosis activator *reaper. Cell* **23,** 457–466.

Mittmann, B., and Scholtz, G. (2001). *Distal-less* expression in embryos of *Limulus polyphemus* (Chelicerata, Xiphosura) and *Lepisma saccharina* (Insecta, Zygentoma) suggests a role in the development of mechanoreceptors, chemoreceptors, and the CNS. *Dev. Genes Evol.* **211,** 232–243.

Mizunuma, T. (1999). "Giant Beetles." ESI Publishers, Tokyo, Japan.

Moczek, A. P. (2003). The behavioral ecology of threshold evolution in a polyphenic beetle. *Behav. Ecol.* **14,** 831–854.

Moczek, A. P. (2005). The evolution and development of novel traits, or how beetles got their horns. *Bioscience* **11,** 935–951.

Moczek, A. P. (2006a). Integrating micro- and macroevolution of development through the study of horned beetles. *Heredity* **97,** 168–178.

Moczek, A. P. (2006b). Pupal remodeling and the development and evolution of sexual dimorphism in horned beetles. *Am. Nat.* **168,** 711–729.

Moczek, A. P. (2007). Pupal remodeling and the evolution and development of alternative male morphologies in horned beetles. *BMC Evol. Biol.* **7,** 151.

Moczek, A. P., and Emlen, D. J. (1999). Proximate determination of male horn dimorphism in the beetle *Onthophagus taurus* (Coleoptera: Scarabaeidae). *J. Evol. Biol.* **12,** 27–37.

Moczek, A. P., and Emlen, D. J. (2000). Male horn dimorphism in the scarab beetle *Onthophagus taurus*: Do alternative tactics favor alternative phenotypes? *Anim. Behav.* **59,** 459–466.

Moczek, A. P., and Nagy, L. M. (2005). Diverse developmental mechanisms contribute to different levels of diversity in horned beetles. *Evol. Dev.* **7,** 175–185.

Moczek, A. P., and Nijhout, H. F. (2002). Developmental mechanisms of threshold evolution in a polyphenic beetle. *Evol. Dev.* **4,** 252–264.

Moczek, A. P., and Nijhout, H. F. (2003). Rapid evolution of a polyphenic threshold. *Evol. Dev.* **5,** 259–268.

Moczek, A. P., and Nijhout, H. F. (2004). Trade-offs during the development of primary and secondary sexual traits in a horned beetle. *Am. Nat.* **163,** 184–191.

Moczek, A.P., and Rose, D. J. (2009). Differential recruitment of limb patterning genes during development and diversification of beetle horns (unpublished data).

Moczek, A. P., Hunt, J., Emlen, D. J., and Simmons, L. W. (2002). Threshold evolution in exotic populations of a polyphenic beetle. *Evol. Ecol. Res.* **4,** 587–601.

Moczek, A. P., Rose, D., Sewell, W., and Kesselring, B. R. (2006). Conservation, innovation, and the evolution of horned beetle diversity. *Dev. Genes Evol.* **216,** 655–665.

Nagy, L. M., and Williams, T. A. (2001). Comparative limb development as a tool for understanding the evolutionary diversification of limbs in arthropods: Challenging the modularity paradigm. *In* "The Character Concept in Evolutionary Biology" (G. Wagner, Ed.), pp. 457–490. Academic Press, San Diego, CA.

Nijhout, H. F., and Emlen, D. J. (1998). Competition among body parts in the development and evolution of insect morphology. *Proc. Natl. Acad. Sci. USA* **95,** 3685–3689.

Oliver, R. H., Albury, A. N., and Mousseau, T. A. (2007). Programmed cell death in flight muscle histolysis of the house cricket. *J. Insect Physiol.* **53,** 30–39.

Palmer, T. J. (1978). A horned beetle which fights. *Nature* **274,** 583–584.

Parzer, H. F., Moczek, A. P. (2008). Rapid antagonistic coevolution between primary and secondary sexual characters in horned beetles. *Evol.* **62,** 2423–2428.

Paulian, R. (1935). Le polymorphisme des males de coléopteres. *In* "Exposés de biométrie et statistique biologique IV" (G. Tessier, Ed.), pp. 1–33. Actualités scientifiques et industrielles 255. Hermann and Cie, Paris, France.

Paulian, R. (1945). "Coléoptère Scarabéides de l'Indochine. Première partie." Faune de l'Empire Français III, Paris, France.

Potten, C., and Wilson, J. (2004). "Apoptosis: The Life and Death of Cells." Cambridge University Press, Cambridge, MA.

Prpic, N. M., and Tautz, D. (2003). The expression of the proximo-distal patterning genes *Distal-less* and *dachshund* in the appendages of *Glomeris marginata* (Myriapoda, Diplopoda) suggest a special role of these genes in patterning head appendages. *Dev. Biol.* **260,** 97–112.

Prpic, N. M., Wigand, B., Damen, W. G., and Klingler, M. (2001). Expression of *dachshund* in wild-type and *Distal-less* mutant *Tribolium* corroborates serial homologies in insect appendages. *Dev. Genes Evol.* **211,** 467–477.

Raff, R. (1996). "The Shape of Life: Genes Development, and the Evolution of Animal Form." University of Chicago Press, Chicago, IL.

Ryoo, H. D., Marty, T., Casares, F., Affolter, M., and Mann, R. S. (1999). Regulation of Hox target genes by a DNA bound Homothorax/Hox/Extradenticle complex. *Development* **126,** 5137–5148.

Schoppmeier, M., and Damen, W. G. M. (2001). Double-stranded RNA interference in the spider *Cupiennius salei*: The role of *Distal-less* is evolutionarily conserved in arthropod appendage formation. *Dev. Genes Evol.* **211,** 76–82.

Shelby, J. A., Madewell, R., and Moczek, A. P. (2007). Juvenile hormone mediates sexual dimorphism in horned beetles. *J. Exp. Zool. B* **308,** 417–427.

Shepherd, B. L., Prange, H. D., and Moczek, A. P. (2008). Some like it hot: Body and weapon size affect thermoregulation in horned beetles. *J. Insect Physiol.* **54,** 604–611.

Simmons, L. W., and Emlen, D. J. (2006). Evolutionary trade-off between weapons and testes. *Proc. Natl. Acad. Sci. USA* **103,** 16346–16351.

Simmons, L. W., Tomkins, J. L., and Hunt, J. (1999). Sperm competition games played by dimorphic male beetles. *Proc. R. Soc. Lond. B* **266,** 145–150.

Simmons, L. W., Emlen, D. J., and Tomkins, J. L. (2007). Sperm competition games between sneaks and guards: A comparative analysis using dimorphic male beetles. *Evolution* **61,** 2684–2692.

Siva-Jothy, M. T. (1987). Mate securing tactics and the cost of fighting in the Japanese horned beetle, *Allomyrina dichotoma* L. (Scarabaeidae). *J. Ethol.* **5,** 165–172.

Suzuki, Y., and Palopoli, M. F. (2001). Evolution of insect abdominal appendages: Are prolegs homologous or convergent traits? *Dev. Genes Evol.* **211,** 486–492.

Svácha, P. (1992). What are and what are not imaginal discs: Reevaluation of some basic concepts (Insecta, Holometabola). *Dev. Biol.* **154,** 101–117.

West-Eberhard, M. J. (2003). "Developmental Plasticity and Evolution." Oxford University Press, New York.

Yao, L. C., Liaw, G. J., Pai, C. Y., and Sun, Y. H. (1999). A common mechanism for antenna-to-leg transformation in *Drosophila*: Suppression of homothorax transcription by four HOM-C genes. *Dev. Biol.* **211,** 268–276.

Ziani, S. (1994). Un interessante caso di teraologia simmetrica in *Onthophagus (Paleonthophagus) fracticornis* (Coleoptera, Scarabaeidae). *Boll. Ass. Romana Entomol.* **49,** 165–167.

CHAPTER SEVEN

Axis Formation and the Rapid Evolutionary Transformation of Larval Form

Rudolf A. Raff*,† *and* Margaret Snoke Smith*,1

Contents

Abstract

Marine invertebrate embryos and larvae are diverse and can evolve rapidly, providing a link between early developmental and evolutionary mechanisms. We here discuss the role of evolutionary changes in axis formation, which is a crucial part of the patterning of marine embryos and larvae. We focus on sea urchin embryos, where axial features are well defined and subject to active current investigation. The genetic control of processes of formation of the three axial systems, animal–vegetal, dorsal–ventral, and left–right, is becoming established for species that undergo development via the feeding pluteus larva. These species represent the primitive condition among living sea urchins. We compare their developmental processes to the highly modified development of a species that has evolved a nonfeeding larva. This derived form has

* Department of Biology, Indiana University, Bloomington, Indiana, USA
† School of Biological Sciences, University of Sydney, Sydney, Australia
1 Current Address: Department of Entomology, University of Georgia, Athens, Georgia, USA

Current Topics in Developmental Biology, Volume 86
ISSN 0070-2153, DOI: 10.1016/S0070-2153(09)01007-2

accelerated some elements of axis formation, and eliminated or modified others. Three features of embryonic/larval evolution stand out (1) evolution of developmental features occurs rapidly over geological time; (2) upstream gene regulatory systems of axis formation are conserved, whereas downstream features evolve rapidly; and (3) heterochronies play an important role.

1. Introduction: Sea Urchins and the Evolution of Marine Larval Development

Many marine invertebrates have a complex life history, meaning that they develop via a larva that possesses a body plan and an ecological niche distinct from the adult and undergo a dramatic metamorphic process that releases the developing adult. For example, most echinoderms have a bilaterally symmetric, planktonic, feeding larva, in which a pentameral, benthic, motile, or sessile juvenile adult develops. This biphasic life history is frequently referred to as indirect development. There are competing hypotheses of how a biphasic life history may have evolved in metazoan history (Raff, 2008). The first hypothesis suggests that the current larval form represents the original adult, and biphasic ontogeny evolved by the addition of a new "adult" phase to the end of development, accompanied by evolution of a corresponding transition from larval gene regulation to that of adult development (Davidson *et al.*, 1995). However, phylogenetic studies indicate that the last common bilaterian ancestor of marine phyla did not exhibit such a biphasic life history and that feeding larval forms arose subsequently (Jenner, 2000). A competing hypothesis suggests that the larval phase was an intercalation into the development of an ancestor in which the bilaterian adult form developed directly from the embryo (Peterson, 2005; Peterson *et al.*, 2005; Raff, 2008; Sly *et al.*, 2003). Evolution of indirect-developing larvae took place during the Cambrian expansion of metazoan life, yielding a diversity of bilaterian marine phyla with a biphasic life history along with persisting direct-developing bilaterian phyla (Nützel *et al.*, 2006; Peterson, 2005; Peterson *et al.*, 2005). Remarkably, the evolution of larval forms has not ceased, with novel larval forms evolving among marine groups like annelids, mollusks and echinoderms beyond the Cambrian (Allen and Pernet, 2007; Collin, 2004; Jeffery *et al.*, 2003; Love *et al.*, 2007; Raff and Byrne, 2006; Rouse, 2000). A number of clades rapidly evolved derived direct-developing, nonfeeding forms within the past 0.5–4 plus million years (Hart *et al.*, 1997; Jeffery *et al.*, 2003; Zigler *et al.*, 2003). These provide important experimental models for the study of larval evolution.

The observation of larval and adult forms that possess distinct body plans raises interesting questions about how developmental mechanisms were modified during the evolution of a complex life history mode (Raff, 2008). Sea urchins are a particularly useful group for studying the evolution of development associated with the evolution of a larval phase because they have been used over a century as experimental model systems in developmental biology. Sea urchin embryos are amenable to experimental manipulations, and there is substantial variation among larval forms across clades (Wray, 1996).

There are two major modes of development among sea urchins, each with a characteristic larval form and life history. The ancestral mode of development of sea urchins and other echinoderms is indirect development, which is characterized by the production of many small eggs that develop into elaborate, swimming and feeding larvae. The pluteus larvae of sea urchins must feed in the water column for several weeks to amass the resources for adult development and metamorphosis. While in the water column, they are subject to high levels of larval predation and other sources of mortality (Morgan, 1995; Rumrill, 1990). In contrast, the other major mode of development in sea urchins, direct development, minimizes larval predation by producing larger eggs that develop into nonfeeding larvae that metamorphose after only a few days. However, given a finite amount of resources for reproduction, fewer eggs are produced and more resources are invested into each egg. Although this second mode of sea urchin development has been termed direct development because it lacks a feeding larval phase, it is a secondarily evolved kind of direct development, and still involves formation of a larva. It should not be confused with the use of direct development to describe the developmental mode of the bilaterian ancestor, which did not include a larval phase.

Within the species pair, *Heliocidaris tuberculata* (which develops via a small egg and a pluteus larva) and *Heliocidaris erythrogramma* (a direct developer with a large egg), there is an egg volume differential of 100-fold resulting from a modified process of oogenesis that late in the process vastly increases egg content in *H. erythrogramma* (Byrne *et al.*, 1999). The outcome of a pentameric adult sea urchin is conserved across developmental modes.

Pluteus larvae of indirect-developing sea urchins have a characteristic morphology generally defined by bilateral symmetry, the development of eight arms, a mouth, a tripartite gut, and a complete ciliary band involved in food capture (Fig. 7.1). The larvae of direct developers morphologically retain bilateral symmetry but usually are simplified relative to pluteus larvae, meaning that they have fewer or often no arms, no functional gut, no mouth, and have an incomplete ciliary band. Many larvae of direct developers are barrel-shaped (Figs. 7.1 and 7.2). Direct development has evolved independently multiple times within several sea urchin clades (Emlet *et al.*, 1987; Jeffery *et al.*, 2003; Strathmann, 1978; Wray, 1996). The factors affecting the evolution of direct development are complex, but the ovoid larvae of many direct developers likely resulted from selection for rapid

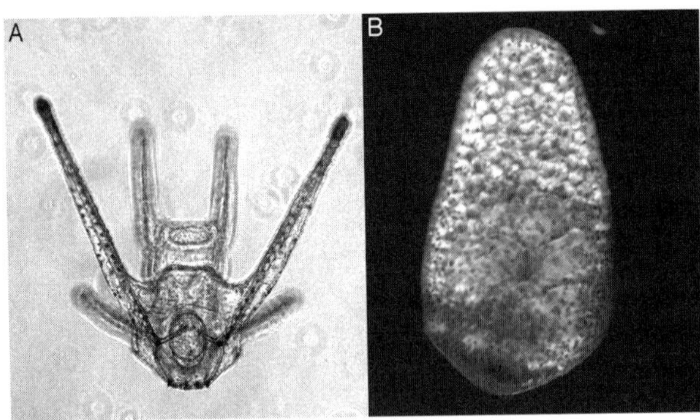

Figure 7.1 Photographs of representative live larvae of (A) *H. tuberculata* and (B) *H. erythrogramma* show the major morphological differences in larval form across developmental modes. The *H. tuberculata* pluteus (2 weeks old, ventral view) has six arms. The large larval mouth opens upward. The oval gut shows in the lower center. The arms each have a skeletal rod. Two more arms and the rudiment of the juvenile adult will develop in the next month, and metamorphosis will then occur. The *H. erythrogramma* larva (2 days old, left side view) has no mouth or gut. The juvenile with its five primary tube feet is forming in the lower center. It will metamorphose in 1 or 2 days more to yield the juvenile sea urchin (photo by E. C. Raff). (See Color Insert.)

metamorphosis (Smith *et al.*, 2007). Evolution of highly derived direct-developing larvae has taken place in 4–7 million years (Jeffery *et al.*, 2003; Zigler *et al.*, 2003). The evolution of these distinct larval morphologies raises the question of how developmental mechanisms underlying larval form changed during the evolution of direct development.

We have discovered modifications to early development in the direct developer *H. erythrogramma*, including changes in oogenesis, sperm, cell lineages, gene expression, cell fates, and downstream changes in cell differentiation (Byrne *et al.*, 1999; Love and Raff, 2006; Raff *et al.*, 1990; Smith *et al.*, 2008c, 2009b; Wray and Raff, 1989, 1990a). These reflect a spectrum of changes in developmental gene regulation, indicating that during the 4 million years since divergence from the indirect-developing last common ancestor with *H. tuberculata*, *H. erythrogramma* has evolved a number of new features. However, crucial initial steps in establishing each of the larval body plans are the events of symmetry breaking and specification of larval body axes. To understand the evolution of larval form, it is important to compare the larvae of indirect- and direct-developing sea urchins in the proper axial framework. Sea urchin embryos differentiate three axes, in the following order: animal–vegetal (A–Vg), dorsal–ventral (D–V), and left–right (L–R). Here, we briefly review the emerging literature on how larval axes are established in both indirect- and direct-developing sea urchins as a framework for highlighting how changes in axial formation are associated with the evolution of larval form.

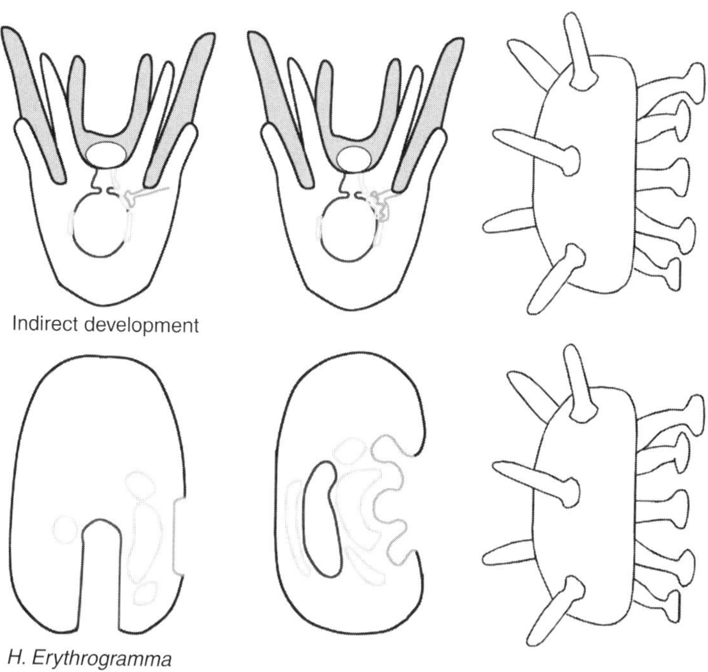

Figure 7.2 Schematic comparing internal features of indirect developers and *H. erythrogramma*. Top row: a pluteus and juvenile. On the left side of the larva, the vestibular ectoderm, derived from oral ectoderm (orange) makes contact with the middle coelomic compartment, the hydrocoel (green) to form the rudiment. Larval gut in blue. The juvenile at right is oriented in the same way as the developing rudiment in the pluteus. Lower row: *H. erythrogramma* is shown in ventral view. The coelom forms within about 30 h, and by 36 h the left middle compartment (hydrocoel) is interacting with the vestibule (orange) to form the juvenile. The postmetamorphic juvenile (about 4 days postfertilization) is shown on the right. (See Color Insert.)

2. Axis Formation

To relate changes in axis formation to the evolution of larval form, we define three distinct phases of axis formation that result in the final polarity along an axis—specification, determination, and execution. Specification refers to the initiation of the axis formation process, where the trajectory for setting up the poles of an axis has begun, but the process is still reversible. Determination is defined as the point at which formation of an axis is irreversible (Gilbert, 2003). The standard embryological demonstration that an axis is determined is to separate embryo blastomeres and show that the axis cannot be re-established during the development of the resulting half-embryos; that is, the two halves are not equivalent. We introduce the word execution to mean the events downstream of axis determination, in which morphological asymmetries along an axis come into being. Execution thus encompasses the morphogenetic, cell differentiation, molecular and timing events that lead to visible polarity becoming manifest along an axis. Specification, determination, and execution necessarily occur in this order. As they describe a continuum of axis establishment, they are not necessarily mutually exclusive processes.

2.1. Animal–vegetal axis

The first axis to be specified in both indirect developers and direct developers is the animal–vegetal axis, the primary axis, which runs from top to bottom of the egg. In both developmental modes, it is specified and determined maternally (Henry and Raff, 1990; Henry *et al.*, 1990; Horstadius, 1973). The classic experiment of Horstadius showed that when unfertilized eggs of indirect-developing sea urchins were cut in half with a glass needle at the equator, the resulting halves when fertilized did not establish a normal A–Vg axis. The vegetal half could gastrulate, but that the animal half could form only a ciliated hollow ball of cells. Animal-half blastomeres separated from eight-cell stage embryos likewise differentiate only animal pole-like ciliated balls of cells. The experimental addition of vegetal pole micromeres to animal pole cells restores a vegetal fate and normal development (Horstadius, 1973; Ransick and Davidson, 1993).

The earliest morphological evidence of execution of the A–Vg axis in indirect-developing sea urchins arises during fourth cleavage. The blastomeres divide so that animal pole daughter cells are all medium size (mesomeres), the next most vegetal cells are large (macromeres), and the most vegetal cells are small (micromeres) (Fig. 7.3, top row). Micromeres give rise to the primary mesenchyme cells (PMCs) that deposit the larval skeleton. The result of fourth cleavage is segregation of cell size and cell fate along the A–Vg axis. Later in development, the vegetal plate flattens followed by gastrulation at the vegetal end and not the animal end of the embryo. The animal–vegetal asymmetry arises from a maternal determinant, the localization of Disheveled (Dsh) protein at the vegetal pole cortex of the unfertilized sea urchin egg (Ettensohn, 2006). Dsh protein inhibits the degradation of β-catenin mediated by GSK3b. Dsh localization thus provides the first detectable asymmetry marking a pole of the A–Vg axis in sea urchin eggs.

The canonical Wnt signaling pathway plays a key role in execution of the A–Vg axis in indirect developers (Emily-Fenouil *et al.*, 1998; Ettensohn, 2006; Vonica *et al.*, 2000; Wikramanayake *et al.*, 1998, 2004). Wnt signaling confers a vegetal fate to cells in the embryo, whereas its absence results in an animal fate. Briefly, signaling through the canonical Wnt pathway is initiated by binding of the Wnt ligand to a Frizzled receptor. This binding

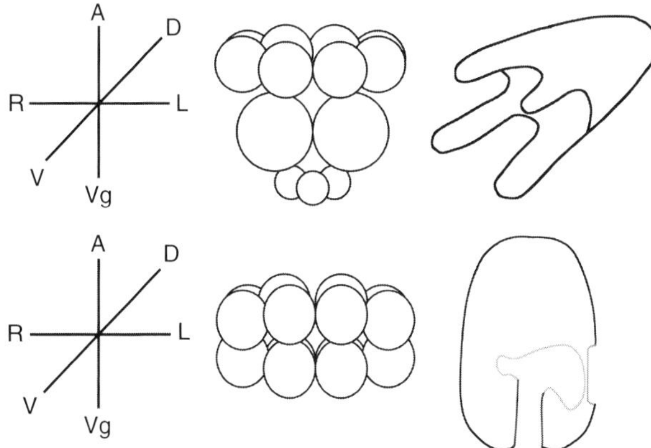

Figure 7.3 Schematic of axial relationships in sea urchin embryos. Left column shows axes of indirect-developing (top) and direct-developing embryos (lower). Animal–vegetal axis (A–Vg), dorsal–ventral (D–V), and left–right axis (L–R). Middle column shows the 16-cell stage larvae (indirect, top). Execution of the A–Vg axis in indirect developers is first obvious after fourth cleavage when there is asymmetric cleavage resulting in blastomere size segregation along the A–Vg axis (top). The blastomeres of the direct developer *H. erythrogramma* are the same size regardless of position (bottom). Right column shows indirect- and direct-developing larvae with ventral side toward the viewer.

results in activation of Disheveled, which prevents GSK3 from phosphorylating β-catenin, thus allowing β-catenin to enter the nucleus. Nuclear β-catenin acts in coordination with TCF/Lef to transcribe specific genes that in sea urchins confer a vegetal fate (reviewed by Croce and McClay, 2006). The function of β-catenin in establishment of vegetal fates and establishing the operation of the gene regulatory network (GRN) of PMCs has both a maternal and a zygotic phase.

The GRN that establishes micromere differentiation into downstream cell fates including PMCs in indirect-developing sea urchins has been defined by Oliveri *et al.* (2008). The network includes maternal inputs, including β-catenin, that lead to activation of the gene *Pmar1*, which blocks a repressor, HesC, allowing activation of a set of regulatory genes that stabilize the micromere differentiation stage and regulate downstream differentiation genes. These downstream genes include those involved in secretion of the larval skeleton. Then, zygotic transcription of *Wnt8* and Wnt8 signaling becomes part of a feedback loop. Two other signal systems, ES, and the Notch ligand, Delta, are also activated in PMCs and signal to endomesoderm and mesoderm. The specification of the pattern of endomesoderm cell fates at the vegetal pole of the early embryo involves a dynamically expanding torus of regulatory gene expression dependent on expression of Wnt8 (Smith *et al.*, 2008a). By the blastula stage, the animal pole acquires its first localized gene transcript, for the transcription factor *foxQ2* (Tu *et al.*, 2006). It functions in animal pole differentiation and has a further role to play in D–V axis formation.

Direct development has been studied almost exclusively in the Australian sea urchin *H. erythrogramma*, so we rely heavily on data from this species when discussing direct development. The first evidence of a maternal A–Vg axis is the "hillock" on the animal pole of the unfertilized egg and the greater density of the animal pole (Henry *et al.*, 1990; Wray and Raff, 1989). In *H. erythrogramma*, all blastomeres are the same size, with no micromeres formed at any point in cleavage (Fig. 7.3, lower row). As *H. erythrogramma* is evolutionarily derived from an ancestor with a pluteus larva, micromeres were present in the ancestor and subsequently lost (Strathmann, 1978; Wray, 1996; Wray and Raff, 1989, 1990a). There is no visible evidence of execution of the A–Vg axis until the vegetal plate flattens during the late blastula stage just preceding gastrulation. However, as in indirect developers, animal halves isolated at the eight-cell stage develop into hollow ciliated balls of cells, indicating that the polarity of cell fate along the A–Vg axis is present much earlier (Henry and Raff, 1990). Gastrulation proceeds by involution of cells at the vegetal end of embryos as in the embryos of indirect-developing sea urchins. Although cell cleavage patterns differ, *H. erythrogramma* forms broadly similar cell fates along the A–Vg axis as indirect developers. However, cell lineages differ and timing of specification of fates occurs later for mesenchyme and endomesodermal cell types (Wray and Raff, 1990b).

The maternal inputs into A–Vg axis formation are not defined in *H. erythrogramma*, but regardless of the difference in timing of execution of the A–Vg axis, signaling through the canonical Wnt pathway is responsible for early zygotic events establishing the A–Vg axis (Kauffman and Raff, 2003). Interference with Wnt8 signaling or with TCF function causes animalization in *H. erythrogramma*. Overexpression of Wnt8 or TCF causes vegetalization. These studies are comparable to those carried out in indirect-developing sea urchins and indicate that the role of Wnt8 signaling along the A–Vg axis is conserved. Similar experiments with an independently evolved direct developer from a different sea urchin clade, *Holopneustes purpurescens*, showed that Wnt signaling is conserved in this lineage as well (Kauffman and Raff, 2003).

2.2. Dorsal–ventral axis

The dorsal–ventral axis is the second larval axis to be determined in indirect-developing sea urchins. Experimental separation of embryo blastomeres at the two-cell stage results in half-embryos that can each re-establish a D–V axis, indicating that the D–V axis is not determined maternally or by fertilization. The D–V axis is determined sometime between (early) cleavage and the mesenchyme blastula stage (Cameron *et al.*, 1990; Hardin *et al.*, 1992). Execution of the D–V axis occurs toward the end of gastrulation when the ventral side flattens, the mouth forms on the ventral side, and the archenteron bends toward the ventral side. Larval ectoderm differentiates along the D–V axis of indirect developers to form two sharply distinct ectodermal domains. The ventral (oral) ectoderm of the pluteus is distinct from dorsal (aboral) ectoderm in cell shape and gene expression (Angerer and Davidson, 1984; Hardin and Armstrong, 1997; Love and Raff, 2006) (Fig. 7.2, top). The ciliary band runs along the arms, defining the ventral (oral) ectoderm within which is the larval mouth that lies between the arms, and the dorsal (aboral) ectoderm covering the remainder of the pluteus.

Specification of the D–V axis may be maternal because asymmetric positioning of mitochondria creates a redox gradient in the egg that directly affects the D–V axis (Coffman and Davidson, 2001), and there is asymmetric gene expression in the egg along the future D–V axis (Bradham and McClay, 2006). The ventral (oral) pole of the D–V axis is specified by a mechanism that links it to the A–Vg axis. β-Catenin-dependent signaling from the macromeres has been shown to restrict expression of the *foxQ2* gene to the animal pole (Yaguchi *et al.*, 2008). In early cleavage, *foxQ2* mRNA is widely distributed in the animal half of the embryo, but becomes restricted to the animal pole by the blastula stage. FoxQ2 blocks expression of *nodal* and, along with Lefty, prevents *nodal* expression in the animal plate. If β-catenin is prevented from being produced by vegetal cells, no D–V axis can form. Thus, in the normal course of events, vegetal Wnt signaling of A–Vg axis execution limits FoxQ2 to the animal pole and coordinates the two first axes by allowing *nodal* to be expressed on the future ventral side but not in the animal plate (Yaguchi *et al.*, 2008).

Events downstream of specification are affected by TGF-β family signaling pathway members. Nodal itself is expressed at the ventral pole of the presumptive D–V axis and, with other TGF-β family members expressed there, is required for proper execution of the D–V axis (Duboc *et al.*, 2004; Flowers *et al.*, 2004). A number of inputs feeding into the regulation of Nodal have been reported recently, including Univin and a redox gradient operating through p38 MAPK (Bradham and McClay, 2006; Nam *et al.*, 2007; Range *et al.*, 2007). Overexpression of *nodal* results in embryos that form increased ventral ectoderm at the expense of dorsal ectoderm (Duboc *et al.*, 2004; Flowers *et al.*, 2004). Inhibition of Nodal signaling results in a highly ciliated ball of cells lacking expression of both ventral (oral) and dorsal (aboral) molecular markers (Duboc *et al.*, 2004). BMP2/4, another TGF-β family member, although transcribed on the ventral side, acts on the dorsal side to antagonize the formation of the ventral (oral) fate (Angerer *et al.*, 2000). Angerer *et al.* (2001) further demonstrated that the execution of ventral (oral) and dorsal (aboral) fates were governed by the transcription factor Goosecoid. *Gsc* is expressed in the ventral (oral) ectoderm and represses the transcription factor Otx in the ventral (oral) domain. *Gsc* is under the positive regulation of Nodal (Duboc *et al.*, 2004). Thus, the execution of the D–V axis, with its corresponding oral and aboral ectodermal fates, depends on the operation of Nodal and BMP2/4 proteins at the two poles of the axis to create two distinct ectodermal transcriptional domains.

Direct-developing sea urchin larvae lack mouths, distinct, ventral (oral) and dorsal (aboral) ectoderms, larval arms, and a complete ciliary band. Despite the drastic differences in morphology, the ventral and dorsal sides of the direct-developing larvae of *H. erythrogramma* can be identified from the position of the relict ciliary band and larval skeleton (Emlet, 1995). Expression of the pluteus skeletogenic marker MSP130 in *H. erythrogramma* suggests that the downstream PMC-expressed genes are conserved despite the great evolutionary change in morphogenesis (Parks *et al.*, 1988). In *H. erythrogramma* the D–V axis is maternally specified and, unlike indirect developers, also maternally determined (Henry and Raff, 1990; Henry *et al.*, 1990). There are extensive differences in the execution of the D–V axis. The embryos of *H. erythrogramma* do not flatten their ventral side toward the end of gastrulation. They form no mouth, and the archenteron does not bend toward the ventral side or even fully extend across the blastocoel. Additionally, *H. erythrogramma* larvae exhibit no distinction of larval ectoderms, but have a novel ectoderm that covers the entire larva surrounding a domain of vestibular ectoderm on the left side of the larva that becomes the ectoderm of the juvenile adult (Haag and Raff, 1998; Love and Raff, 2006). The vestibular ectoderm that arises late in pluteus development is derived from the left side of the oral ectoderm of the pluteus (Ferkowicz and Raff, 2001). The large vestibular ectoderm of the *H. erythrogramma* larva is all that remains of the pluteus oral ectoderm, and it arises differently than in the pluteus.

Execution of the D–V axis in *H. erythrogramma* larvae occurs at a similar time in development relative to pluteus larvae but is truncated with respect to differentiation of ventral and dorsal features. Midstage larvae nucleate two pairs of skeletal spicules in locations orthogonal to the D–V axis, but they do not elongate to protrude from the larva to form arms. Around the same time as skeletal nucleation, a horseshoe-shaped ciliary band forms where the open end of the horseshoe marks the dorsal side of the larvae, and the closed end indicates the ventral side (Emlet, 1995).

Regardless of extensive differences in morphology expressed along the D–V axis, Nodal signaling is required for events downstream of maternal determination of the D–V axis in *H. erythrogramma*. Figure 7.4 shows the effects on larval ectoderm and mesoderm when *nodal* is mis/overexpressed or inhibited early in the development of *H. erythrogramma*. Controls (Fig. 7.4A, A′, D) differentiate larval ectoderm that covers the entire larva

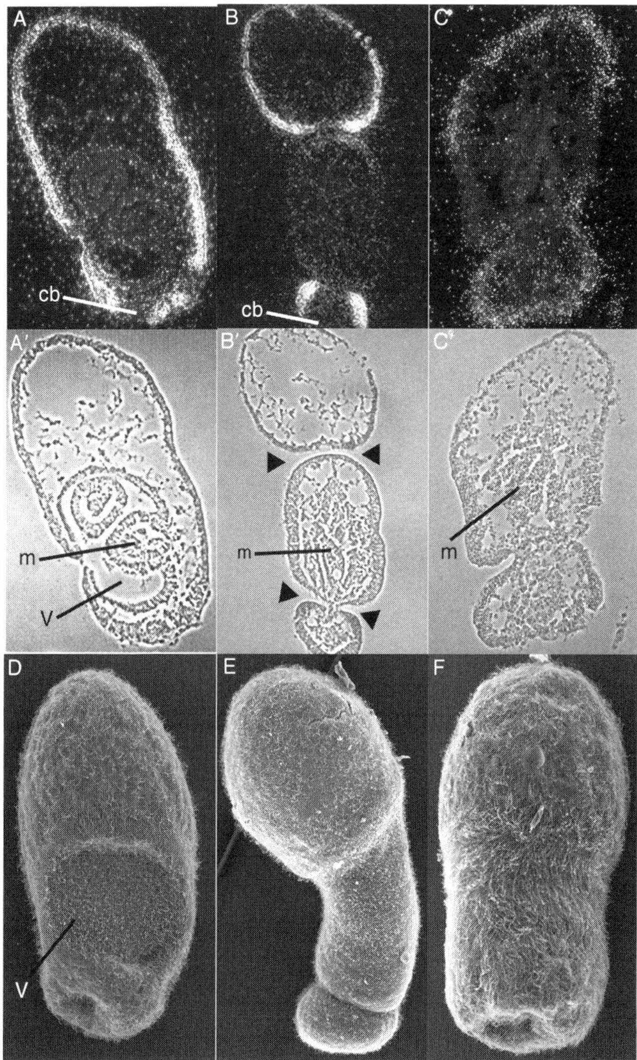

Figure 7.4 Effects of *nodal* mis/overexpression, and knockdown in *H. erythrogramma*. *H. erythrogramma* control larvae (A, A′, D); *H. erythrogramma* larvae from eggs into which *nodal* transcripts were injected (B, B′, E); and *H. erythrogramma* larvae of embryos exposed to 5 uM SB431542 at 5 h (C, C′, F). A, B, C are images of sections of embryos showing expression of the larval ectoderm marker *HeET1* detected using radioactive *in situ* hybridization. A′, B′, C′ are brightfield images of the sections above showing morphology. D, E, F are SEMs (250×). Panel D shows a 29-h control larva in which vestibular ectoderm (ve) is obviously ingressing. Panel E shows a 35-h larva in which *nodal* was mis/overexpressed by injection of *nodal* mRNA. Panel F shows a 35-h larva in which Nodal signaling was inhibited by exposure to SB431542 at 5 h. *Abbreviations*: m, mesoderm; v, vestibule; cb, ciliary band (from Smith *et al.*, 2008c, used by permission of John Wiley & Sons, Inc.).

until vestibular ectoderm differentiates in a patch on the left side of the larva (Fig. 7.4D). *HeET1* expression is a marker that differentiates between larval and vestibular ectoderm. *In situ* hybridization shows that it is expressed in the larval ectoderm and not in vestibular ectoderm, mesoderm, or ciliary band (Fig. 7.4A, A′). In addition to differences in *HeET1* expression, larval and vestibular ectoderms differ in the morphology of their cilia. The cilia of vestibular ectoderm are shorter than the cilia of larval ectoderm and have looped ends (Smith *et al.*, 2008c). Mis/overexpression of *nodal* via injection of *nodal* mRNA into eggs results in larvae that have the same ectoderm 360° around, lacking any D–V polarity. The animal and vegetal regions of these larvae express *HeET1* suggesting that these territories are larval ectoderm, but the middle band of ectoderm appears to be vestibular ectoderm based on lack of HeET1 expression and cilia morphology (Fig. 7.4B, B′, E). The increase in vestibular ectoderm represents an expansion of the ventral domain, but *H. erythrogramma* manifests this hypertrophy as vestibular ectoderm, which in the pluteus arises from oral ectoderm. In addition, *Gsc* gene expression in the ventral ectoderm lies downstream of and is positively regulated by Nodal in *H. erythrogramma* (Smith *et al.*, 2008c). In normal larvae, *goosecoid* (*gsc*) is first expressed in the presumptive ventral ectoderm of *H. erythrogramma* (Wilson *et al.*, 2005a) and later is localized to the left side of the larva. However, in sections of in larvae in which *nodal* was mis/overexpressed, *gsc* is expressed around the larva.

Nodal signaling can also be perturbed in sea urchin embryos by exposure to the drug SB431542 that interferes with the binding of ligands to ALK 4, 5, and 7 receptors. Although this reagent inhibits Nodal signaling, its inhibitory effects are not specific to only Nodal so these results must be interpreted with caution (Duboc *et al.*, 2005; Inman *et al.*, 2002; Range *et al.*, 2007). In *H. erythrogramma*, inhibition of Nodal signaling early in development also abolishes any D–V polarity (Fig. 7.4C, C′, F). Early Nodal-inhibited larvae express *HeET1* in the animal and vegetal regions and have cilia in these areas consistent with control larval ectoderm. However, the band of ectoderm in the middle of the embryo lacks *HeET1* expression but have very long cilia lacking looped ends (Smith *et al.*, 2008c). These long cilia indicate that this ectoderm is not vestibular ectoderm as is produced in larva mis/overexpressing *nodal*, but are consistent with the extremely long cilia of the ciliary band. Inhibition of Nodal signaling in the indirect developer *Paracentrotus lividus* also results in the formation of highly ciliated ectoderm (Duboc *et al.*, 2004). Additionally, inhibiting Nodal results in a decrease in *Gsc* expression (Smith *et al.*, 2008c).

These results indicate an overall conservation of the upstream regulatory systems for D–V axis execution in *H. erythrogramma*, but there clearly have been changes downstream of Nodal signaling responsible for the formation of a novel larval ectoderm in *H. erythrogramma* rather than differentiation of ventral and dorsal ectoderm territories as in pluteus larvae. The expression of downstream markers of ectoderm shows pronounced differentiation of ventral versus dorsal ectoderm in the pluteus. These markers are still expressed in *H. erythrogramma*, but now in a common larval ectoderm that does not respect the dorsal–ventral geography or partitioning of ectodermal markers observed in the pluteus (Love and Raff, 2006).

Within the general picture of conserved upstream D–V regulation by Nodal, combined with pronounced downstream rewiring of ectoderm gene expression, there are other apparent gene regulatory changes as well. Wilson *et al.* (2005a,b) observed that *Gsc* is expressed transiently in the ventral ectoderm of *H. erythrogramma*. Smith *et al.* (2008c) observed that *Gsc* is positively regulated by Nodal in *H. erythrogramma* during D–V axis formation. However, *H. erythrogramma* never differentiates oral ectoderm as seen in the pluteus. A knockdown of *Gsc* expression has little effect on *H. erythrogramma*. Overexpression of injected *Gsc* mRNA restores a substantial degree of pluteus-like oral morphogenesis in *H. erythrogramma* embryos (Wilson *et al.*, 2005b). Together, these observations suggest that *Gsc* is still regulated by Nodal, but *Gsc* has a more transient and limited effect that encompasses only the earliest stages of ventral differentiation when normally expressed in the derived direct developer.

Finally, we have evidence of another regulatory change in control of the location of the relict larval skeleton in *H. erythrogramma*. The bilateral position of the larval skeleton is established in the pluteus by a signal from the oral ectoderm. Di Bernardo *et al.* (1999) discovered that a single-copy zygotic gene, *Otp*, which encodes a homeodomain transcription factor, is expressed in the presumptive oral ectoderm at the midgastrula stage in two pairs of oral ectoderm cells located in a ventrolateral position. These cells lie over the PMC clusters that secrete the skeletal primordia. Inhibition of *Otp* expression was found to abolish skeletal secretion. Conversely, expression of an *Otp–GFP* fusion gene construct driven in the oral ectoderm by the hatching enzyme promoter induces ectopic and abnormal spiculogenesis. Cavalieri *et al.* (2003) concluded that Otp triggers a signal from the ectoderm that promotes skeletogenesis in PMCs.

Zhou *et al.* (2003) compared the role of *Otp* in larvae of the indirect-developing sea urchin *H. tuberculata* to *H. erythrogramma*. *Otp* is a single-copy gene with an identical protein sequence in these species. Expression of *Otp* is initiated by late gastrula, initially in two cells of the oral ectoderm in *H. tuberculata*. These cells are restricted to oral ectoderm and exhibit left–right symmetry. Approximately 266 copies of *Otp* mRNA were present per *Otp*-expressing cell in *H. tuberculata*. As in other indirect developers, misexpression of *Otp* mRNA in *H. tuberculata* radialized the embryos and caused defects during larval skeletogenesis. Misexpression of *Otp* mRNA in *H. erythrogramma* embryos did not affect skeleton formation. This is consistent with the observation by *in situ* hybridization of no concentration of *Otp* transcript in any particular cells or region of the *H. erythrogramma* larva, and measurement of a level of less than one copy of *Otp* mRNA per cell in *H. erythrogramma*. *Otp* plays an important conserved role in patterning the larval skeleton of *H. tuberculata*, but this role apparently has been lost in the evolution of the *H. erythrogramma* larva as part of a de-emphasis on differentiation of morphological features of the D–V axis.

2.3. Left–right axis

The final axis formed in indirect-developing sea urchins is the L–R axis, which is determined in the embryo by the late blastula stage (McCain and McClay, 1994). The L–R axis may be specified during early cleavage because signals from the small micromeres, formed after the fifth cleavage, are required for proper establishment of the L–R axis (Kitazawa and Amemiya, 2007). Execution of the L–R axis does not occur in the pluteus until weeks after fertilization when the hydrocoel enlarges only on the left side (Smith *et al.*, 2008b). The hydrocoel is a derivative of the left coelom. Toward the end of gastrulation, small right and left coelomic pouches pinch of from the sides of the archenteron. They are developmentally quiescent for a few weeks while the pluteus larva feeds but eventually each divide into three compartments, an axocoel, somatocoel, and hydrocoel (Smith *et al.*, 2008b) (Fig. 7.2, top). The left hydrocoel enlarges, and this event is followed by differentiation of the overlying ventral (oral) ectoderm into vestibular ectoderm. Contact between the left hydrocoel and vestibular ectoderm initiates adult development on the left side of the larva with the development of a pentameral pattern of primary tube feet (Ferkowicz and Raff, 2001) (Fig. 7.2, lower). However, there is evidence that the L–R axis is established before morphological differences are obvious. Fluorescent labeling indicates that micromeres contribute asymmetrically to the coelomic pouches, so that more micromeres are apart of the left coelom than the right coelom (Pehrson and Cohen, 1986).

In addition to its role along the D–V axis, Nodal signaling has been shown to be crucial to the execution of the L–R axis (Duboc *et al.*, 2005). Following its initial expression on the ventral side of the embryo, *nodal* is expressed again but this time on the right side of the embryo where it represses development of the hydrocoel and vestibular ectoderm on the right so that these events occur only on the left side of the pluteus. Inhibition of Nodal signaling results in formation of adult rudiments on both the left and right sides of the larval gut rather than just on the left side as in normal embryos (Fig. 7.5).

Heliocidaris erythrogramma specifies and determines the L–R axis maternally (Henry and Raff, 1990; Henry *et al.*, 1990). However, unlike the D–V axis, morphological differences along the L–R axis are obvious in *H. erythrogramma* larvae long before they are in pluteus larvae, in both developmental and absolute time. The first evidence of execution of the

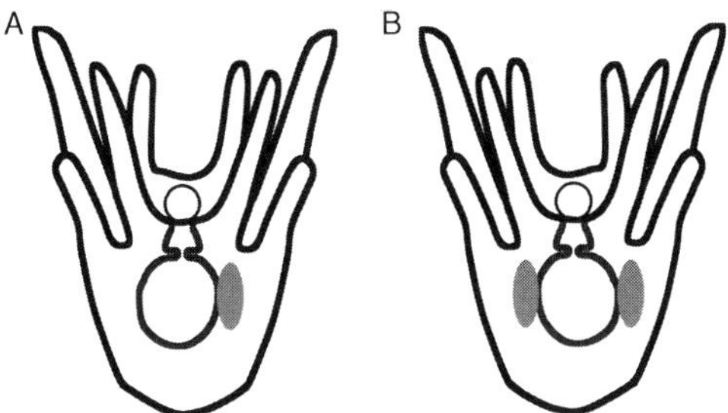

Figure 7.5 Schematic of the effect of inhibiting Nodal after the prism stage in indirect developers. This treatment results in a larva with a duplicated rudiment. In control embryos, the rudiment normally forms only on the left side (A), but when Nodal is inhibited, rudiments develop on both the right and left sides (B) (based on data of Duboc *et al.*, 2005). (See Color Insert.)

L–R axis occurs when *H. erythrogramma* forms a left coelom. Rather than pinching off small, dormant right and left coelomic pouches, a relatively large left coelomic pouch forms from the left-top of the archenteron late in gastrulation, and this event is followed by the formation of the right coelom (Fig. 7.3, lower). Shortly after its formation, the left coelom in *H. erythrogramma* divides into left axocoel, hydrocoel, and somatocoel. The left hydrocoel grows, and a large patch of the overlying larval ectoderm invaginates and differentiates into vestibular ectoderm (Ferkowicz and Raff, 2001) (Fig. 7.2, lower). As in the pluteus, contact between the left hydrocoel and vestibular ectoderm initiates adult development on the left side of the larva. Soon after, the hydrocoel forms the pentameral pattern of the adult and initiates primary tube foot formation with the vestibular ectoderm (Ferkowicz and Raff, 2001) (Fig. 7.2). Left hydrocoel formation, vestibular ectoderm differentiation, and adult development initiation all occur within hours of fertilization (~24–36 h) rather than after a few weeks as occurs in pluteus larvae. There are other significant changes along the L–R axis in *H. erythrogramma* preceding left coelom formation. During gastrulation, more cells ingress over the left lip of the blastopore than the right (Wray and Raff, 1991). Rapid formation of the coelomic pouches in *H. erythrogramma* takes place by rearrangement of archenteron cells rather than cell division (Smith *et al.*, 2009a).

Despite the difference in timing and the developmental course of events associated with execution of the L–R axis, Nodal signaling is involved in establishing the L–R axis of *H. erythrogramma* larvae. As in the pluteus, *nodal* is expressed on the right side of the larvae, and inhibition of Nodal signaling by exposure to SB431542 after ~16 h, after the effect of Nodal on the D–V axis, results in hypertrophy of the adult rudiment (Smith *et al.*, 2008c) (Fig. 7.6). However, unlike pluteus larvae, elimination of the Nodal signaling along the L–R axis does not cause duplication of the adult rudiment but

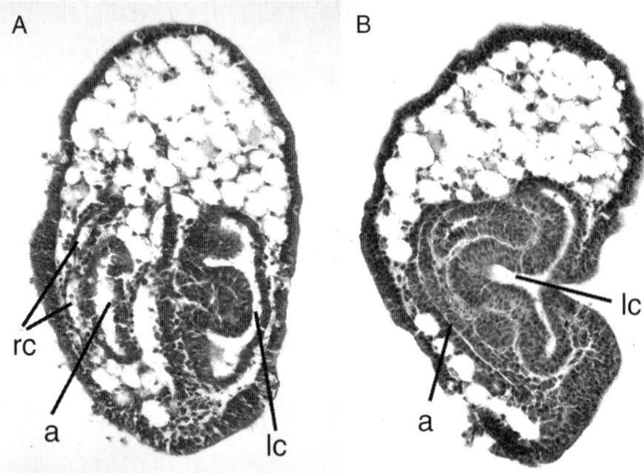

Figure 7.6 Inhibiting Nodal later in the development of *H. erythrogramma* (after ~16 h) results in a hypertrophied rudiment and lack of any right coelom. The large left coelom derivative likely results from a conversion of right-sided fates to left side as occurs in indirect developers (from Smith *et al.*, 2008c, used by permission of John Wiley & Sons, Inc.). (See Color Insert.)

rather results in the development of one giant rudiment. This difference in outcome of inhibition of Nodal action reflects the lack of a distinct, functioning gut in *H. erythrogramma* larvae, which is present in pluteus larvae and physically separates the cells of the left and right coeloms. With a shift in identity of right coelomic cells to left, these cells are incorporated into a double-sized, but correctly patterned adult rudiment.

2.4. Cross-species hybrids and axes

Cross-species hybrids between eggs of *H. erythrogramma* and sperm of *H. tuberculata* (*H. e.* × *H. t.*) have revealed that combining the genomes of these two disparately developing species directs development via a novel ontogeny through metamorphosis (Raff *et al.*, 1999). A harmonious but distinct ontogeny occurs that sets up the three larval axes and allows development despite heterochronies with respect to either parent. Maternal effects are revealed in the opposite hybrid made by fertilizing eggs of *H. tuberculata* by sperm of *H. erythrogramma*. D–V and L–R axes fail to form. This observation confirms our earlier observations that the determination of D–V and L–R axes have undergone a profound switch from embryonic to maternal in the evolution of *H. erythrogramma* (Henry and Raff, 1990; Henry *et al.*, 1990). *H. e.* × *H. t.* hybrids have pronounced oral ectoderm development, and the hybrids express *gsc* similarly to indirect developers. When *gsc* mRNA is overexpressed in *H. erythrogramma*, a hybrid-like phenotype is produced. These results support a role for modification of the expression of the D–V axis by downregulation of *gsc* expression in the evolution of the *H. erythrogramma* larva (Wilson *et al.*, 2005a,b).

3. Heterochrony and Modularity in the Evolution of Larval Form

Heterochrony refers to shifts in the relative timing of events in ontogeny and has been suggested to be a general factor affecting the evolution of development. Before the advent of an effective discipline of developmental genetics, evolutionary developmental biologists were limited in effective hypothetical mechanisms of developmental evolution to heterochrony (deBeer, 1958; Gould, 1977). In studies of morphological evolution, heterochronies are often important. One potential problem is that heterochrony might explain too much, as all events in development take place along a time axis (Raff, 1996). In tetrapod limb reduction and loss, apparent heterochronies are really the byproducts of degeneration of expression of limb developmental regulatory genes (Bejder and Hall, 2002; Cohn and Tickle, 1999). Perhaps, heterochrony is in many cases not *per se* a mechanism, but a result, although this may not be of great significance if the heterochrony itself is under selection. Studies of heterochrony at the levels of gene networks or signaling are still few. Kim *et al.* (2000) have reported heterochronies of expression of the developmental regulatory gene *hairy* among species of *Drosophila*.

An elegant heterochronic gene system controls developmental events, including timing of cuticle differentiation, in larval stages of the nematode *Caenorhabditis elegans* (Ambros and Horwitz, 1984). This pathway consists of a cascade of genes where mutations cause timing changes in development. These genes lie upstream of *let-7*, which is a switch gene that regulates timing through the action of the *let-7* microRNA (Reinhart *et al.*, 2000). *Let-7* is conserved and is involved in the regulation of stem cell differentiation by miRNAs in mammals (Ibarra *et al.*, 2007; Rybak *et al.*, 2008). In *Drosophila*, *let-7* microRNA is required for the remodeling of neuromusculature in metamorphosis (Sokol *et al.*, 2008). Among nematodes, there is variation in number of juvenile stages and larval heterochronies in moults and cell lineages in *Pristionchus pacificus* relative to *C. elegans* (Felix *et al.*, 1999). The molecular basis was not reported. The heterochronic system in *C. elegans* in which *let-7* is involved is not only extraordinary for its broad importance, but also as an (the) example of the extreme rarity of detailed studies of heterochronies at the level of gene action.

Although the upstream gene regulatory machinery for the molecular specification of each axis is conserved in *H. erythrogramma*, heterochronic shifts in aspects of formation of all three axes are dramatic and appear to be crucial for the evolution of direct development. The nature of heterochronies can be examined at the levels of axis specification, determination, and execution as well as in aspects of the genes regulating axis formation such as gene expression and execution of gene function. Comparisons reveal a general trend toward earlier onset and acceleration of events in *H. erythrogramma* (Table 7.1). The most drastic difference in axis formation between indirect and direct developers is in the timing of the onset of execution of the L–R axis which occurs hours after fertilization in *H. erythrogramma* rather than days after fertilization as in most pluteus larvae. The heterochronic shift in events along the L–R axis is associated with a truncation of developmental events along the D–V axis, including the lack of arm and mouth development. The elimination of these developmental events are responsible for much of the morphological differences between pluteus larvae and the barrel-shaped larvae of direct developers, so the shift in timing of execution of the L–R axis has had drastic effects on the evolution of larval morphology, likely driven by selection for rapid metamorphosis (Smith *et al.*, 2007).

Heterochrony also appears to be an effective dissociation mechanism between levels of developmental gene regulation. The observation that the effect of the second *nodal* pathway has been shifted earlier in developmental and in absolute time in *H. erythrogramma* allows comparisons of how changes in timing affect L–R axis formation at the levels of axis determination, gene

Table 7.1 Timing of specification, determination, and execution for all three larval axis for direct and indirect developers showing the shifts in timing of components of axis formation

Axis	Event	Indirect developers	*H. erythrogramma*
Animal–vegetal	Specification	Maternal	Maternal
	Determination	Maternal	Maternal
	Execution	Embryonic	Embryonic
Dorsal–ventral	Specification	Maternal*	Maternal
	Determination	Embryonic	Maternal
	Execution	Embryonic/early larval	Embryonic/early larval
Left–right	Specification	Embryonic	Maternal
	Determination	Embryonic	Maternal
	Execution	Late larval	Embryonic/early larval

An asterisk occurs next to specification of the D–V axis in indirect developers to represent some uncertainty in the literature as to when the D–V axis is specified. Evidence from gene expression and experiments perturbing the mitochondrial gradient suggest that it is maternally specified (based on the studies cited in the text).

expression, and axis execution (Fig. 7.7). Determination of the L–R axis, the axis that the second *nodal* pathway later affects, occurs maternally in *H. erythrogramma* but not until cleavage (or later) in indirect-developing species (Cameron *et al.*, 1990; Duboc *et al.*, 2005; Henry and Raff, 1990; Henry *et al.*, 1990; McCain and McClay, 1994).

However, despite the earlier determination of L–R axes in *H. erythrogramma*, there appears to be little difference in the timing of onset of *nodal* expression in *H. erythrogramma* embryos as compared to indirect developers (Duboc *et al.*, 2004; Flowers *et al.*, 2004; Smith *et al.*, 2008c). Additionally, the shift from an initial, presumptive ventral domain of expression of *nodal* to later expression on the right side of the larva occurs at approximately the same time even though *H. erythrogramma* larvae form a large hydrocoel and initiate adult development long before these events occur in pluteus larvae. This observation indicates that changes in the timing of gene expression are not always a good proxy for understanding differences in the timing of gene function. The execution of the function of Nodal along the L–R axis must require other developmental signals because otherwise in pluteus larvae the expression of *nodal* would coincide with execution of Nodal function. These other signals may occur earlier in *H. erythrogramma*, or have been lost, preventing the delay of coelom development observed in the pluteus.

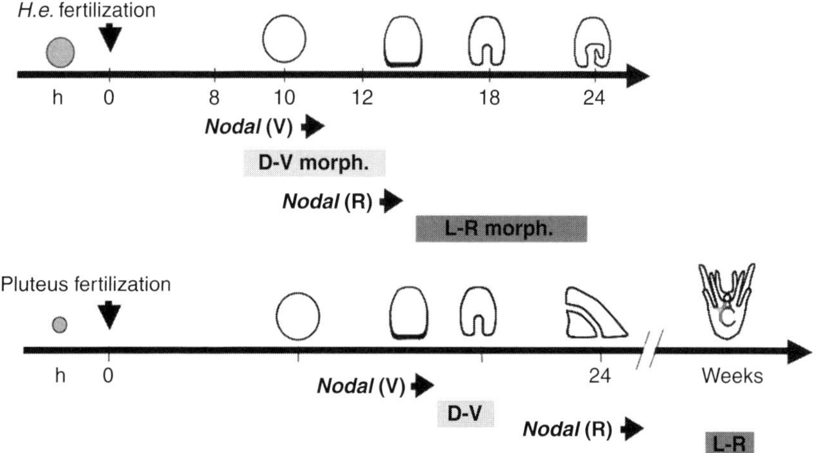

Figure 7.7 Comparison of timing of *nodal* expression (arrows) and function (colored blocks) along the D–V ((V) and green block) and L–R ((R) and blue block) for *H. erythrogramma* (top) and indirect developers (bottom). Expression of *nodal* is at roughly similar stages across developmental modes, but execution of the function of *nodal* along the L–R axis has been shift drastically earlier in development (hours vs weeks) (based on the data of Duboc *et al.*, 2004, 2005; Ferkowicz and Raff, 2001; Flowers *et al.*, 2004; Smith *et al.*, 2008c). (See Color Insert.)

Development of the left hydrocoel is the earliest morphogenetic asymmetry along the L–R axis and first obvious sign of L–R axis execution in *H. erythrogramma* (Smith *et al.*, 2008c). The formation in *H. erythrogramma* of the large hydrocoel occurs at about 24 h after fertilization rather than weeks after fertilization as occurs in indirect developers. The heterochronies in execution of the L–R are thus dramatic in timing and in offset with respect to other developmental events. Therefore, although determination and execution of the L–R axis occur earlier in the larvae of *H. erythrogramma* relative to pluteus larvae, the expression of *nodal* remains basically the same. This observation suggests that the role of heterochronies in the evolution of development can differ across developmental levels and produce major changes in morphology.

At the cellular level, the Nodal pathway, or circuit, as defined by Carroll (2005) to include necessarily connected elements such as those required for intracellular signaling, appears to be the same between both Nodal signal events in sea urchins. Signal transduction initiated through ALK 4/5/7 receptors is conserved because SB431542, which interferes with Nodal binding to an ALK receptor, inhibits Nodal regardless of the time it is applied in both *H. erythrogramma* and larvae of the indirect developer *P. lividus* (Duboc *et al.*, 2005).

The next higher level of organization, a gene network, includes the gene targets affected by signaling of pathways (Davidson, 2006). Evidence from *gsc* expression indicates that there are differences between *nodal* networks within *H. erythrogramma*. *Gsc* appears to be expressed in the same domain as *nodal* and is positively regulated by *nodal* in the D–V axis. In contrast, *gsc* is likely not a part of the second (L–R) *nodal* network because it does not appear to be directly downstream or regulated by *nodal*. Additionally, there have been changes relative to pluteus larvae in the gene networks affected by *nodal*, a broader level of organization encompassing downstream events. This phenomenon is seen along the D–V axis where, associated with Nodal signaling, there have been downstream changes in the gene networks affecting ectoderm differentiation, because *H. erythrogramma* larvae no longer differentiate distinct ventral and dorsal ectoderms territories like pluteus larvae but produce a novel larval ectoderm with distinct patterns of downstream gene expression and cellular morphology (Love and Raff, 2006). The changes in this ectoderm network occur downstream of Nodal because Nodal is still capable of promoting a ventral fate. Changes early in the network are suggested by functional studies of *gsc*. *Gsc* is downstream of *nodal* in (D–V) formation in indirect and direct developers and it retains the capability of affecting a ventral (oral) fate in both developmental modes. However, in *H. erythrogramma* knocking down *gsc* has no obvious effect on the development, suggesting that it is no longer essential for axis differentiation in *H. erythrogramma* larval development (Wilson *et al.*, 2005b).

The observations, that *nodal* is expressed in the ventral ectoderm and functions in establishing the D–V axis in *H. erythrogramma* despite the lack of distinct ventral and dorsal ectoderms and obvious *gsc* function, raise the question of why Nodal signaling along the D–V axis is conserved. One possibility is that early Nodal signaling is maintained through indirect selection via the second *nodal* network, which underlies the L–R axis. Proper execution of the L–R axis is likely under strong selection given that the adult, reproductive form develops on the left side of the larva. This idea that pleiotropic interactions among developmental modules affect the evolution of morphology has long been recognized (Frankino and Raff, 2004; Kleinenberg, 1886; Wolf, 2000), and Nodal signaling appears to supply a striking example in *H. erythrogramma*. Downstream components of the second *nodal* network, which affect the L–R axis, are not known, but the network is likely conserved because the function of Nodal along this axis is conserved.

At this broadest level of organization, the two *nodal* gene networks can be thought of as separate modules, as defined by Raff (1996), because, relative to each other, they are deployed at different times, have different functions, and express distinct downstream genes. Although the second *nodal* network has been conserved in function, the observation that Nodal is involved in execution of the L–R axis in *H. erythrogramma* and that this event occurs much earlier in time indicates that this second *nodal* network must be deployed at an earlier time relative to pluteus larvae, which could involve upstream regulatory changes in this module.

4. An Evolutionary Intermediate in Heterochronic Evolution of Direct-Developing Larvae

To examine how axial changes play a role in the evolution of direct development and the associated modification of larval form, we can examine the few sea urchin species intermediate between indirect and direct development. Although such intermediate forms are only proxies for evolutionary intermediates, these species allow us to examine basic premises of how direct development evolved. One hypothesis is that the ovoid larvae of direct developers are a result of the loss of larval structures associated with feeding, such as arms, a mouth, and a functional gut, due to relaxed selection after the requirement to feed has been lost. *Clypeaster rosaceus* makes an intermediate sized egg (~275 uM) that develops into a pluteus larva that can, but is not required to feed in order to metamorphose 5–7 days after fertilization. By this hypothesis, *C. rosaceus*, which still has the ability to feed, should still develop like indirect developers.

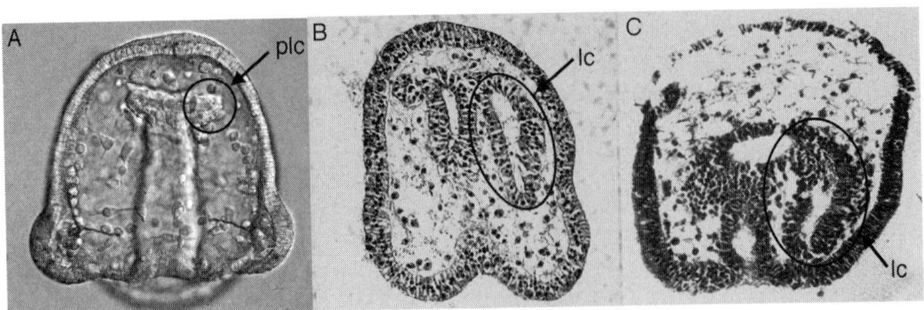

Figure 7.8 Early larval stages of *Clypeaster subdepressus* (A), *C. rosaceus* (B), and *H. erythrogramma* (C) show that *C. rosaceus* forms a large left coelom early in development more consistent with coelom development in *H. erythrogramma* (direct developer) than *C. subdepressus* (indirect developer) (from Smith *et al.*, 2007, used by permission of John Wiley & Sons, Inc.). (See Color Insert.)

Surprisingly, although *C. rosaceus* maintains an outward pluteus morphology, there has been a heterochronic acceleration in the formation of a large left coelom reminiscent in form, size and timing of direct-developing larvae, and drastically different than the development of typical indirect-developing larvae (Smith *et al.*, 2007) (Fig. 7.8). Early onset and acceleration of the formation of a large left coelom in *C. rosaceus* suggests two major insights into how direct development evolved. First, major developmental changes occur before the trophic transition to nonfeeding is complete. Likely other changes, such as egg content, have also occurred given that *C. rosaceus* eggs and plutei are opaque, reminiscent of direct developers, whereas in *Clypeaster subdepressus*, a smaller egged indirect-developing relative of *C. rosaceus*, eggs and plutei are clear as usual with indirect developers. Secondly, this observation suggests that the transition to a barrel-shaped larva may be driven by selection for streamlined development of the adult and rapid metamorphosis. Subsequent loss of some feeding structures may be due to selection for allocation of the embryo resources to rapid development of adult structures rather than to larval structures, which lets us infer that adaptive selection may play a much stronger role than previously thought in shaping the evolution of the direct-developing larval form.

5. Conclusions

Formation of the left coelom and development of adult structures only on the left side of larvae are obvious markers of asymmetry along the L–R axis. Timing of development of these structures is a key difference between indirect- and direct-developing sea urchins and controls much of the larval form of each developmental mode. Based on *C. rosaceus*, selection for early formation of these structures (left coelom/adult) appears to be early and important steps in the evolution of direct development. Therefore, major developmental changes driven by selection for early execution of the L–R axis are important for the evolution of direct development and larval form.

Work with the gene *nodal*, important for execution of the L–R axis in indirect- and direct-developing sea urchins, shows that at least in the case of *H. erythrogramma* changes in execution along the L–R axis are not a result of evolution of the genes regulating axis execution but rather heterochronic shifts in controls of timing of Nodal execution and likely in deployment of the whole *nodal* network. This contrasts with differences along the D–V axis where the timing of initiation of events is relatively conserved, but changes in genes downstream of Nodal signaling are responsible for the lack distinct differentiation of the ventral and dorsal ectoderms in *H. erythrogramma* that is so prominent in pluteus larvae.

Therefore, studies of *H. erythrogramma* development indicate that the evolution of larval forms associated with the evolution of direct development in sea urchins is highly dynamic. Changes in the timing of developmental events played a major role in the evolution of direct development, but modification of specific gene functions (*Otp*, *Gsc*) and possibly downstream networks were also involved.

ACKNOWLEDGMENTS

Our work was supported by the National Science Foundation through a grant to RAR, and a predoctoral fellowship to MSS. Work in Australia was made possible by assistance and facilities provided by the School of Biological Sciences, University of Sydney, and the Sydney Aquarium.

REFERENCES

Allen, J. D., and Pernet, B. (2007). Intermediate modes of larval development: Bridging the gap between planktotrophy and lecithotrophy. *Evol. Dev.* **9,** 643–653.

Ambros, V., and Horwitz, H. R. (1984). Heterochronic mutants of the nematode *Caenorhabditis elegans*. *Science* **226,** 409–416.

Angerer, R. C., and Davidson, E. H. (1984). Molecular indices of cell lineage specification in sea urchin embryos. *Science* **226,** 1153–1160.

Angerer, L. M., Oleksyn, D. W., Logan, C. Y., McClay, D. R., Dale, L., and Angerer, R. C. (2000). A BMP pathway regulates cell fate allocation along the sea urchin animal–vegetal embryonic axis. *Development* **127,** 1105–1114.

Angerer, L. M., Oleksyn, D. W., Levine, A. M., Li, X., Klein, W. H., and Angerer, R. C. (2001). Sea urchin goosecoid function links specification along the animal–vegetal and oral–aboral embryonic axes. *Development* **128,** 4393–4404.

Bejder, L., and Hall, B. K. (2002). Limbs in whales and limblessness in other vertebrates: Mechanisms of evolutionary and developmental transformation and loss. *Evol. Dev.* **4,** 445–458.

Bradham, C. A., and McClay, D. R. (2006). p38 MAPK is essential for secondary axis specification and patterning in sea urchin embryos. *Development* **133,** 21–32.
Byrne, M., Villinski, J. T., Cisternas, P., Popodi, E., and Raff, R. A. (1999). Maternal factors and the evolution of developmental mode: Evolution of oogenesis in *Heliocidaris erythrogramma. Dev. Genes Evol.* **209,** 275–283.
Cameron, R. A., Fraser, S. E., Britten, R. J., and Davidson, E. H. (1990). Segregation of oral from aboral ectoderm precursors is completed at fifth cleavage in the embryogenesis of *Strongylocentrotus purpuratus. Dev. Biol.* **137,** 77–85.
Carroll, S. B., Grenier, J. K., and Weatherbee, S. D. (2005). "From DNA to Diversity: Molecular Genetics and the Evolution of Animal Diversity." Blackwell, Malden, MA.
Cavalieri, V., Spinelli, G., and Di Bernardo, M. (2003). Impairing Otp homeodomain function in oral ectoderm cells affects skeletogenesis in sea urchin embryos. *Dev. Biol.* **262,** 107–118.
Coffman, J. A., and Davidson, E. H. (2001). Oral–aboral axis specification in the sea urchin embryo. I. Axis entrainment by respiratory asymmetry. *Dev. Biol.* **230,** 18–28.
Cohn, M. J., and Tickle, C. (1999). Developmental basis of limblessness and axial patterning in snakes. *Nature* **399,** 474–479.
Collin, R. (2004). Phylogenetic effects, the loss of complex characters, and the evolution of development in calyptraeid gastropods. *Evolution* **58,** 1488–1502.
Croce, J. C., and McClay, D. R. (2006). The canonical Wnt pathway in embryonic axis polarity. *Semin. Cell Dev. Biol.* **17,** 168–174.
Davidson, E. H. (2006). "The Regulatory Genome." Academic Press, Amsterdam.
Davidson, E. H., Peterson, K. J., and Cameron, R. A. (1995). Origin of bilaterian body plans: Evolution of developmental regulatory mechanisms. *Science* **270,** 1319–1325.
deBeer, G. (1958). "Embryos and Ancestors" 3rd Edn. Clarendon Press, Oxford.
Di Bernardo, M., Castagnetti, S., Bellomonte, D., Oliveri, P., Melfi, R., Palla, F., and Spinelli, G. (1999). Spatially restricted expression of PlOtp, a *Paracentrotus lividus* orthopedia-related homeobox gene, is correlated with oral ectodermal patterning and skeletal morphogenesis in late-cleavage sea urchin embryos. *Development* **126,** 2171–2179.
Duboc, V., Rottinger, E., Besnardeau, L., and Lepage, T. (2004). Nodal and BMP2/4 signaling organizes the oral–aboral axis of the sea urchin embryo. *Dev. Cell* **6,** 397–410.
Duboc, V., Rottinger, E., Lapraz, F., Besnardeau, L., and Lepage, T. (2005). Left–right asymmetry in the sea urchin embryo is regulated by nodal signaling on the right side. *Dev. Cell* **9,** 147–158.
Emily-Fenouil, F., Ghiglione, C., Lhomomd, G., Lepage, T., and Gache, C. (1998). GSK3beta/shaggy mediates patterning along the animal–vegetal axis of the sea urchin embryo. *Development* **125,** 2489–2498.
Emlet, R. B. (1995). Larval spicules, cilia and symmetry and remnants of indirect development in the direct developing sea urchin *Heliocidaris erythrogramma. Dev. Biol.* **167,** 405–415.
Emlet, R. B., McEdward, L. R., and Strathmann, R. R. (1987). Echinoderm larval ecology viewed from the egg. *In* "Echinoderm Studies" (M. Jangoux and J. M. Lawrence, Eds.), Vol. 2, pp. 55–136. Balkema, Rotterdam.
Ettensohn, C. A. (2006). The emergence of pattern in embryogenesis: Regulation of β-catenin localization during early sea urchin development. *Sci. STKE* **2006**(361), pe48.
Felix, M.-A., Hill, R. J., Schwarz, H., Sternberg, P. W., Sudhaus, W., and Sommer, R. J. (1999). *Pristionchus pacificus*, a nematode with only three juvenile stages, displays major heterochronic changes relative to *Caenorhabditis elegans. Proc. R. Soc. Lond. B* **266,** 1617–1621.
Ferkowicz, M. J., and Raff, R. A. (2001). Wnt gene expression in sea urchin development: Heterochronies associated with the evolution of developmental mode. *Evol. Dev.* **3,** 24–33.

Flowers, V. L., Courteau, G. R., Poustka, A. J., Weng, W., and Venuti, J. M. (2004). Nodal/activin signaling establishes oral–aboral polarity in the early sea urchin embryo. *Dev. Dyn.* **231,** 727–740.

Frankino, W. A., and Raff, R. A. (2004). Evolutionary importance and pattern of phenotypic plasticity. *In* "Phenotypic Plasticity, Functional and Conceptual Approaches" (T. J. DeWitt and S. M. Scheiner, Eds.), pp. 64–81. Oxford University Press, New York.

Gilbert, S. F. (2003). "Developmental Biology", 7th Edn. Sinauer Associates, Sunderland, MA.

Gould, S. J. (1977). "Ontogeny and Phylogeny." The Belknap Press of Harvard University Press, Cambridge, MA.

Haag, E. S., and Raff, R. A. (1998). Isolation and characterization of three mRNAs enriched in embryos of the direct-developing sea urchin *Heliocidaris erythrogramma*: Evolution of larval ectoderm. *Dev. Genes Evol.* **208,** 188–204.

Hardin, J., and Armstrong, N. (1997). Short-range cell–cell signals control ectodermal patterning in the oral region of the sea urchin embryo. *Dev. Biol.* **182,** 134–149.

Hardin, J., Coffman, J. A., Black, S. D., and McClay, D. R. (1992). Commitment along the dorsoventral axis of the sea urchin embryo is altered in response to $NiCl_2$. *Development* **116,** 671–685.

Hart, M. W., Byrne, M., and Smith, M. J. (1997). Molecular phylogenetic analysis of life-history evolution in asterinid starfish. *Evolution* **51,** 1848–1861.

Henry, J. J., and Raff, R. A. (1990). Evolutionary change in the process of dorsoventral axis determination in the direct developing sea urchin, *Heliocidaris erythrogramma. Dev. Biol.* **141,** 55–69.

Henry, J. J., Wray, G. A., and Raff, R. A. (1990). The dorsoventral axis is specified prior to first cleavage in the direct developing sea urchin *Heliocidaris erythrogramma. Development* **110,** 875–884.

Horstadius, S. (1973). "Experimental Embryology of Echinoderms." Clarendon Press, Oxford.

Ibarra, I., Erlich, Y., Muthuswamy, S. K., Sachidanandam, R., and Hannon, G. J. (2007). A role for microRNAs in maintenance of mouse mammary epithelial progenitor cells. *Genes Dev.* **21,** 3238–3243.

Inman, G. J., Nicolas, F. J., Callahan, J. F., Harling, J. D., Gaster, L. M., Reith, A. D., Laping, N. J., and Hill, C. S. (2002). SB-431542 is a potent and specific inhibitor of transforming growth factor-β superfamily type 1 activin receptor-like kinase (ALK) receptors ALK4, ALK5, and ALK7. *Mol. Pharmacol.* **62,** 65–74.

Jeffery, C. H., Emlet, R. B., and Littlewood, D. T. J. (2003). Phylogeny and evolution of developmental mode in temnopleurid echinoids. *Mol. Phylogenet. Evol.* **28,** 99–118.

Jenner, R. A. (2000). Evolution of body plans: The role of metazoan phylogeny at the interface between pattern and process. *Evol. Dev.* **2,** 208–221.

Kauffman, J. S., and Raff, R. A. (2003). Patterning mechanisms in the evolution of derived developmental life histories: The role of Wnt signaling in axis formation of the direct-developing sea urchin *Heliocidaris erythrogramma. Dev. Genes Evol.* **213,** 612–624.

Kim, J., Kerr, J. Q., and Min, G. S. (2000). Molecular heterochrony in the early development of *Drosophila. Proc. Natl. Acad. Sci. USA* **97,** 212–216.

Kitazawa, C., and Amemiya, S. (2007). Micromere-derived signal regulates larval left–right polarity during sea urchin development. *J. Exp. Zool. Part A Ecol. Genet. Physiol.* **307,** 249–262.

Kleinenberg, N. (1886). Die entstehung des annelids aus der larve von lopadorhynchus. Nebst bemarkungen uber die entwicklunh anderer polychaten. *Ztschr. Wiss. Zool.* **44,** 1–227.

Love, A., and Raff, R. A. (2006). Larval ectoderm, organizational homology, and the origins of evolutionary novelty. *J. Exp. Zool. B Mol. Dev. Evol.* **306,** 18–34.

Love, A., Andrews, M., and Raff, R. A. (2007). Pluteus larval arm morphogenesis and evolution: Gene expression patterns in a novel animal appendage and their transformation in the origin of direct development. *Evol. Dev.* **9,** 51–68.

McCain, E. R., and McClay, D. R. (1994). The establishment of bilateral asymmetry in sea urchin embryos. *Development* **12,** 395–404.

Morgan, S. G. (1995). Life and death in the plankton: Larval mortality and adaptation. *In* "Ecology of Marine Invertebrate Larvae" (L. R. McEdward, Ed.), pp. 279–321. CRC Press, Boca Raton, FL.

Nam, J., Su, Y. H., Lee, P. Y., Robertson, A. J., Coffman, J. A., and Davidson, E. H. (2007). *Cis*-regulatory control of the nodal gene, initiator of the sea urchin oral ectoderm gene network. *Dev. Biol.* **306,** 860–869.

Nützel, A., Lehnert, O., and Fryda, J. (2006). Origin of planktotrophy—Evidence from early mollusks. *Evol. Dev.* **8,** 325–330.

Oliveri, P., Tu, Q., and Davidson, E. H. (2008). Global regulatory logic for specification of an embryonic cell lineage. *Proc. Natl. Acad. Sci. USA* **105,** 5955–5962.

Parks, A. L., Parr, B. A., Chin, J. E., Leaf, D. S., and Raff, R. A. (1988). Molecular analysis of heterochrony in the evolution of direct development in sea urchins. *J. Evol. Biol.* **1,** 27–44.

Pehrson, J. R., and Cohen, L. H. (1986). The fate of the small micromeres in sea urchin development. *Dev. Biol.* **113,** 522–526.

Peterson, K. J. (2005). Macroevolutionary interplay between planktonic larvae and benthic predators. *Geology* **33,** 929–932.

Peterson, K. J., McPeek, M. A., and Evans, D. A. D. (2005). Tempo and mode of early animal evolution: Inferences from rocks, Hox, and molecular clocks. *Paleobiology* **31** (Suppl.), 36–55.

Raff, R. A. (1996). "The Shape of Life." University of Chicago Press, Chicago, IL.

Raff, R. A. (2008). Origins of the other metazoan body plans: The evolution of larval forms. *Philos. Trans. R. Soc. B* **363,** 1473–1480.

Raff, R. A., and Byrne, M. (2006). The active evolutionary lives of echinoderm larvae. *Heredity* **97,** 244–252.

Raff, R. A., Herlands, L., Morris, V. B., and Healy, J. (1990). Evolutionary modification of echinoid sperm correlates with developmental mode. *Dev. Growth Differ* **32,** 283–291.

Raff, E. C., Popodi, E. M., Sly, B. J., Turner, F. R., Villinski, J. T., and Raff, R. A. (1999). A novel ontogenetic pathway in hybrid embryos between species with different modes of development. *Development* **126,** 1937–1945.

Range, R., Lapraz, F., Quirin, M., Marro, S., Besnardeau, L., and Lepage, T. (2007). Cis-regulatory analysis of nodal and maternal control of dorsal–ventral axis formation by Univin, a TGF-beta related to Vg1. *Development* **134,** 3649–3664.

Ransick, A., and Davidson, E. H. (1993). A complete second gut induced by transplanted micromeres in the sea urchin embryo. *Science* **259,** 1134–1138.

Reinhart, B. J., Slack, F. J., Basson, M., Pasquinelli, A. E., Bettinger, J. C., Rougvie, A. E., Horvitz, H. R., and Ruvkun, G. (2000). The 21-nucleotide *let*-7 RNA regulates developmental timing in *Caenorhabditis elegans. Nature* **403,** 901–906.

Rouse, G. W. (2000). The epitome of hand waving? Larval feeding and hypotheses of metazoan phylogeny. *Evol. Dev.* **2,** 222–233.

Rumrill, S. S. (1990). Natural mortality of invertebrate larvae. *Ophelia* **32,** 163–198.

Rybak, A., Fuchs, H., Smirnova, L., Brandt, C., Pohl, E. E., Nitsch, R., and Wulczyn, F. G. (2008). A feedback loop comprising lin-28 and let-7 controls pre-let-7 maturation during neural stem-cell commitment. *Nat. Cell Biol.* **10,** 987–993.

Sly, B. J., Snoke, M. S., and Raff, R. A. (2003). Who came first? Origins of bilaterian metazoan larvae. *Int. J. Dev. Biol.* **47,** 623–632.

Smith, M. S., Zigler, K. S., and Raff, R. A. (2007). Evolution of direct-developing larvae: Selection versus loss. *Bioessays* **29,** 566–571.

Smith, J., Kraemer, E., Liu, H., Theodoris, C., and Davidson, E. (2008a). A spatially dynamic cohort of regulatory genes in the endomesodermal gene network of the sea urchin embryo. *Dev. Biol.* **313,** 863–875.

Smith, M. M., Cruz Smith, L., Cameron, R. A., and Urry, L. A. (2008b). The larval stages of the sea urchin, *Strongylocentrotus purpuratus*. *J. Morphol.* **269,** 713–733.

Smith, M. S., Turner, F. R., and Raff, R. A. (2008c). Nodal expression and heterochrony in the evolution of dorsal–ventral and left–right axis formation in the direct-developing sea urchin *Heliocidaris erythrogramma*. *J. Exp. Zool. B Mol. Dev. Evol.* **310,** 609–622.

Smith, M. S., Collins, S., and Raff, R. A. (2009a). Morphogenetic mechanisms in the direct developing sea urchin *Heliocidaris erythrogramma*. *Dev. Genes Evol.* **219,** 21–29.

Smith, M. S., Wray, G. A., and Raff, R. A. (2009b). Larval axes and cell fates in the evolution of the direct-developing larva of *Heliocidaris erythrogramma* (submitted).

Sokol, N. S., Xu, P., Jan, Y.-N., and Ambros, V. (2008). *Drosophila let-7* microRNA is required for remodeling of the neuromusculature during metamorphosis. *Genes Dev.* **22,** 1591–1596.

Strathman, R. R. (1978). The evolution and loss of feeding larval strategies of marine invertebrates. *Evolution* **32,** 894–906.

Tu, Q., Brown, C. T., Davidson, E. H., and Oliveri, P. (2006). Sea urchin forkhead gene family: Phylogeny and embryonic expression. *Dev. Biol.* **300,** 49–62.

Vonica, A., Weng, W., Gumbiner, B. M., and Venuti, J. M. (2000). TCF is the nuclear effector of the beta-catenin signal that patterns the sea urchin animal–vegetal axis. *Dev. Biol.* **217,** 230–243.

Wikramanayake, A., Huang, L., and Klein, W. (1998). B-catenin is essential for patterning the maternally specified animal–vegetal axis in the sea urchin embryo. *Proc. Natl. Acad. Sci. USA* **95,** 9343–9348.

Wikramanayake, A., Peterson, R., Chen, J., Huang, L., Bince, J., McClay, D., and Klein, W. (2004). Nuclear B catenin dependent Wnt-8 signaling in vegetal cells of early sea urchin embryos regulates gastrulation and differentiation of endoderm and mesoderm cell lineages. *Genesis* **39,** 194–205.

Wilson, K., Andrew, M. A., and Raff, R. A. (2005a). Dissociation of expression patterns of homeodomain transcription factors in the evolution of developmental mode in the sea urchins *Heliocidaris tuberculata* and *H. erythrogramma*. *Evol. Dev.* **7,** 401–415.

Wilson, K., Andrews, M. A., Turner, F. R., and Raff, R. A. (2005b). Major regulatory factors in the evolution of development: The roles of goosecoid and Msx in the evolution of the direct-developing sea urchin *Heliocidaris erythrogramma*. *Evol. Dev.* **7,** 416–428.

Wolf, S. B. (2000). Gene interactions from maternal effects. *Evolution* **54,** 1882–1898.

Wray, G. (1996). Parallel evolution of nonfeeding larvae in echinoids. *Syst. Biol.* **45,** 308–322.

Wray, G. A., and Raff, R. A. (1989). Evolutionary modification of cell lineage in the direct-developing sea urchin *Heliocidaris erythrogramma*. *Dev. Biol.* **132,** 458–470.

Wray, G. A., and Raff, R. A. (1990a). Novel origins of lineage founder cells in the direct-developing sea urchin *Heliocidaris erythrogramma*. *Dev. Biol.* **141,** 41–54.

Wray, G. A., and Raff, R. A. (1990b). Pattern and process heterochronies in the early development of sea urchins. *Semin. Dev. Biol.* **1,** 245–251.

Wray, G. A., and Raff, R. A. (1991). Rapid evolution of gastrulation mechanisms in a direct-developing sea urchin. *Evolution* **45,** 1741–1750.

Yaguchi, S., Yaguchi, J., Angerer, R. C., and Angerer, L. M. (2008). A Wnt–FoxQ2–Nodal pathway links primary and secondary axis specification in sea urchin embryos. *Dev. Cell* **14,** 97–107.

Zhou, N., Wilson, K. A., Andrews, M. E., Kauffman, J. S., and Raff, R. A. (2003). Evolution of OTP-independent larval skeleton patterning in the direct-developing sea urchin, *Heliocidaris erythrogramma. J. Exp. Zool. B Mol. Dev. Evol.* **300,** 58–71.

Zigler, K. S., Raff, E. C., Popodi, E., Raff, R. A., and Lessios, H. A. (2003). Adaptive evolution of bindin in the genus *Heliocidaris* is correlated with the shift to direct development. *Evolution* **57,** 2293–2302.

CHAPTER EIGHT

Evolution and Development in the Cavefish *Astyanax*

William R. Jeffery

Contents

Abstract

The teleost *Astyanax mexicanus* is a single species consisting of two radically different forms: a sighted pigmented surface-dwelling form (surface fish) and a blind depigmented cave-dwelling form (cavefish). The two forms of *Astyanax* have favorable attributes, including descent from a common ancestor, ease of laboratory culture, and the ability to perform genetic analysis, permitting their use as a model system to explore questions in evolution and development. Here, we review current research on the molecular, cellular, and developmental mechanisms underlying the loss of eyes and pigmentation in *Astyanax* cavefish. Although functional eyes are lacking in adults, cavefish embryos begin to develop eye primordia, which subsequently degenerate. The major cause of eye degeneration appears to be apoptotic cell death of the lens, which prevents

Department of Biology, University of Maryland, College Park, Maryland, USA

Current Topics in Developmental Biology, Volume 86
ISSN 0070-2153, DOI: 10.1016/S0070-2153(09)01008-4

the growth of other optic tissues, including the retina. Ultimately, the loss of the eye is the cause of craniofacial differences between cavefish and surface fish. Lens apoptosis is induced by enhanced activity of the Hedgehog signaling system along the cavefish embryonic midline. The absence of melanin pigmentation in cavefish is due to a block in the ability of undifferentiated melanoblasts to accumulate L-tyrosine, the precursor of L-DOPA and melanin, in melanosomes. Genetic analysis has shown that this defect is caused by a hypomorphic mutation in the *p/oca2* gene encoding an integral melanosomal membrane protein. We discuss how current studies of eye and pigment regression have revealed some of the mechanisms in which cavefish development has been changed during evolution.

1. Introduction

Many studies in evolutionary developmental biology have been centered exclusively on the generation of novel traits. Although molecular and developmental analysis of trait loss is often more tractable than analysis of gain, considerably less attention has been focused on the reduction and loss of traits. It can be argued that trait modification or loss is just as important as gain in providing a complete understanding of evolution as a developmental process, and may be one of the first steps in the cascade of events leading to evolutionary innovations. For example, during the evolution of flippers in marine mammals, significant changes, including reductions and losses, must have occurred in the limbs of their terrestrial ancestors prior to their conversion to perform a swimming function. Therefore, it is important to study the evolution of novelties, referred to here as constructive traits, within the context of reduced or lost traits, referred to here as regressive traits. One of the most important animal models for studying regressive and constructive traits in the same context is the teleost *Astyanax mexicanus* (Jeffery, 2001, 2008).

Astyanax mexicanus consists of two conspecific forms: a surface-dwelling form (surface fish) and a cave-dwelling form (cavefish). Surface fish adults have large eyes and three different types of pigment cells, whereas cavefish have reduced or lost both these traits (Fig. 8.1), a phenotype shared with a diverse community of cave animals (Culver, 1982). Cavefish have also gained constructive features, larger jaws, more taste buds, larger cranial neuromasts, fat reserves, and possibly a more sensitive olfactory system than their surface fish counterparts. At least 30 different populations of *Astyanax* cavefish are present in limestone caverns in Mexico, having been isolated from their surface fish conspecifics for the past few million years (Porter *et al.*, 2007). Each cavefish population is named after their cave of origin (Mitchell *et al.*, 1977). For example, Pachón, Chica, Los Sabinos, and Río Subterráneo, and Molino cavefish are found in La Cueva de El Pachón,

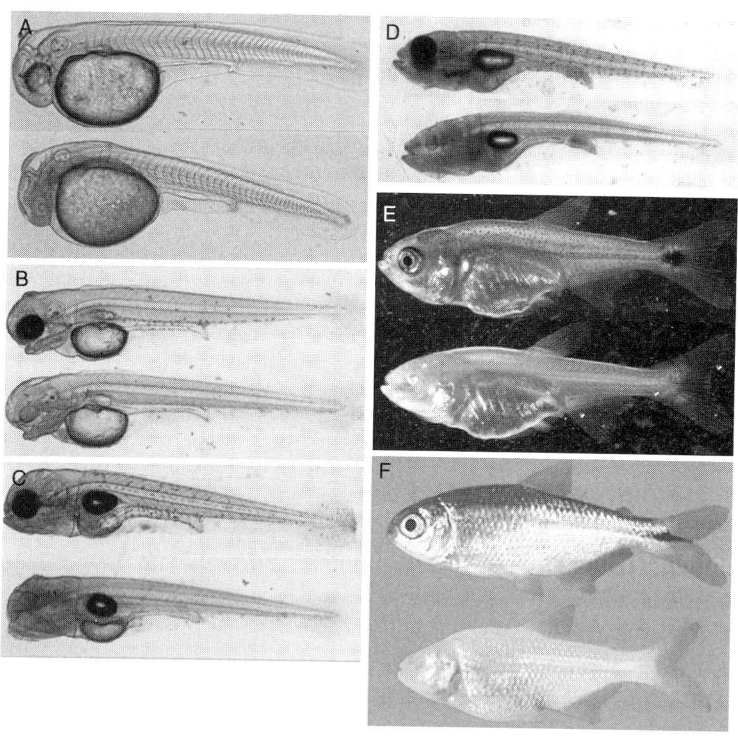

Figure 8.1 Gradual loss of eyes and absence of body pigment development in cavefish. Surface fish (above) and cavefish (below) are shown in each frame. (A) One-day postfertilization (dpf). (B) Three dpf. (C) One-week postfertilization (wpf). (D) Two wpf. (E) One-month postfertilization (mpf). (F) Adults. Note developmental arrest and progressive loss of eyes in cavefish, development and rapid growth of eyes in surface fish, body pigment cell development in surface fish, and absence of eye and body pigmentation in cavefish. Scale bars are 62.5 mm (A), 125 mm (B), 250 mm (C, D), and 500 mm (F) [(F) from Yamamoto and Jeffery, 2000].

La Cueva Chica, La Cueva de Los Sabinos, La Cueva de la Río Subterráneo, and Sótano de El Molino, respectively. In this chapter, unless another cavefish population is named specifically, reference to cavefish will imply the Pachón population, which has been the subject of the most extensive developmental analysis. Genetic (Borowsky, 2008; Wilkens, 1971) and phylogenetic (Dowling *et al.*, 2002; Strecker *et al.*, 2003, 2004) evidence suggests that some of these cavefish populations have evolved regressive and constructive phenotypes independently. Thus, *Astyanax* cavefish are an excellent model system to study parallel evolution of developmental mechanisms.

The *Astyanax* system has many advantages for studies in evolutionary developmental biology. First and perhaps foremost is its value as a laboratory model. Surface fish and cavefish are easy to maintain in the laboratory, spawn frequently, and produce fairly large and robust embryos. Second, the polarity of evolutionary changes in this system is known with certainty: cavefish lacking eyes and pigment have evolved from surface fish ancestors that exhibited both of these traits. Evolutionary polarity is must be inferred by phylogenetic analysis and is rarely understood with such confidence in other cases. Third, the similarity of present-day surface fish to the historical source of cavefish provides an excellent comparative system in which an evolutionary product can be compared to a prototype of its ancestral form. Finally, cavefish and surface fish are completely interfertile, allowing the power of genetic analysis to be applied to the evolution of constructive and regressive traits.

It has been said that evolution is the effect of ecology on development (Van Valen, 1973). Accordingly, not only do *Astyanax* cavefish provide an excellent model system to study the evolution of development, but they also provide a context in which evolutionary events can be understood with respect to the environmental conditions that forged them. In most instances, the ecological effects that led to the emergence of new phenotypes are difficult to discern because they occurred in the distant past and are no longer in existence. In contrast, perpetual darkness, the ecological cue leading to evolutionary changes in *Astyanax* cavefish and other cave animals has remained constant through time. Thus, it is likely that present conditions in the caves harboring cavefish are the same as they were when surface fish first entered and began the process of evolutionary change leading to cavefish. This chapter reviews the molecular, cellular, and developmental mechanisms responsible for loss of eyes and pigmentation, which have occurred in a background of constructive changes in *Astyanax* cavefish.

2. Eye Development and Degeneration

Vertebrate eyes develop from three different parts of the ectoderm. The ocular lens is formed from a thickening in the surface ectoderm, known as the lens placode. The retina and retinal pigment epithelium (RPE) are formed from bilateral protrusions of forebrain neuroectoderm, which are called optic vesicles. The optic vesicles arise from bilaterally symmetric optic fields in the anterior neural plate. Each optic vesicle rotates about 90° and then buckles inward to form the optic cup, with the future RPE on the convex side and the retina on the concave side. The connection of the optic cup to the forebrain will become the optic stalk, which encases the optic nerve fibers. As the optic vesicles are forming, the lens placode reorganizes into a vesicle, which detaches from the surface epithelium and enters the concave opening of the optic cup. The neural crest is the third part of the ectoderm responsible for eye development. Cranial neural crest cells migrate from the anterior neural tube region into spaces between the lens and surface epithelium to form the inner parts of the cornea, between the lens and the distal edges of the retina to contribute to the iris and ciliary body, and into the areas surrounding the RPE to form the choroid and sclera. Neural crest cells probably also contribute to the ocular dermal bones that develop much later around the orbit, forming a part of the adult craniofacial skeleton.

The eye primordium also consists of three major parts—the lens, retina, and RPE—which differentiate in concert. The lens vesicle produces fiber cells, which synthesize crystallin proteins, and becomes transparent, leaving behind a layer of undifferentiated stem cells. The retina differentiates into several layers. From distal to proximal, they consist of (1) the ganglion cell layer, which transmits neural signals to the brain via axons extending through the optic stalk into the optic tectum; (2) the intermediate layers, which consist of interneurons and glial cells; and (3) the photoreceptor layer, where rod and cone cells translate photons into neural signals. The RPE forms tight connections with the photoreceptor layer and produces melanin pigment. Pigment cells also become organized around the RPE, but outside the eye proper. These body pigment cells have a different origin from those of the RPE, and will be discussed later in this chapter.

The sequence of events during surface fish and cavefish eye development are compared in Fig. 8.2A (Cahn, 1958). The cavefish eye primordium is slightly smaller than its surface fish counterpart (Fig. 8.2B and C). This difference in size is due to a smaller lens and optic cup, which appears to be missing its ventral sector. In contrast to the surface fish eye, cavefish optic tissues either fail to be induced (cornea, iris, and ciliary body) or begin to differentiate and then degenerate (lens, retina, and probably the RPE). However, the most important flaw in the cavefish eye primordium is the absence of net optic growth after the conclusion of the embryonic stages (Fig. 8.2A). Eventually, the arrested cavefish eye primordium, which has not markedly increased in size during the larval stages, is overgrown by head epidermis and connective tissue, and disappears into the orbit, making adult cavefish appear eyeless.

Because cavefish eye development involves growth arrest, it is important to consider the possible effects on the origin of new cells. Stem cells in the epithelial layer are the source of new lens fiber cells. The source of most new retinal and all new RPE cells is a stem cell niche at the edge of the optic cup, a region known as the ciliary marginal zone (CMZ). As the eye enlarges during larval development, it is surrounded by orbital bones, which form a part of the craniofacial skeleton. The orbital bones presumably differentiate from mesenchyme of neural crest origin and their number, size, and organization are distinct between surface fish and cavefish, and even among different cavefish populations (Alvarez, 1947; Yamamoto *et al.*, 2003). As described below, the presence or absence of a functional eye is critical in the morphogenesis of the orbital bones and organization of the craniofacial skeleton.

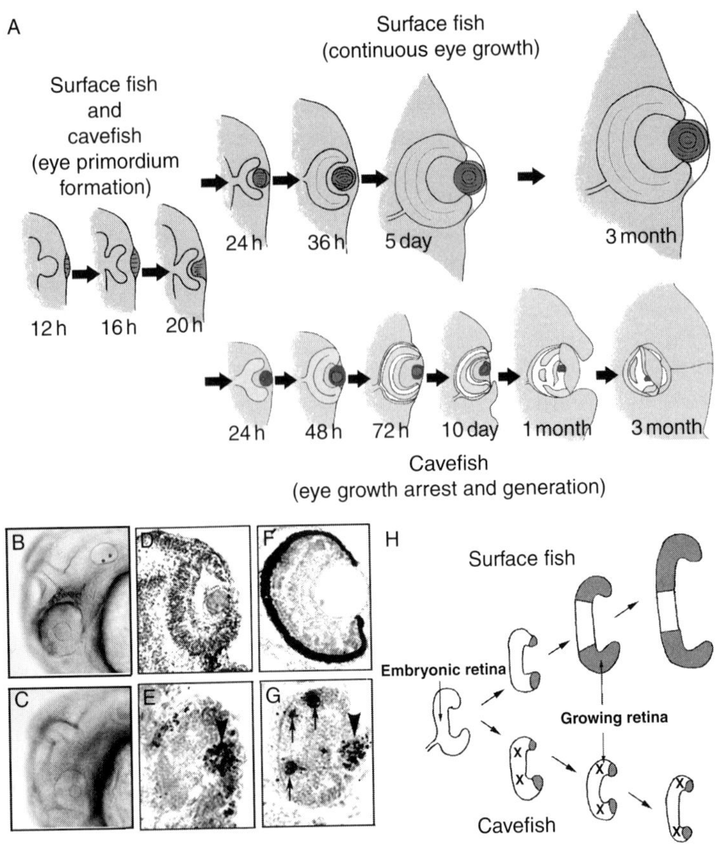

Figure 8.2 Eye development and degeneration in *Astyanax*. (A) Development of the eye primordium from up to about 12 h postfertilization (hpf) in cavefish and surface fish (left). The surface fish eye differentiates and rapidly increases in size (top) from 1 dpf to 1 mpf, whereas the cavefish eye arrests in growth, degenerates, and gradually sinks into the orbit. (B, C). Size differences in the 24 hpf surface fish (B) and cavefish (C) eye primordia. (D–G) Sections of 2 (D, E) and 3 (F, G) dpf surface fish (D, F) and cavefish (E, G) eye primordia showing apoptosis (dark-stained spots) in various eye tissues. In cavefish, apoptosis begins in the lens (arrowheads) and spreads to the retina (arrows). There is no apoptosis in these tissues in surface fish. (H) The roles of cell proliferation and apoptosis during retina/RPE growth in surface fish cavefish. Clear retinal areas: embryonic retina and central part of growing retina derived from embryonic retina. Shaded area: part of retina derived from cell proliferation at the CMZ after the embryonic stages. X: apoptotic areas. The surface fish retina grows continuously due to cell proliferation at the CMZ, whereas the cavefish retina is arrested in growth because the products of cell proliferation at the CMZ die before they contribute to the differentiated retina [(B, C) from Yamamoto and Jeffery, 2000; (D–G) from Strickler *et al.*, 2007a].

3. Cellular Mechanisms of Eye Degeneration

A block in cell proliferation, an increase in programmed cell death, or a combination of these processes could be the cause of arrested eye development in cavefish. Current evidence suggests that cell death has a major role in this process (Jeffery and Martasian, 1998; Yamamoto and Jeffery, 2000). If cell death is restricted to a single eye tissue, or starts in one tissue and later spreads to others, then the tissue that dies first is a candidate to initiate the entire degeneration process. The early cavefish eye primordium is largely free of cell death, except for one tissue: the lens (Fig. 8.2E). Apoptosis is not detected in the surface fish lens (Fig. 8.2D). Although cell proliferation does not cease in the cavefish lens, the rate of apoptosis is very high, eventually obliterating the lens, or reducing it to a tiny vestige in the adult (Soares *et al.*, 2004). A few days after the initiation of lens apoptosis, the cavefish retina also begins to undergo apoptosis (Fig. 8.2G) (Alunni *et al.*, 2007; Strickler *et al.*, 2007a). Retinal cell death is restricted to the intermediate layers and regions adjacent to the CMZ. Later in development, the cavefish RPE also shows dying cells (Strickler *et al.*, 2007a). As for the lens, cell death is not observed in the surface fish retina (Fig. 8.2F) or RPE. Clearly, the lens is the first tissue to undergo cell death in the cavefish eye primordium, suggesting that its absence may be the trigger for eye degeneration.

In contrast to cell death, there is no evidence that cell proliferation stops in the degenerating cavefish eye. In the surface fish retina the primary zone of cell proliferation is the CMZ, in which proliferating cells can be detected by labeling replicating DNA or the presence of DNA replication enzymes, such as proliferating cell nuclear antigen (PCNA). Continuous cell proliferation in the CMZ displaces newly born cells into the adjacent retinal layers and RPE, where they differentiate and increase the general mass of the retina. The profile of cell proliferation in the cavefish retina is not changed compared to surface fish. The cell proliferation markers BrdU and PCNA are expressed normally in the cavefish CMZ during the period in which the retina does not markedly increase in size (Alunni *et al.*, 2007; Strickler *et al.*, 2002, 2007a). The reason that the cavefish retina does not show net growth is that the new cells are quickly removed by apoptosis, which persists during cavefish larval development and into adult life (Strickler *et al.*, 2007a). Thus, the cavefish eye is arrested in growth because newly born cells die before they are able to differentiate and join the retinal layers. The relationship between growth, cell proliferation, and apoptosis in the surface fish and cavefish retina is illustrated in Fig. 8.2H.

Does the absence of a functional lens play a role in the survival of newly born cavefish retinal cells? This possibility has been tested by transplantation of embryonic lenses between surface fish and cavefish (Yamamoto and Jeffery, 2000). The lens transplantation method is illustrated in Fig. 8.3A.

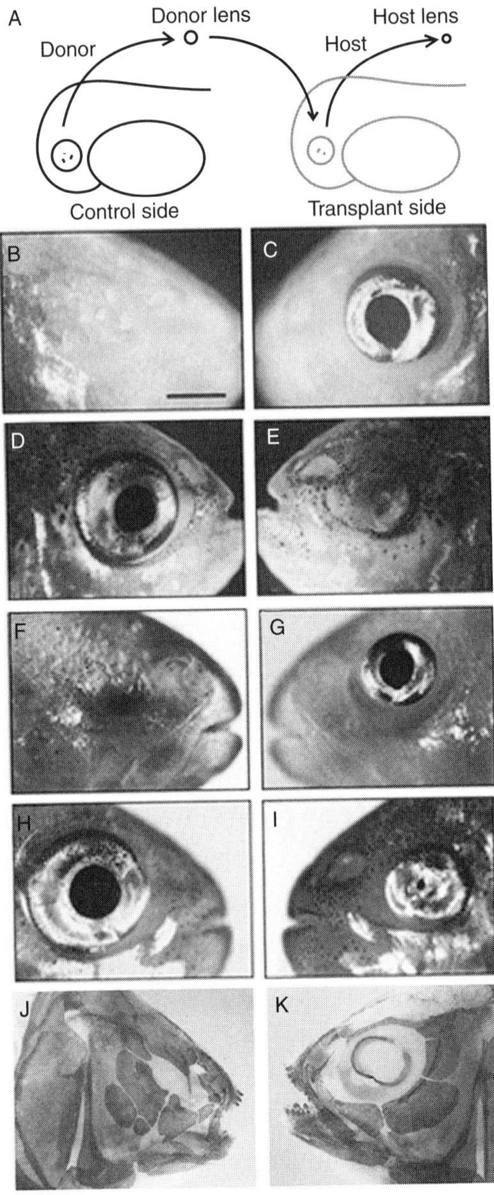

Figure 8.3 Lens transplantation. (A) Diagram showing the transplantation method in which a donor lens is removed from the optic cup of one form of *Astyanax* embryo and transplanted unilaterally into the optic cup of another form after the host lens is removed. This operation is carried out at about 1 dpf. (B–K) Changes in eye development after lens transplantation during embryogenesis. (B, C, F, G) Surface fish lens was transplanted into a Pachón (B, C) or Los Sabinos (F, G), cavefish host. (D, E, H, I)

The embryonic lens is removed from a donor embryo shortly after its formation and transplanted into the optic cup of a host embryo. Lens transplantation is done unilaterally, with the unoperated eye of the host serving as a control, and reciprocally: a surface fish lens is transplanted into a cavefish optic cup and *vice versa*. When a cavefish lens was transplanted into a surface fish optic cup it died on schedule, just as if it had not been removed from the donor embryo. In contrast, when a surface fish lens was transplanted into a cavefish optic cup it continued to grow and differentiate, just as it would have done in the surface fish host. These results indicate that the cavefish lens is autonomously fated for apoptosis.

Cavefish with a transplanted surface fish lens show a dramatic restoration of eye development. The eye primordium of Pachón or Los Sabinos cavefish containing a surface fish lens begins to grow (Fig. 8.3C and G) (Jeffery *et al.*, 2003; Yamamoto and Jeffery, 2000). Eventually, the cornea and iris appear, and the enlarged retina is more highly organized. Further growth results in the presence of a highly developed eye containing a cornea, iris, and photoreceptor cells. In contrast to the eye with a transplanted lens, the unoperated eye of the cavefish host degenerates and disappears into the orbit (Fig. 8.3B and F). Likewise, after obtaining a cavefish lens, development of the surface fish eye is retarded, the cornea and iris do not differentiate, and the size and organization of the retina are reduced. The degenerate surface fish eye eventually disappears into the orbit, mimicking the cavefish eye (Fig. 8.3E and I), whereas the unoperated eye develops normally, producing a one-eyed surface fish (Fig. 8.3D and H).

Several important conclusions can be made from the lens transplantation experiments. First, the lens is required for normal development of the retina, cornea, and iris. Second, as a result of apoptosis the cavefish lens has lost the ability to organize eye development. Third, the cavefish optic cup (RPE/retina) has retained the ability to respond to signals generated by a normal surface fish lens. Fourth, the lens has a role in promoting the survival of retinal cells: a transplanted surface fish lens can protect the cavefish retina from apoptosis (Strickler *et al.*, 2007a). Finally, the lens has an indirect role in determining craniofacial morphology. When a surface fish lens is transplanted into a cavefish optic cup, an orbital bone phenotype is obtained resembling surface fish rather than cavefish (Fig. 8.3J and K) (Yamamoto *et al.*, 2003). The cavefish host develops with a hybrid

---

Changes in eye development after a Pachón (D, E) or Los Sabinos (H, I) cavefish lens was transplanted into a surface fish host. (J, K) Changes in orbital bone structure after a surface fish lens was transplanted into a cavefish optic cup unilaterally. (B, D, F, H, J) Control (unoperated) side. (C, E, G, I, K) Transplant side [(B–E) from Yamamoto and Jeffery, 2000; (F–I) from Jeffery *et al.*, 2003; (J, K) from Yamamoto *et al.*, 2003].

craniofacial morphology, one side (the lens transplant side) resembling surface fish and the other (the control side) cavefish.

Clearly, the lens has a major role in regulating cavefish eye degeneration. Whether the death of the lens is the only cause of eye degeneration, or other optic alterations, such as independent changes in the retina or RPE, are also involved (Strickler *et al.*, 2007a), remains to be determined.

4. Molecular Mechanisms of Eye Degeneration

Understanding the molecular mechanisms of eye degeneration requires identification of the genes involved in this process and how they function during development. Many eye development genes are known in vertebrates, allowing a candidate gene approach to be used for gene identification (Jeffery, 2005). In addition, many genes that are differentially expressed in cavefish have been identified by a microarray-based approach (Strickler and Jeffery, 2009). A list of some of the differentially expressed genes is provided in Table 8.1. These genes encode transcription factors that function near the top of eye gene hierarchies, as well as structural genes encoding proteins that function at the bottom of these cascades. In many cases, *in situ* hybridization or staining with specific antibodies was used to determine their expression patterns.

Most of the genes surveyed by candidate gene analysis do not show expression changes in surface fish and cavefish embryos. For example, the Prox1 transcription factor is expressed normally in the developing lens and retina of cavefish until after the eye begins to degenerate (Jeffery *et al.*, 2000). Likewise, prior to lens degeneration, genes encoding the membrane proteins MIP and MP19 are expressed normally (Strickler *et al.*, 2007b). Many genes also show the same or similar expression patterns in the developing surface and cavefish retinas (Jeffery *et al.*, 2000; Menuet *et al.*, 2007; Strickler *et al.*, 2002). However, some genes are downregulated or upregulated in cavefish (Table 8.1). For example, *gamma M crystallin* and *rhodopsin* expression are reduced in the cavefish lens and retina, respectively (Strickler and Jeffery, 2009; see also Langecker *et al.*, 1993 for *rhodopsin*). These downregulated genes are consistent with respective lack of lens fiber cell differentiation and degeneration of the retinal photoreceptor layer in cavefish.

Among the upregulated genes is one related to human *ubiquitin-specific protease 53* and several other genes (not shown in Table 8.1) encoding factors related to apoptotic cell death. Two genes related to apoptosis are especially interesting. First, the *hsp90α* gene is specifically activated in the cavefish lens vesicle just prior to apoptosis (Hooven *et al.*, 2004). Outside of the lens, *hsp90α* expression remains unchanged, and expression of its close relative *hsp90β* remains unchanged between cavefish and surface fish

Table 8.1 Differentially expressed genes in cavefish embryos relative to surface fish embryos

Gene	Status	Expression	Identification[a]	References
hsp90α	Upregulated	Lens	Candidate analysis	Hooven *et al.* (2004)
shhA	Upregulated	Midline, brain	Candidate analysis	Yamamoto *et al.* (2004)
shhB	Upregulated	Midline	Candidate analysis	Yamamoto *et al.* (2004)
patched 1, 2	Upregulated	Midline	Candidate analysis	Yamamoto *et al.* (2004)
pax2.1a	Upregulated	Optic vesicles	Candidate analysis	Yamamoto *et al.* (2004)
nkx2.1a, b	Upregulated	Midline, brain	Candidate analysis	Menuet *et al.* (2007) and Yamamoto *et al.* (2004)
vax1	Upregulated	Optic vesicles	Candidate analysis	Yamamoto *et al.* (2004)
	Downregulated	Retina	Candidate analysis	Alunni *et al.* (2007)
lhx6, 7	Upregulated	Brain	Candidate analysis	Menuet *et al.* (2007)
ubiquitin-specific protease 53	Upregulated	Unknown	Microarray analysis	Strickler and Jeffery (2009)
pax6	Downregulated	Optic vesicles	Candidate analysis	Strickler *et al.* (2001)
gamma M crystallin	Downregulated	Lens	Microarray analysis	Strickler and Jeffery (2009)

(*continued*)

Table 8.1 *(continued)*

Gene	Status	Expression	Identification[a]	References
gamma B crystallin	Downregulated	Unknown	Microarray analysis	Strickler and Jeffery (2009)
αA-crystallin	Downregulated	Lens	Candidate analysis	Behrens *et al.* (1998) and Strickler *et al.* (2007b)
rhodopsin	Downregulated	Retina	Microarray analysis	Strickler and Jeffery (2009)
neurofilament protein M	Downregulated	Unknown	Microarray analysis	Strickler and Jeffery (2009)
guanosine nucleotide-binding proteins 1, 2	Downregulated	Unknown	Microarray analysis	Strickler and Jeffery (2009)

[a] In microarray analysis, only genes with at least a 10-fold difference are included.

(Hooven *et al.*, 2004). Pharmacological inhibition of Hsp90α suppresses lens apoptosis and rescues lens differentiation. Second, the α *A-crystallin* gene, which encodes a potent antiapoptotic factor, is strongly downregulated in the lens vesicles of Piedras (Behrens *et al.*, 1998) and Pachón cavefish (Strickler *et al.*, 2007b). αA-crystallin may normally protect the lens from apoptosis and is a required chaperone for the normal function of other crystallins in the lens. It is possible that αA-crystallin and Hsp90α interact in a cascade leading to lens apoptosis.

Some of the changes in expression detected by *in situ* hybridization are more subtle than those described above. The *pax6* gene encodes a transcription factor that is expressed in the lens, retina, RPE, and their precursors early in teleost eye development (Krauss *et al.*, 1991; Püschel *et al.*, 1992). Later, *pax6* expression becomes restricted to the lens epithelial cells, some of the retinal layers, and the corneal epithelium. In surface fish embryos, the *pax6* expression domains in the bilateral optic fields connect across the midline at their anterior margins (Fig. 8.4A). In cavefish embryos, however, the corresponding *pax6* domains are diminished in size and show a large gap across the midline (Strickler *et al.*, 2001) (Fig. 8.4B). The division of the optic vesicle into the optic cup and stalk is controlled by reciprocal antagonistic interactions between the Pax6, Pax2, and Vax1 transcription factors (Schwarz *et al.*, 2000; Take-uchi *et al.*, 2003). Pax6 directs optic cup development, whereas Pax2 and Vax1 control optic stalk development. Accordingly, a reduction of *pax6* levels (or an increase in *pax2* and *vax1* levels) increases the optic stalk at the expense of the optic cup. The reduction of *pax6* expression coupled with the overexpression of *pax2a* (Fig. 8.4C–F) and *vax1* (Fig. 8.4G–L) genes accounts for the ventrally reduced optic cup in cavefish embryos (Yamamoto *et al.*, 2004). The *vax1* gene is also expressed on the ventral side of the developing retina in surface fish and other teleosts. However, vax1 expression is missing in the ventral portion of the cavefish retina (Alunni *et al.*, 2007), showing that this gene is either upregulated or downregulated in cavefish depending on developmental stage.

The wider gap between *pax6*-expressing optic fields in the cavefish neural plate provides further insight into how eye degeneration is controlled. During vertebrate development, the presumptive optic cup is initially determined as a single medial optic field, which is subsequently split into two bilateral eye domains by Hedgehog (Hh) signals emanating from the underlying prechordal plate (Ekker *et al.*, 1995; Macdonald *et al.*, 1995). Hh signaling inhibits *pax6* expression along the midline to divide the original eye domain into bilateral eyes. Teleosts have at least two-*hh* midline-signaling genes, *sonic hedgehog A* (*shhA*) and *shhB*, which show overlapping expression patterns (Ekker *et al.*, 1995). Yamamoto *et al.* (2004) compared *shhA* and *shhB* expression patterns during surface fish and cavefish development and demonstrated that the midline expression domains of both genes are expanded in cavefish relative to surface fish (see Fig. 8.5A and B for

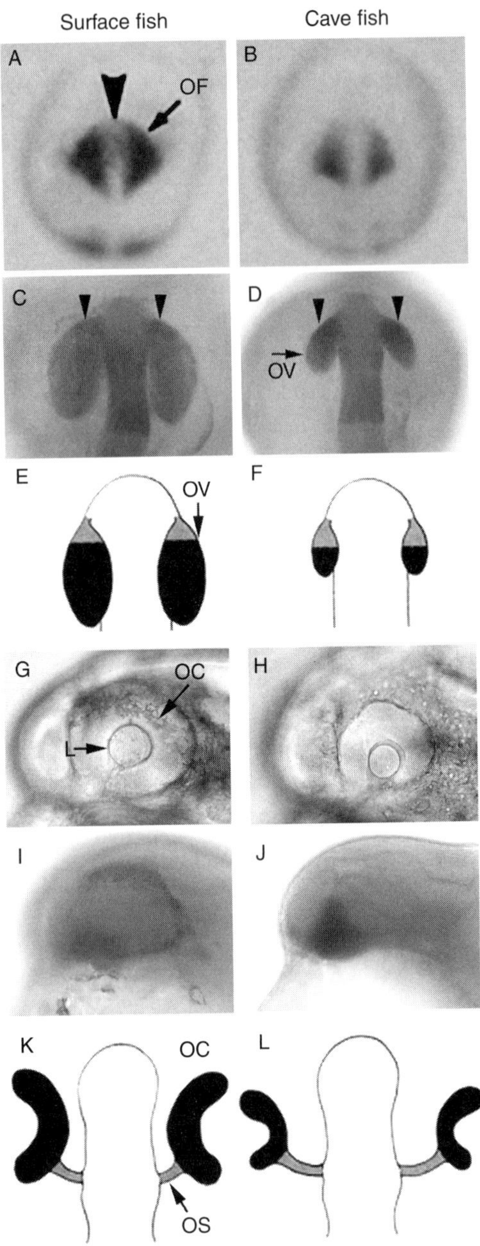

Figure 8.4 Optic vesicle (A–F) and optic cup (G–L) development in surface fish and cavefish. (A, B) Neural plate stage embryo showing differences in *pax6* expression in the surface fish and cavefish optic fields (OF). Arrowhead shows the midline *pax6* expression gap, which is wider in cavefish. (C, D) Optic vesicles (OV) showing size and *pax2a* expression (arrowheads) differences in surface fish and cavefish. In (A–D),

shhA). Later in cavefish development, *shhA* expression is also expanded anteriorly, curling around the rostrum in the presumptive oral area (Fig. 8.5C and D). The expression patterns of genes acting downstream of *shhA* and *shhB* in the Hh midline-signaling pathway, such as *patched 1* and *patched 2*, encoding Shh receptors, and *nkx2.1a* and *nkx2.1b*, encoding Shh-dependent transcription factors, are also expanded (Yamamoto *et al.*, 2004), suggesting that a general increase in midline Hh signaling has evolved in cavefish.

The *shh* genes are expressed in many places in vertebrate embryos. Does Shh expansion also occur in these places in cavefish? Although further studies are needed to completely investigate this important question, the answer appears to be yes and no. Although *shhA* is overexpressed early in the notochord as well as the anterior midline, at later stages of development the notochord expression domain appears normal (Y. Yamamoto and W. R. Jeffery, unpublished data), suggesting that compensatory mechanisms are active during later cavefish development. Likewise, there appear to be no differences in the size or intensity of *shhA* expression domains in cavefish and surface fish fin buds (Y. Yamamoto and W.R. Jeffery, unpublished data). In contrast, early *shhA* expansion is continued in various regions of the cavefish embryonic forebrain, where working through downstream transcription factors such as Nkx2.1a and Lhx6/Lhx7, it seems to be instrumental in increasing the size of the cavefish hypothalamus and ventral forebrain (Menuet *et al.*, 2007). The conclusion is that persistent expansion of Shh signaling is restricted to the anterior midline, a region known as the prechordal plate, as well as the developing forebrain immediately dorsal to this region.

5. Role of Hedgehog Signaling in Eye Degeneration

The epicenter of expanded *shh* expression along the cavefish anterior midline is a critical position with respect to eye development. The role of enhanced Hh signaling in cavefish eye development was investigated by

embryos are viewed dorsally with anterior at the top. (E, F) Diagram showing size differences in the surface fish and cavefish optic vesicles. Territories fated to form optic stalk are lightly shaded and those fated to form retina/RPE are darkly shaded. (G, H) Surface fish and cavefish optic cups (OC) showing ventral size reduction in the latter. L: lens. (I, J) The *vax1* gene is overexpressed ventrally in the cavefish optic cup relative to surface fish. In (G–J), embryos are viewed laterally with dorsal at the top. (K, L) Diagram showing size and relative optic cup territorial differences between cavefish and surface fish. The optic stalk (OS) is lightly shaded and the optic cup is darkly shaded [(A–D, G–J) from Yamamoto *et al.*, 2004; (E, F, K, L) from Strickler *et al.*, 2001].

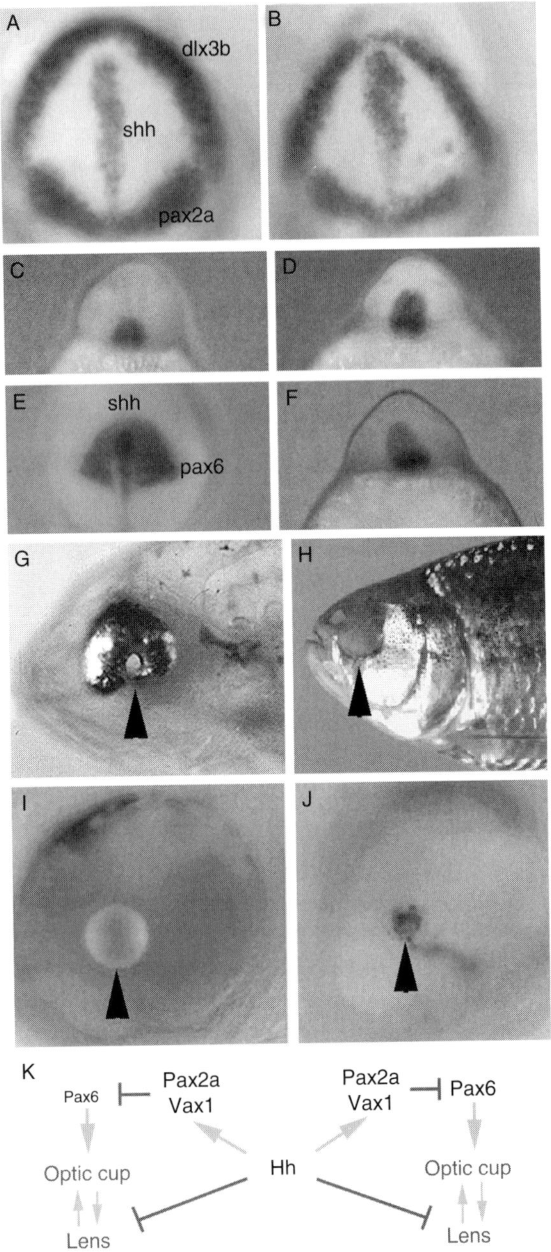

Figure 8.5 Role of Hh midline signaling in cavefish eye degeneration. (A–D) The cavefish embryonic midline shows a wider *shh* expression domain than its surface fish counterpart. The expression of *dlx3* and *pax2a* marker genes does not change. (A, B) Tailbud stage. (C, D) Ten somite stage. (E–J) Effects of *shh* overexpression in

increasing *shh* expression in surface fish embryos (Yamamoto *et al.*, 2004). When *shhA* mRNA was injected into one side of a cleaving embryo, *shhA* expression was expanded along that side of the prechordal plate (anterior embryonic midline), and *pax6* expression was downregulated unilaterally in the corresponding optic field (Fig. 8.5E and F). Surface fish larvae that developed from embryos overexpressing Shh were missing an eye on one side of the head (Fig. 8.5G and H). Thus, blind cavefish were phenocopied by increasing the levels of *shh* gene expression in surface fish, demonstrating a key role for Shh signals in eye degeneration. Importantly, lens apoptosis is also induced by *shh* overexpression in surface fish embryos (Yamamoto *et al.*, 2004) (Fig. 8.5I and J). A diagram of the proposed gene network leading to eye degeneration via hyperactive Shh signaling, reduction of the optic vesicle, and lens cell death is shown in Fig. 8.5K.

In summary, a sequence of regulatory events beginning with expanded midline signaling, proceeding through reduction in size of the eye primordia, lens apoptosis, and retinal apoptosis, and resulting in arrested eye growth and alteration of craniofacial morphology, is responsible for cavefish optic degeneration (Fig. 8.6). Alterations in the activity of many different genes and their upstream regulators are likely to control these changes. Early genetic studies showed that eye degeneration is a multigenic trait (Wilkens, 1988). More recently, it has been determined that at least 12 quantitative trait loci (QTL) are involved in the loss of eyes in Pachón cavefish (Protas *et al.*, 2007). None of these QTL are near the locations of *shhA* or *shhB* on the *Astyanax* genetic map, showing that *shh* genes themselves are not mutated to cause eye degeneration. One possibility for further consideration is that some or all of these QTL may act upstream in the pathway leading to *shh* overexpression in cavefish.

6. Pigment Cell Regression

Astyanax surface fish have three types of body pigment cells: light-reflecting iridophores, yellow-orange xanthophores, and black melanophores. Pigmentation normally functions in protection from the

surface fish. (E, F) Increased *shh* expression (compare F with C) and reduced *pax6* expression (E) on one side of the midline of an embryo injected with *shh* mRNA. (G, H) As a result of *shh* overexpression, the optic cup (retina/RPE) is missing its ventral sector (G) and the adult eye has degenerated (H). Arrowhead in (G): missing ventral sector of the retina. Arrowhead in (H): missing eye. (I, J) Lens apoptosis (J) after injection of an embryo with *shh* mRNA. Arrowheads: lens. (K) Diagram showing antagonistic relationship between Pax6, Pax2, and Vax1 transcription factors, ventralization of the optic cup, and lens apoptosis in cavefish. Arrows: activations. Blocked lines: inhibitions [(A–J) from Yamamoto *et al.*, 2004].

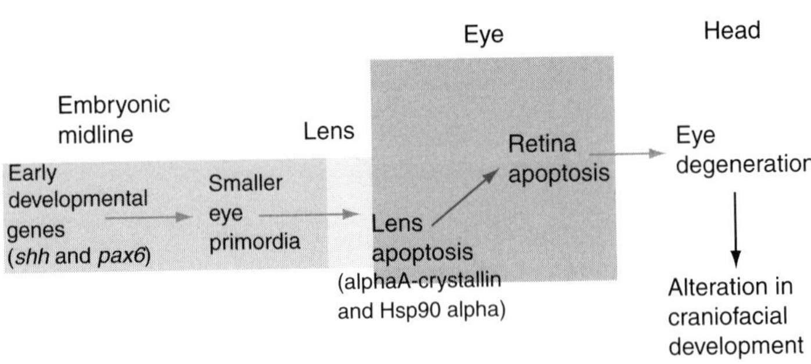

Figure 8.6 Summary of early and late events in cavefish eye degeneration and consequences on craniofacial development.

damaging effects of sunlight, in camouflage, and in species and sex recognition. Selective pressure for retaining these functions is relaxed in the absence of light. What are the consequences in *Astyanax* cavefish?

The early studies of Rasquin (1947) showed that melanophores are decreased in numbers although xanthophores seem to be present at the same levels in Chica cavefish. Very little is known about changes in iridophores. Of the three types of pigment cells, most is known about melanophores (McCauley *et al.*, 2004; Wilkens, 1988). Río Subterráneo cavefish show a modest reduction in melanophore pigmentation, Chica, Curva, Los Sabinos, and Tinaja cavefish show substantial decreases in melanophore pigmentation, and most Molino and Pachón cavefish show little if any melanophores. In addition to changes in the number of melanophores, cavefish also show defects in the ability to produce melanin, the pigment found in melanophores. In Pachón cavefish, melanin pigment seems to be entirely absent, both in body pigment cells (including those surrounding the eye) and in the pigment-containing layer of the RPE. Loss of pigmentation is a typical feature of a diverse assemblage of cave animals and may indeed represent one of the most broad examples of evolutionary convergence in nature. What are the mechanisms of pigment cell regression?

All types of body pigment cells are derived from the neural crest, a unique class of migratory cells derived from the border of the neural tube and surface ectoderm (Erickson, 1993; Le Douarin and Kalcheim, 1999). Vertebrate neural crest cells produce a myriad of different cell types, including sensory ganglia, the peripheral nervous system, cranial cartilage and bone, endocrine and fat cells, as well as body pigment cells. Considering the diversity of their derivatives, it is unlikely that neural crest cells could be modified without inducing lethality. However, a subset of neural crest cells involved in pigment cell development could be missing in cavefish. To test this possibility, cell tracing and tissue culture methods have been used to follow neural crest development in *Astyanax* (Jeffery, 2006; McCauley *et al.*, 2004). In DiI-labeling experiments, neural crest cells migrate into the epidermis (Fig. 8.7A–D), suggesting that there is no defect in neural crest cells during cavefish development.

Another possibility to explain the regression of pigment cells would be cell death. We have already seen how lens cell death mediated by Shh overexpression along the embryonic midline has major effects on cavefish eye regression. Neural crest cells that do not migrate properly or receive normal differentiation signals often die by apoptosis (Morales *et al.*, 2005). Therefore, apoptosis could remove neural crest-derived precursors in cavefish embryos before they differentiate into pigment cells. When this possibility was tested, only a few dying neural crest cells were observed in cavefish embryos, and their number was about the same as in surface fish embryos (Jeffery, 2006). Therefore, melanophores or their progenitor cells do not undergo massive apoptosis during cavefish embryogenesis.

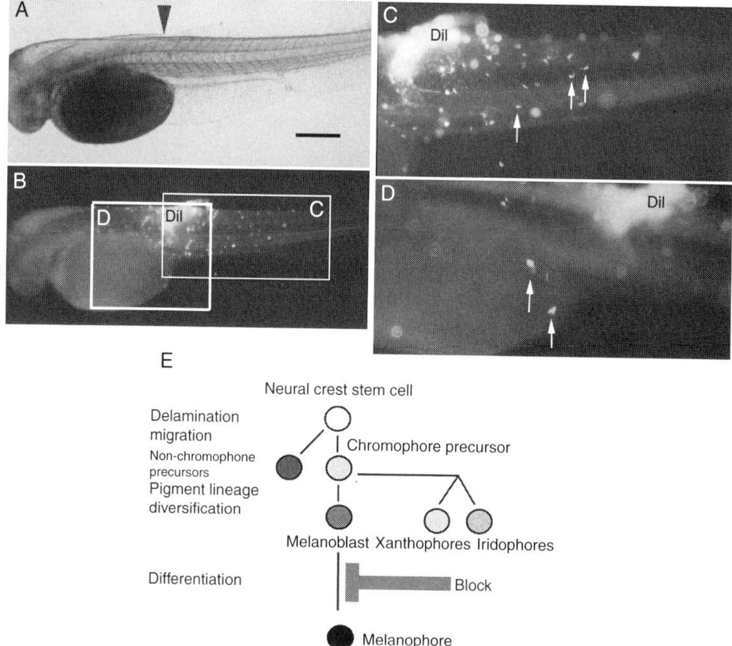

Figure 8.7 Neural crest development in cavefish. (A–C) Detection of migratory neural crest cells in cavefish embryos by DiI injection and subsequent tracing of labeled cells. (A) A 1.5 dpf cavefish embryo showing the site of DiI injection (arrowhead). (B) Fluorescence image of the embryo in (A) showing migration of DiI-injected cells. DiI: original injection site. (C, D) Higher magnification images of insets in (B) showing morphology of injected cells (arrows). (E) Diagram of pigment cell development from the neural crest-derived precursor cells showing the location of the pigmentation block in cavefish [(A–D) from McCauley *et al.*, 2004].

Cavefish pigmentation defects must arise downstream of the generation, migration, and divergence of pigment cell types. This conclusion is supported by the fact that iridophores and xanthophores, which are also products of the migratory neural crest, are apparently present in cavefish that are completely lacking melanophores.

7. Defective Melanogenesis and Undifferentiated Melanoblasts in Cavefish

The early events of pigment cell formation and diversification are not completely understood in vertebrates. However, the fates of iridophores, xanthophores, and melanophores, which are derived from the same neural crest cell lineage, may be somewhat interchangeable (Fig. 8.7E). The presence of appreciable numbers of other pigment cell types in cavefish lacking melanophores (McCauley *et al.*, 2004; Rasquin, 1947) suggests that the lesion in melanophore development lies downstream of the split between the pigment cell progenitors.

Melanophore differentiation involves the initial formation of colorless melanoblasts, which subsequently synthesize black melanin pigment and become functional melanophores. The biochemical steps involved in melanin synthesis during the transition from melanoblast to melanophore are well known and conserved throughout the vertebrates (Fig. 8.8C). First, cytoplasmic L-tyrosine is transported into the melanosome, where it is converted to L-DOPA by the multifunctional enzyme tyrosinase. Next, L-DOPA is converted into melanin by a series of enzymatic reactions, the first of which is also catalyzed by tyrosinase. Most subsequent reactions in this biosynthetic pathway are spontaneous. If adequate L-tyrosine is available and the tyrosinase, tyrosinase-related protein 1 (TRP-1), and TRP-2 enzymes are active, then melanin will be produced. This series of reactions have been investigated to determine the lesion in cavefish melanin synthesis.

Tyrosinase is the limiting enzyme in melanogenesis. Do cavefish pigment progenitor cells have functional tyrosinase? Tyrosinase activity was determined by the L-DOPA assay, in which melanin production is determined after exogenous L-DOPA is provided to fixed specimens. The L-DOPA assay showed that Pachón, Chica, Los Sabinos, Tinaja, and Curva cavefish all exhibit active tyrosinase in cells resembling melanoblasts in their morphology and location within the embryo (Fig. 8.8A and B). Tyrosinase-positive melanoblasts were also observed in the scales and fins of adult cavefish in the positions in which differentiated melanophores are found in their surface fish counterparts (Fig. 8.8D and E). The results of this experiment show that the inability to synthesize melanin in cavefish is due

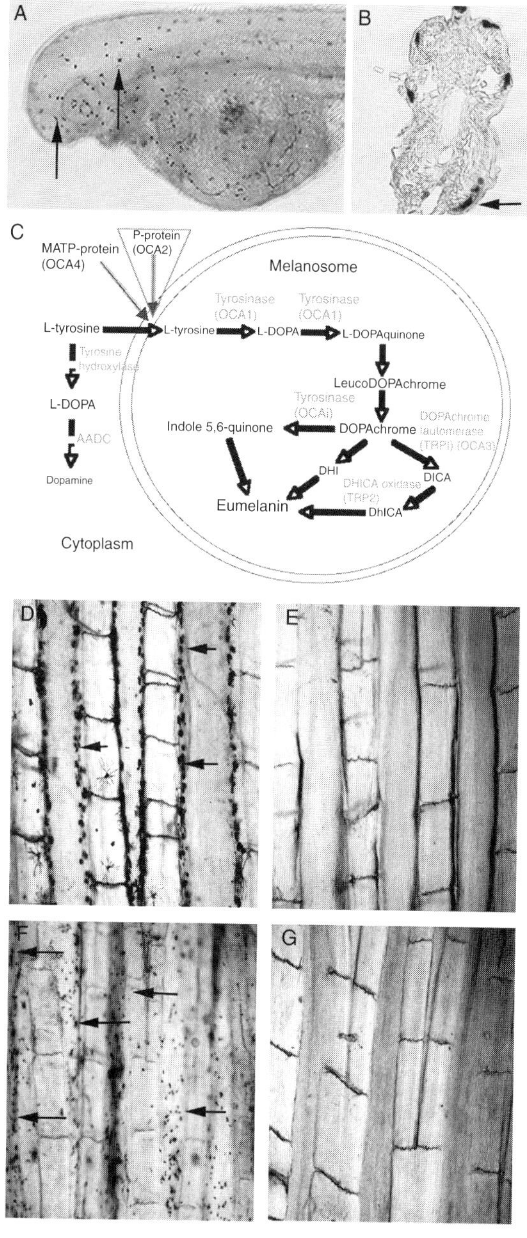

Figure 8.8 Block in melanogenesis in cavefish. (A, B) Cavefish embryos after L-DOPA assay showing tyrosinase-positive melanoblasts (arrows). (A) Whole mount viewed laterally. (B) Section through the trunk. (C) Eumelanin and dopamine synthesis from L-tyrosine in the melanosome and cytoplasm, respectively. Substrates and products, enzymes, and melanosome membrane proteins involved in the reactions are

to a block in the melanogenic pathway immediately upstream of the tyrosinase-dependent steps.

The first step in melanin synthesis is the conversion of L-tyrosine to L-DOPA, which is also catalyzed by tyrosinase (Fig. 8.8C). Cavefish must have L-tyrosine itself because it is required for protein synthesis. However, because cavefish seem to lack endogenous melanosomal L-DOPA, there may be a deficiency in the ability of L-tyrosine to be converted to L-DOPA. This possibility was investigated by a modified L-DOPA assay in which exogenous L-tyrosine was provided to fixed specimens instead of L-DOPA (McCauley *et al.*, 2004). If cavefish can convert L-tyrosine to L-DOPA, then black pigment would be deposited in the same cells that have active tyrosinase. However, melanin deposition was not detected in cavefish exposed to excess L-tyrosine (Fig. 8.8E and G). The results indicate that cavefish melanoblasts are unable to convert L-tyrosine to melanin, implying that melanogenesis is blocked because cytoplasmic L-tyrosine cannot be transported into cavefish melanosomes.

8. Genetic Basis of Cavefish Albinism

A single gene controls cavefish albinism (Borowsky and Wilkens, 2002; Sadoglu, 1957). Accordingly, all F1 progeny of surface fish × Pachón cavefish cross are pigmented and their F2 progeny show a 3:1 ratio of pigmented to unpigmented fishes. Further, crosses between Pachón and Molino cavefish (Wilkens and Strecker, 2003) generate albino F1 offspring, suggesting that mutations in the same gene underlie albinism in many different *Astyanax* cavefish populations.

Using crosses between surface fish and Pachón or Molino cavefish, Protas *et al.* (2006) determined the location of the albinism gene on a microsatellite map of the *Astyanax* genome. The albinism gene was mapped to the same linkage group in both cavefish populations. This result could be explained either by the same mutation in the same gene, different mutations in the same gene, or mutations in different but very closely linked genes. To address this issue, a complementation test was performed in which Pachón and Molino cavefish were crossed and pigmentation was examined in the

indicated at their position(s) in the pathways. Inverted triangle indicates the lesion in cavefish melanogenesis involving P/OCA2. (D–G) Whole mounts of tail fins of adult surface fish (D) and cavefish (E–G). (D) The surface fish fin has melanophores (arrows). (E) The cavefish fin lacks melanophores. (F) The cavefish fin has melanoblasts (arrows) that can convert exogenously supplied L-DOPA to melanin. (G) Cavefish melanoblasts lack the ability to convert L-tyrosine to L-DOPA and melanin [(A, B, D–G) from McCauley *et al.*, 2004].

offspring. If the progeny are pigmented this would suggest that different genes are responsible for albinism, whereas if they are colorless the same gene locus would be implicated. Colorless progeny were obtained showing that the same gene is responsible for albinism in Pachón and Molino cavefish.

Human tyrosinase-positive albinisms have been classified as OCA1, OCA2, OCA3, and OCA4, which are defined by mutations in different genes (Oeting and King, 1999). OCA1 albinism is caused by mutations in the multifunctional enzyme tyrosinase, which acts at three different points in the melanin biosynthetic pathway (Fig. 8.8C). As described above, cavefish can convert L-DOPA to melanin. This means that functional tyrosine must be present in melanoblasts and that cavefish are not OCA1 albinos (McCauley *et al.*, 2004). OCA3 albinism is due to mutations in the gene encoding tyrosinase-related protein 1 (DOPAChrome tautomerase), which functions downstream of the initial tyrosinase-catalyzed steps. By the same reasoning as applied immediately above, this enzyme is also likely to be functional in cavefish, which are therefore not OCA3 albinos. OCA2 and OCA4 albinisms are caused by mutations in the *pink-eyed dilution/oca2* (*p/oca2*) (Rinchik *et al.*, 1993) and *matp* (Baxter and Pavan, 2002) genes, respectively, which encode melanosome membrane proteins. Mutations in *p/oca2* also cause albinism in mice, in which the mutant gene was originally named *pink-eyed dilution* (*p*), and in a teleost, the Medaka (Fukamachi *et al.*, 2004). The *matp* gene is responsible for hypopigmentation in the mouse *underwhite* mutant, where it encodes a putative membrane transporter (Newton *et al.*, 2002). Protas *et al.* (2006) compared the positions of three candidate genes, tyrosinase (OCA1), *p/oca2* (OCA2), and tyrosinase-related protein 1 (OCA3) to the albinism locus on the *Astyanax* genome map. These studies identified *p/oca2* as the cavefish albinism gene. These results suggest that cavefish are OCA2 albinos, which is also the most common form of albinism in humans.

The mammalian *p/oca2* gene contains 24 exons encoding a putative 12-pass membrane protein (Brilliant *et al.*, 1994; Rosenblatt *et al.*, 1994) (Fig. 8.9A). Several functions have been proposed. One possibility is that P/OCA2 transports L-tyrosine into the melanosome (Toyofuku *et al.*, 2002), thus explaining why cavefish melanosomes can use exogenous L-DOPA, but not L-tyrosine as a tyrosinase substrate. Another possibility is that P/OCA2 modulates the processing and transport of tyrosinase (Toyofuku *et al.*, 2002). However, the conservation of tyrosinase activity in cavefish is inconsistent with this possibility. Finally, it has been proposed that P/OCA2 is a proton transporter responsible for regulating melanosomal pH, a key factor in melanogenesis (Brilliant, 2001). Further studies are needed to define the molecular function of P/OCA2 and the physiological lesion it mediates in cavefish melanosomes.

The molecular basis of loss of function was determined by identifying cavefish *p/oca2* mutations. Protas *et al.* (2006) isolated and compared surface

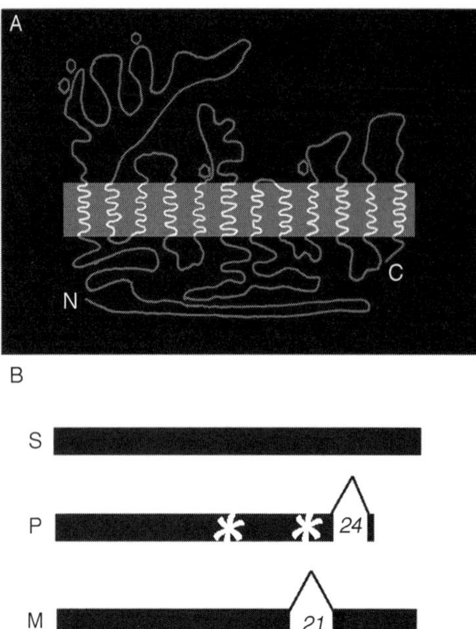

Figure 8.9 Mutations in *p/oca2* responsible for albinism in Pachón and Molino cavefish. (A) The predicted structure of the human P/OCA2 protein showing 12 membrane spanning domains. N: N-terminus. C: C-terminus. Thick bar: melanosome membrane. Thin line: P/OCA2 protein (after Brilliant *et al.*, 1994). (B) Diagram showing the positions of single amino acid changes (asterisks) and deletions (peaked thin lines) in the Pachón (P) and Molino (M) cavefish P/OCA2 proteins. S: the intact surface fish P/OCA2 protein consisting of 24 exons of the *p/oca2* gene. P: the nonfunctional Pachón cavefish P/OCA2 protein showing loss of a major part of exon 24. For clarity, additional translated sequence in Pachón cavefish P/OCA2 protein corresponding to part of intron 23 (see text) is not indicated in the diagram. M: the nonfunctional P/OCA2 protein in Molino cavefish showing the loss of exon 21. Thick black lines: exon sequence. Sequence lengths are not drawn to scale (after Protas *et al.*, 2006).

fish, Pachón cavefish, and Molino cavefish *p/oca2* cDNAs. Three differences from surface fish were discovered in Pachón cavefish *p/oca2*: two point mutations resulting in conserved amino acid substitutions and a large deletion extending from within intron 23 through most of exon 24. Because of this deletion, the Pachón P/OCA2 protein would contain a part of intron 23 as a translated sequence and would be missing most of exon 24 (Fig. 8.9B). In Molino cavefish, there was a single change, another large deletion encompassing exon 21 that would also shorten the P/OCA2 protein. Both deletions are in regions predicted to be parts of membrane spanning domains.

To determine which of these mutations cause *p/oca2* loss of function, Protas *et al.* (2006) examined the ability of DNA constructs containing wild-type surface fish *p/oca2* and the individual polymorphisms in Pachón and Molino cavefish *p/oca2* to rescue the colorless phenotype in a melanocyte cell line derived from a P/OCA2 deficient albino mouse (Sviderskaya *et al.*, 1998). The surface fish *p/oca2* DNA construct and the two Pachón cavefish *p/oca2* DNA constructs with different amino acid polymorphisms rescue melanogenesis in the cell line, indicating that the corresponding point mutations do not prevent melanogenesis. In contrast, *p/oca2* DNA constructs containing the large deletions found in Pachón and Molino *p/oca2* do not induce melanin synthesis, suggesting that they are responsible for loss of function. Although the *p/oca2* gene appears to be responsible for loss of melanin pigment in many different cavefish populations, the mutations are distinct in Pachón and Molino cavefish, suggesting that cavefish albinism evolved by independent changes in *p/oca2*.

9. Evolution of Development

The comparative studies of *Astyanax* provide important insights into the evolution of development in cavefish. In the final section of this chapter, we discuss evolutionary insights gleaned from the studies described above pertaining to cavefish eye and pigment regression.

9.1. Developmental constraints

It is clear that cavefish regressive evolution is channeled to a large extent by developmental constraints, which restrict the amplitude of evolutionary changes, or make them unlikely or impossible, by limiting developmental flexibility. This lack of flexibility appears to have a very important role in cavefish eye and pigment evolution. Consider the following. If these traits are ultimately lost, why is it necessary to construct an eye or produce melanoblasts in the first place? The answer may be that early steps in eye and pigment development are required for other essential steps in development, and the elimination of these steps would be fatal.

Eyes are initially formed and then degraded during larval or adult development in all sightless cave-dwelling vertebrates (Berti *et al.*, 2001; Durand, 1976; Eigenmann, 1908). Indeed, we feel that cave vertebrates lacking embryonic eye primordia will not be discovered because of this strong developmental constraint. Because all vertebrates have bilateral eyes arising from a single medial optic field, the subsequent separation of optic fields is likely to be an ancient vertebrate trait that evolved in concert with other head features. Thus, if the Shh midline-signaling pathway is altered, as we have seen in cavefish, there may be automatic consequences on eye development, in this case leading to degeneration (Jeffery, 2005).

In cavefish, a block in the pigment cell-generating pathway occurs relatively late in the developmental pathway, during the conversion of melanoblasts into melanophores. Earlier steps in this pathway, such as the determination and migration of neural crest cells, the restriction to pigment cell fate, and the diversification of different pigment cell lineages is apparently not changed, even though the usefulness of any pigment cell type is questionable in cavefish. The reason neural crest cells are formed is clear: they have many critical derivatives and their loss would be lethal. Why are any pigment cell types are formed in cavefish? The constraint might be that progression toward making a general set of pigment cells precursors (including melanoblasts) may be required to produce other types of pigment cells (e.g., iridophores and xanthophores), whose function is in some unknown way essential in cave-dwelling teleosts.

The process in which retinal development is arrested in cavefish may be another example of a developmental constraint. We have shown that the arrest of retinal development is not caused by inhibition of cell division at the CMZ, which would seem to be the simplest way to stop growth. Instead, retinal growth is curtailed by apoptosis of newly born cells (Strickler *et al.*, 2007a). This must be a very costly process in terms of energy expenditure, so why has inhibition of cell proliferation, the most parsimonious and least expensive route to preventing retina development, not been taken? The probable answer lies in the fact that the retina is actually a part of the brain. In both retina and brain, stem cells replenish the laminated areas through the same course of action, which may be a fundamental property of nervous system development and difficult to modify. Accordingly, killing new cells after they proliferate in the retina may be more allowable than blocking stem cell division in the CMZ because of an ancient constraint on how different parts of the brain grow in concert during development.

9.2. Developmental amplification

Cavefish show how large-scale changes in the phenotype can occur rapidly during evolution. The differences in craniofacial skeletons between cavefish and surface fish, in particular the ocular bones surrounding the eye, are so extreme that they were formerly used to support their designation as separate genera (Alvarez, 1947). However, the majority of these changes are related to whether or not a large eye punctuates the craniofacial skeleton. When the eye is absent from the surface of the head, as in cavefish, the craniofacial skeleton is patterned differently from when an eye is present. Major changes in the craniofacial skeleton can be elicited by transplanting a surface fish lens into a cavefish optic cup during early development (Yamamoto *et al.*, 2003). The sequence of events is as follows: a normal lens induces anterior eye parts and promotes the growth of a normal retina, producing a large growing eye, which in turn dictates the morphology of surrounding bones in the adult (Fig. 8.8). Cavefish show that slight changes in early development can be amplified to have major impacts in the adult.

9.3. Pleiotropy and tradeoffs

Pleiotropy, the control of multiple, often seemingly unrelated phenotypes, by a single gene is a possible mechanism for the evolution of regressive traits in cave animals (Barr, 1968). Accordingly, if downregulation of genes controlling eye development simultaneously increases the development of a beneficial trait, such as olfaction or another sensory system, the latter might be adaptive and subject to natural selection. The potential for trait linkage is the reason that it is important to study regressive traits in the context of constructive traits. The discovery of enhanced midline signaling mediated by highly pleiotropic *hh* genes (Yamamoto *et al.*, 2004) opens many possibilities that may be able to explain eye degeneration in cavefish. As we have seen, Hh overexpression has a negative effect on eye development, and it is known from studies on other vertebrates that Hh signaling has positive effects on many other developmental traits. Thus, selection for the positive traits would automatically affect the negative ones. In the future, it will be important to determine the identity of positive traits influenced by Hh signaling.

9.4. Evolutionary forces

Why have eyes and pigment been lost in cavefish? No one really knows the answer but the regressive features of cave animals are usually explained by one of two hypotheses (1) the accumulation of selectively neutral (loss of function) mutations and genetic drift (Wilkens, 1988) or (2) indirect selection based on energy conservation and/or antagonistic pleiotropy (Culver, 1982; Jeffery, 2005). Although neither hypothesis has been proved in the case of *Astyanax* cavefish, developmental and genetic studies generally support one or the other for loss of pigmentation and eyes, respectively.

In the case of eye loss, the developmental information seems to support selection over neutral mutation. First, the genes involved in eye development that have been studied thus far do not appear to have mutated to a degree in which they have lost function. In addition, the restoration of eyes by lens transplantation suggests that all genes that act downstream of lens function are present and potentially active in cavefish. Also, supportive of selection is that most genes with modified expression patterns, such as those in the Shh signaling pathway and *hsp90α*, increase rather than decrease their activity in cavefish. Genetic analysis is also consistent with selection (Protas *et al.*, 2007). QTL have only been found that result in a decrease in eye formation; none have been reported that result in an increase, which would be expected if genetic drift were involved.

In contrast to eye regression, developmental studies on loss of pigmentation could support either selection or neutral mutation. On the one hand, the accumulation of neutral mutations resulting in loss of melanophores might be possible, especially if the *oca2* gene is not pleiotropic and its disruption does not affect other important developmental pathways. Genetic analysis, in which individual QTL governing the extent of melanophore development have been shown to either increase or decrease melanophore abundance, supports the role of neutral mutation and genetic drift (Protas *et al.*, 2007). On the other hand, melanogenesis could be disrupted because it is adaptive, allowing pigment cell precursors to be shunted into other, more beneficial differentiation pathways. Some of these possibilities are testable and predict a bright future for the *Astyanax* system in addressing why, as well as how, developmental changes have occurred during evolution.

ACKNOWLEDGMENTS

The research from the Jeffery laboratory described in this chapter was supported by grants from NIH (R01-EY014619) and NSF (IBN-0542384).

REFERENCES

Alunni, A., Menuet, A., Candal, E., Pénigault, J.-B., Jeffery, W. R., and Rétaux, S. (2007). Developmental mechanisms for retinal degeneration in the blind cavefish *Astyanax mexicanus*. *J. Comp. Neurol.* **505,** 221–233.

Alvarez, J. (1947). Descripción de *Anoptichthys hubbsi* caracinindo ceigo de La Cueva de Los Sabinos. *S. L. P. Soc. Mex. Hist. Nat.* **8,** 215–219.

Barr, T. (1968). Cave ecology and the evolution of troglobites. *Evol. Biol.* **2,** 35–102.

Baxter, L. L., and Pavan, W. J. (2002). The oculocutaneous albinism type IV gene Matp is a new marker of pigment cell precursors during mouse embryonic development. *Mech. Dev.* **116,** 209–212.

Behrens, M., Wilkens, H., and Schmale, H. (1998). Cloning of the αA-crystallin genes of the blind cave form and the epigean form of *Astyanax fasciatus*: A comparative analysis of structure, expression and evolutionary conservation. *Gene* **216,** 319–326.

Berti, R., Durand, J. P., Becchi, S., Brizzi, R., Keller, N., and Ruffat, G. (2001). Eye degeneration in the blind cave-dwelling fish *Phreatichthys andruzzii*. *Can. J. Zool.* **79,** 1278–1285.

Borowsky, R. (2008). Restoring sight in blind cavefish. *Curr. Biol.* **18,** R23–R24.

Borowsky, R., and Wilkens, H. (2002). Mapping a cave fish genome. Polygenic systems and regressive evolution. *J. Hered.* **93,** 19–21.

Brilliant, M. H. (2001). The mouse p (pink-eyed dilution) and human P genes, oculocutaneous albinism type 2 (OCA2), and melanosomal pH. *Pigment Cell Res.* **14,** 86–93.

Brilliant, M. H., King, R., Francke, U., Schuffenhauer, S., Meitinger, T., Gardner, J. M., Durham-Pierre, D., and Nakatsu, Y. (1994). The mouse pink-eyed dilution gene: Association with hypopigmentation in Prader-Willi and Angelman syndromes and with human OCA2. *Pigment Cell Res.* **7,** 398–402.

Cahn, P. H. (1958). Comparative optic development in *Astyanax mexicanus* and in two of its blind cave derivatives. *Bull. Am. Mus. Nat. Hist.* **115,** 73–112.

Culver, D. (1982). "Cave Life: Evolution and Ecology." Harvard University Press, Cambridge MA.

Dowling, T. E., Martasian, D. P., and Jeffery, W. R. (2002). Evidence for multiple genetic lineages with similar eyeless phenotypes in the blind cavefish, *Astyanax mexicanus*. *Mol. Biol. Evol.* **19,** 446–455.

Durand, J. P. (1976). Ocular development and involution in the European cave salamander, *Proteus anguinus* Laurenti. *Biol. Bull.* **151,** 450–466.

Eigenmann, C. H. (1908). The eyes of the blind vertebrates of North America. V. The history of the eye of blind *Amblyopsis* from its appearance to its disintegration in old age. *Contrib. Zool. Lab. Indiana Univ.* Mark Anniversary Volume, 167–204.

Ekker, S. C., Ungar, A. R., von Greenstein, P., Porter, J. A., Moon, R. T., and Beachy, P. (1995). Patterning activities of vertebrate hedgehog proteins in the developing eye and brain. *Curr. Biol.* **5,** 944–955.

Erickson, C. A. (1993). From the crest to the periphery: Control of pigment cell migration and lineage segregation. *Pigment Cell Res.* **6,** 336–347.

Fukamachi, S., Asakawa, S., Wakamatsu, Y., Shimizu, N., Mitanti, H., and Shima, A. (2004). Conserved function of Medaka *pink-eyed dilution* in melanin synthesis and its divergent transcriptional regulation in gonads among vertebrates. *Genetics* **168,** 1519–1527.

Hooven, T. A., Yamamoto, Y., and Jeffery, W. R. (2004). Blind cavefish and heat shock protein chaperones: A novel role for hsp90α in lens apoptosis. *Int. J. Dev. Biol.* **48,** 731–738.

Jeffery, W. R. (2001). Cavefish as a model system in evolutionary developmental biology. *Dev. Biol.* **231,** 1–12.

Jeffery, W. R. (2005). Adaptive evolution of eye degeneration in the Mexican blind cavefish. *J. Hered.* **96,** 185–196.

Jeffery, W. R. (2006). Regressive evolution of pigmentation in the cavefish *Astyanax*. *Isr. J. Ecol. Evol.* **52,** 405–422.

Jeffery, W. R. (2008). Emerging systems in evo/devo: Cavefish and mechanisms of microevolution. *Evol. Dev.* **10,** 265–272.

Jeffery, W. R., and Martasian, D. P. (1998). Evolution of eye regression in the cavefish *Astyanax*: Apoptosis and the *Pax-6* gene. *Am. Zool.* **38,** 685–696.

Jeffery, W. R., Strickler, A. G., Guiney, S., Heyser, D., and Tomarev, S. I. (2000). Prox1 in eye degeneration and sensory organ compensation during development and evolution of the cavefish *Astyanax*. *Dev. Genes Evol.* **210,** 223–230.

Jeffery, W. R., Strickler, A. G., and Yamamoto, Y. (2003). To see or not to see: Evolution of eye degeneration in Mexican blind cavefish. *Integr. Comp. Biol.* **43,** 531–541.

Krauss, S., Johannsen, T., Korzh, V., and Fijose, A. (1991). Zebrafish pax[zf-a]: A paired box gene expressed in the neural tube. *EMBO J.* **10,** 3609–3619.

Langecker, T. G., Schmale, H., and Wilkens, H. (1993). Transcription of the opsin gene in degenerate eyes of cave dwelling *Astyanax fasciatus* (Teleostei, Characidae) and its conspecific ancestor during early ontogeny. *Cell Tissue Res.* **273,** 183–192.

Le Douarin, N. M., and Kalcheim, C. (1999). "The Neural Crest", 2nd Edn. Cambridge University Press, New York.

Macdonald, R., Anukampa Barth, K., Xu, Q., Holder, N., Mikkola, I., and Wilson, S. (1995). Midline signalling is required for *Pax6* gene regulation and patterning of the eyes. *Development* **121,** 3267–3278.

McCauley, D. W., Hixon, E., and Jeffery, W. R. (2004). Evolution of pigment cell regression in the cavefish *Astyanax*: A late step in melanogenesis. *Evol. Dev.* **6,** 209–218.

Menuet, A., Alunni, A., Joly, J.-S., Jeffery, W. R., and Rétaux, S. (2007). Shh overexpression in *Astyanax* cavefish: Multiple consequences on forebrain development and evolution. *Development* **134,** 845–855.

Mitchell, R. W., Russell, W. H., and Elliot, W. R. (1977). Mexican eyeless characin fishes, genus *Astyanax*: Environment, distribution, and evolution. *Spec. Publ. Mus. Texas Tech. Univ.* **12,** 1–89.

Morales, A. V., Barbas, J. A., and Nieto, M. A. (2005). How to become neural crest: From segregation to delamination. *Semin. Cell Dev. Biol.* **16,** 655–662.

Newton, J. M., Cohen-Barak, O., Hagiwara, H., Gardner, J. M., Davisson, M. T., King, R. A., and Brilliant, M. H. (2002). Mutations in the human orthologue of the mouse *underwhite* gene (uw) underlie a new form of oculocutaneous albinism, OCA4. *Am. J. Hum. Genet.* **69,** 981–988.

Oeting, W. S., and King, R. A. (1999). Molecular basis of albinism: Mutations and polymorphisms of pigmentation genes associated with albinism. *Hum. Mutat.* **13,** 99–113.

Porter, M. L., Dittmar de la Cruz, K., and Pérez-Losada, M. (2007). How long does evolution of the troglomorphic form take? Estimating divergence times in *Astyanax mexicanus*. *Acta Carsologica* **36,** 173–182.

Protas, M. E., Hersey, C., Kochanek, D., Zhou, Y., Wilkens, H., Jeffery, W. R., Zon, L. T., Borowsky, R., and Tabin, C. J. (2006). Genetic analysis of cavefish reveals molecular convergence in the evolution of albinism. *Nat. Genet.* **38,** 107–111.

Protas, M., Conrad, M., Gross, J. B., Tabin, C., and Borowsky, R. (2007). Regressive evolution in the Mexican cave tetra, *Astyanax mexicanus*. *Curr. Biol.* **17,** 452–454.

Püschel, A. W., Gruss, P., and Westerfield, M. (1992). Sequence and expression pattern of pax-6 are highly conserved between zebrafish and mice. *Development* **114,** 643–651.

Rasquin, P. (1947). Progressive pigmentary regression in fishes associated with cave environments. *Zoologica* **32,** 35–44.

Rinchik, E. M., Bultman, S. J., Horsthemke, B., Lee, S. T., Strunk, K. M., Spritz, R. A., Avidano, K. M., Jong, M. T., and Nicholls, R. D. (1993). A gene for the mouse *pink-eyed dilution* locus and for human type II oculocutaneous albinism. *Nature* **361,** 72–76.

Rosenblatt, S., Durham-Pierce, D., Garner, J. M., Nakatsu, Y., Brilliant, M. H., and Orlow, S. J. (1994). Identification of a melanosomal membrane protein encoded by the pink-eyed dilution (type II oculocutaneous albinism) gene. *Proc. Natl. Acad. Sci. USA* **91,** 12071–12075.

Sadoglu, P. (1957). A Mendelian gene for albinism in natural cave fish. *Experientia* **13,** 394.

Schwarz, M., Cecconi, F., Berneir, G., Andrejewski, N., Kammandel, B., Wagner, M., and Gruss, P. (2000). Spatial specification of mammalian eye territories by reciprocal transcriptional repression of Pax2 and Pax6. *Development* **127,** 4325–4334.

Soares, D., Yamamoto, Y., Strickler, A. G., and Jeffery, W. R. (2004). The lens has a specific influence on optic nerve and tectum development in the blind cavefish *Astyanax*. *Dev. Neurosci.* **26,** 308–317.

Strecker, U., Bernachez, L., and Wilkens, H. (2003). Genetic divergence between cave and surface populations of *Astyanax* in Mexico (Characidae, Teleostei). *Mol. Ecol.* **12,** 699–710.

Strecker, U., Faúndez, V. H., and Wilkens, H. (2004). Phylogeography of surface and cave *Astyanax* (Teleostei) from Central and North America based on cytochrome *b* sequence data. *Mol. Phylogenet. Evol.* **33,** 469–481.

Strickler, A. G., and Jeffery, W. R. (2009). Differentially expressed genes identified by cross species microarray in the blind cavefish *Astyanax*. *Int. Zool.* **4,** 98–109.

Strickler, A. G., Yamamoto, Y., and Jeffery, W. R. (2001). Early and late changes in *Pax6* expression accompany eye degeneration during cavefish development. *Dev. Genes Evol.* **211,** 138–144.

Strickler, A. G., Famuditimi, K., and Jeffery, W. R. (2002). Retinal homeobox genes and the role of cell proliferation in cavefish eye degeneration. *Int. J. Dev. Biol.* **46,** 285–294.

Strickler, A. G., Yamamoto, Y., and Jeffery, W. R. (2007a). The lens controls cell survival in the retina: Evidence from the blind cavefish *Astyanax*. *Dev. Biol.* **311,** 512–523.

Strickler, A. G., Byerly, M. S., and Jeffery, W. R. (2007b). Lens gene expression analysis reveals downregulation of the anti-apoptotic chaperone αA crystallin during cavefish eye degeneration. *Dev. Genes Evol.* **217,** 771–782.

Sviderskaya, E. V., Novak, E. K., Swank, R. T., and Bennent, D. C. (1998). The murine misty mutation: Phenotypic effects on melanocytes, platelets, and brown fat. *Genetics* **148,** 381–390.

Take-uchi, M., Clarke, J. D., and Wilson, S. W. (2003). Hedgehog signalling maintains the optic stalk–retinal interface through the regulation of Vax gene activity. *Development* **130,** 955–968.

Toyofuku, K., Valencia, J. C., Kushimoto, T., Costin, G.-E., Virador, V. M., Viera, W. D., Ferrans, V. J., and Hearing, V. J. (2002). The etiology of oculocutaneous albinism (OCA) type II: The pink protein modulates the processing and transport of tyrosinase. *Pigment Cell Res.* **15,** 217–224.

Van Valen, L. (1973). Festschrift. *Science* **180,** 488.

Wilkens, H. (1971). Genetic interpretation of regressive evolutionary processes: Studies of hybrid eyes of two *Astyanax* cave populations (Characidae, Pisces). *Evolution* **25,** 530–544.

Wilkens, H. (1988). Evolution and genetics of epigean and cave *Astyanax fasciatus* (Characidae, Pisces). *Evol. Biol.* **23,** 271–367.

Wilkens, H., and Strecker, U. (2003). Convergent evolution of the cavefish *Astyanax* (Characidae, Teleostei): Genetic evidence from reduced eye-size and pigmentation. *Biol. J. Linn. Soc.* **80,** 545–554.

Yamamoto, Y., and Jeffery, W. R. (2000). Central role for the lens in cavefish eye degeneration. *Science* **289,** 631–633.

Yamamoto, Y., Espinasa, L., Stock, D. W., and Jeffery, W. R. (2003). Development and evolution of craniofacial patterning is mediated by eye-dependent and -independent processes in the cavefish *Astyanax*. *Evol. Dev.* **5,** 435–446.

Yamamoto, Y., Stock, D. W., and Jeffery, W. R. (2004). Hedgehog signalling controls eye degeneration in blind cavefish. *Nature* **431,** 844–847.

Subject Index

Contents of Previous Volumes

Volume 47

Volume 48

Volume 49

Volume 50

Volume 51

Volume 52

Volume 53

Volume 54

Volume 57

Volume 60

Volume 61

Volume 62

Volume 63

Volume 64

Volume 65

Volume 66

Volume 67

Volume 68

Volume 70

Volume 71

Volume 72

Volume 73

Volume 74

Volume 75

Volume 77

Volume 78

Volume 79

Volume 81

Volume 82

Volume 83

Volume 84

Volume 85

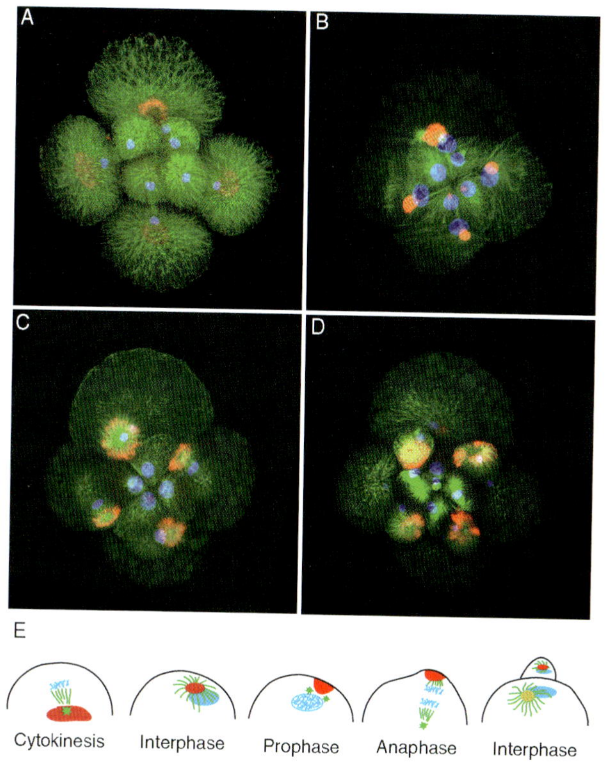

J. David Lambert, Figure 5.2 Centrosomal localization and asymmetric segregation of RNAs in early cleavage. The IoUbiquitin ligase RNA is localized by *in situ* hybridization detected with fluorescent tyramide precipitation (red). This RNA was formerly known as IoLR2 (Kingsley *et al.*, 2007) but significant homology to ubiquitin ligases was identified in additional flanking sequence recovered in an EST sequencing project (unpublished data). DNA is stained with DAPI (blue). Microtubules are stained with an antibody against β-tubulin (green). Images are projections of confocal *Z*-stacks. Yellow indicates that RNA and microtubules are colocalized, or superimposed in different sections. (A) During cytokinesis of the third cleavage cycle, the IoLR2 RNA surrounds the spindle poles in the macromeres. (B) At the interphase–prophase transition of the eight-cell stage, two macromeres are in interphase (1A and 1B, lower and right) and show IoLR2 RNA localization to the large spherical centrosomes. The other two are in prophase: at this stage, the RNA is moving from the centrosomes to the cortex and the prophase asters are visible as two small foci of microtubules under the RNA in 1D (upper macromere). (C) In metaphase of the fourth division, the IoLR2 RNA is on the cortex in all four macromeres, and the spindles are aligned toward the RNA, which changes from a disk-shaped patch to a ring, with the spindle pole at its center. (D) During cytokinesis of the fourth division, the IoLR2 RNA is on the cortex of the second quartet micromeres, which will inherit all of this RNA. (E) Diagram of RNA localization and segregation events during early cleavage cycles, showing the RNA, microtubules and DNA. Data are from Kingsley *et al.* (2007).

Armin P. Moczek, Figure 6.1 Examples of horned beetles illustrating diversity and magnitude of horn expression in adult beetles. Clockwise from top: *Trypoxylus (Allomyrina) dichotoma, Onthophagus watanabei, Golofa claviger*, and *Phanaeus imperator*.

Armin P. Moczek, Figure 6.2 Diversity between and within *Onthophagus* species. (A) Six *Onthophagus* species illustrating the diversity of horn types that exist within the genus. (B) Sexual and male dimorphism in *Onthophagus nigriventris*.

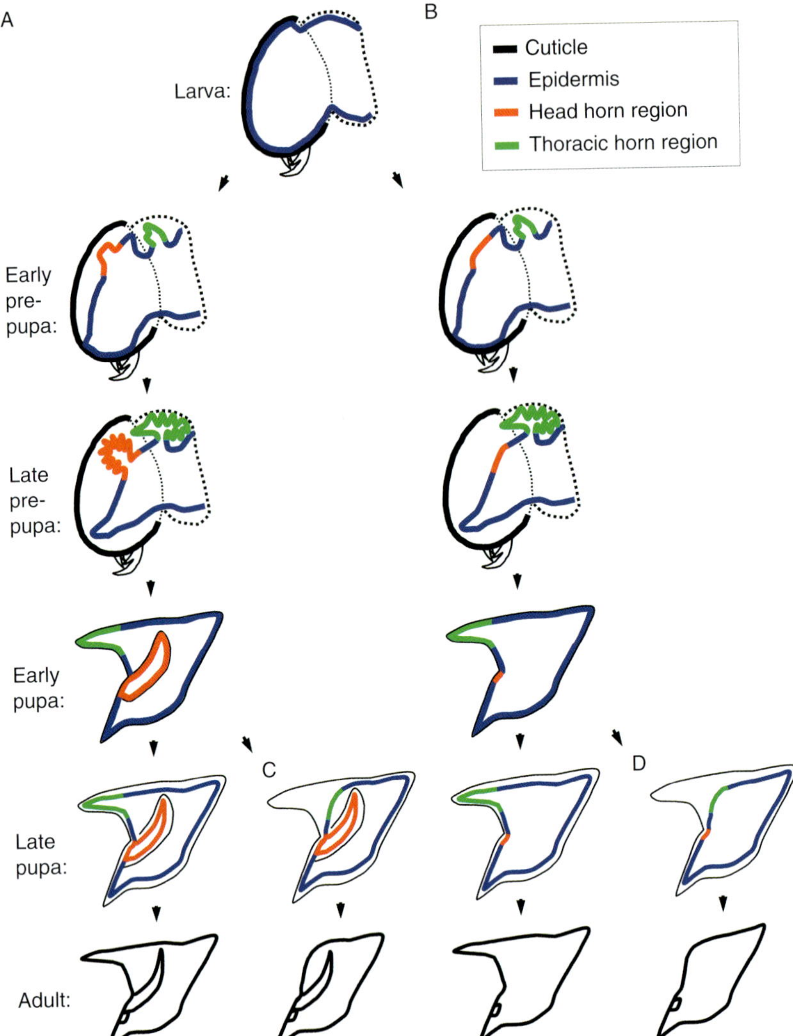

Armin P. Moczek, Figure 6.3 Development of horns and horn dimorphisms in *Onthophagus* beetles. (A) Apolysis is followed by rapid cell proliferation of selected epidermal tissue regions (shown here for a head horn and thoracic horn). Horn primordia expand during the pupal molt and become externally visible. During the pupal stage epidermal cells apolyse once more, followed by remodeling of the pupal epidermis into the final adult shape. The pupa then undergoes one last molt to the final adult stage. (B) Development of horn dimorphisms through differential proliferation of prepupal horn tissue (illustrated here for head horns). During the prepupal stage presumptive horn tissue proliferates little, resulting in the absence of external horns in pupae and resulting adults. (C, D) Development of horn dimorphisms through differential remodeling of pupal horn tissue (illustrated here for thoracic horns). Pupal horn epidermis is resorbed prior to the secretion of the adult cuticle. This mechanism generates sexual dimorphisms for thoracic horns in many species, and can occur in the presence or absence of (differential) head horn development (modified after Moczek, 2005).

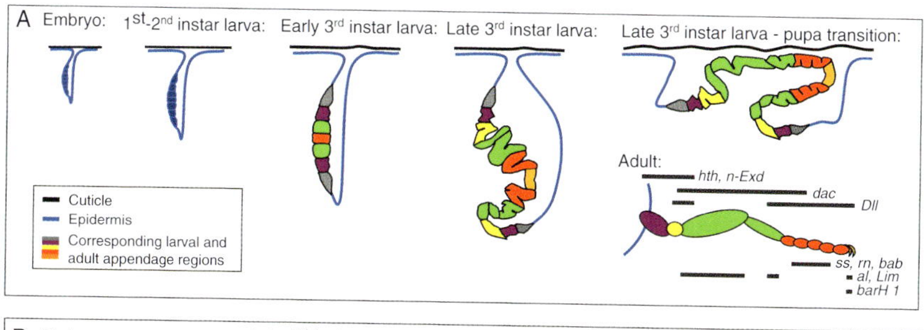

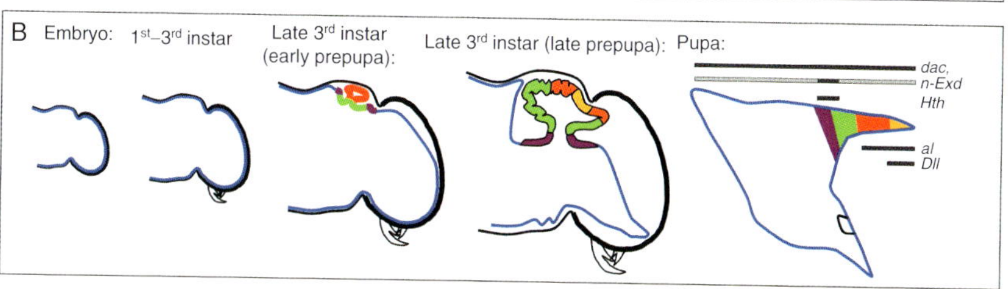

Armin P. Moczek, Figure 6.4 Differences and similarities in the development of the (A) *Drosophila* leg and (B) thoracic horns in beetles (see text for details). Colors indicate tissue types and regional relationships between immature and mature appendage. Also indicated is the approximate relationship between expression domains of common p/d patterning genes during development and the corresponding adult appendage region (modified after Moczek, 2006a,b).

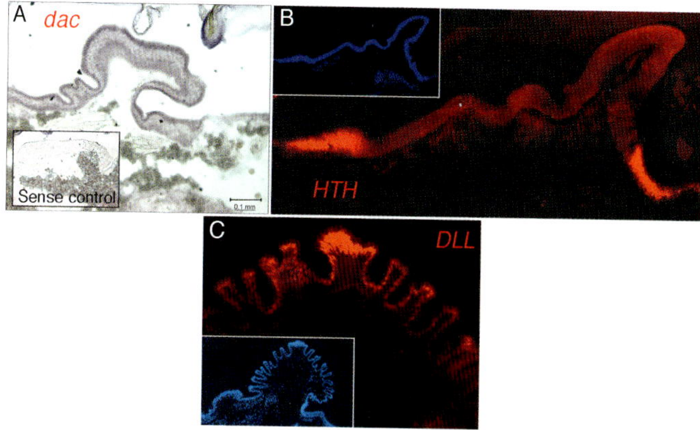

Armin P. Moczek, Figure 6.5 Examples of p/d genes expressed during horn development. (A) *Dachshund in situ* hybridization of the transient thoracic horn primordium in *O. taurus*. (B) Anti-HTH immunostaining of the persisting thoracic horn primordium of *O. binodis*. (C) Anti-DLL immunostaining of one of two head horn primordia in of *O. taurus*.

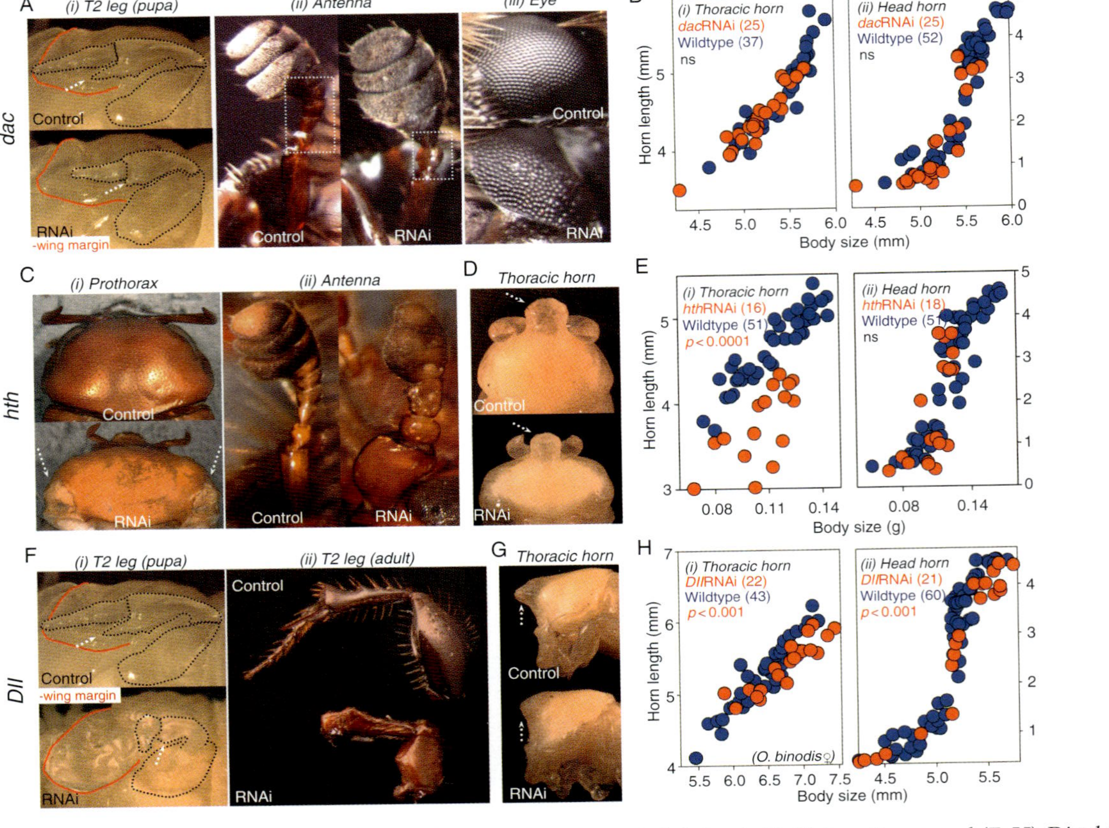

Armin P. Moczek, Figure 6.6 Larval RNAi-mediated transcript depletion of (A, B) *dachshund*, (C–E) *homothorax*, and (F–H) *Distal-less*. Images illustrate typical phenotypes observed in each experiment compared to wild-type phenotypes. Graphs depict scaling relationships between pupal body size and horn length for thoracic horns (i) and head horns (ii). Wild-type is shown in blue and RNAi-treated individuals are shown in red. All data are from male *O. taurus* except H(i) which were collected from female *O. binodis*. Sample sizes are given in parentheses (modified after Moczek and Rose, in preparation).

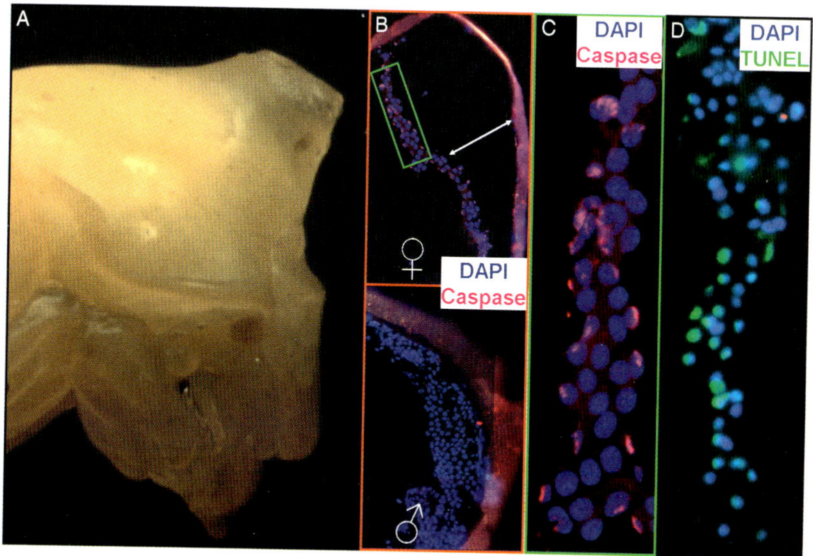

Armin P. Moczek, Figure 6.8 Programmed cell death appears to mediate sex-specific pupal remodeling in *O. binodis*. (A) Pupa indicating distal thoracic horn. (B) Anti-DRICE (activated caspase-3) staining in thoracic horn epidermis on pupal day 1 in (top) and (bottom). (C) Detail of Bè. (D) Corresponding region stained with TUNEL assay to detect PCD-specific DNA fragmentation.

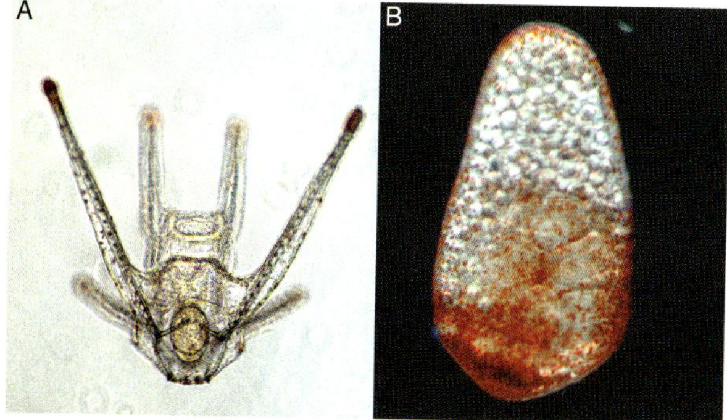

Rudolf A. Raff and Margaret Snoke Smith, Figure 7.1 Photographs of representative live larvae of (A) *H. tuberculata* and (B) *H. erythrogramma* show the major morphological differences in larval form across developmental modes. The *H. tuberculata* pluteus (2 weeks old, ventral view) has six arms. The large larval mouth opens upward. The oval gut shows in the lower center. The arms each have a skeletal rod. Two more arms and the rudiment of the juvenile adult will develop in the next month, and metamorphosis will then occur. The *H. erythrogramma* larva (2 days old, left side view) has no mouth or gut. The juvenile with its five primary tube feet is forming in the lower center. It will metamorphose in 1 or 2 days more to yield the juvenile sea urchin (photo by E. C. Raff).

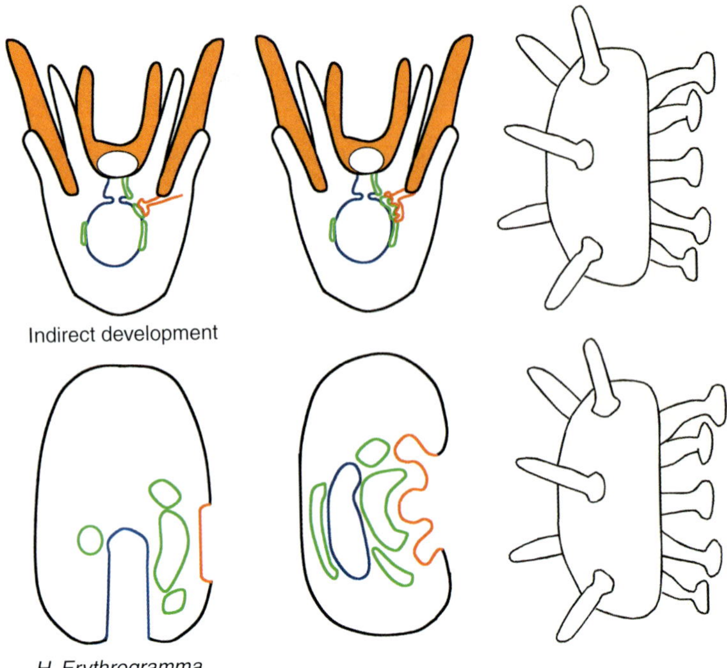

Rudolf A. Raff and Margaret Snoke Smith, Figure 7.2 Schematic comparing internal features of indirect developers and *H. erythrogramma*. Top row: a pluteus and juvenile. On the left side of the larva, the vestibular ectoderm, derived from oral ectoderm (orange) makes contact with the middle coelomic compartment, the hydrocoel (green) to form the rudiment. Larval gut in blue. The juvenile at right is oriented in the same way as the developing rudiment in the pluteus. Lower row: *H. erythrogramma* is shown in ventral view. The coelom forms within about 30 h, and by 36 h the left middle compartment (hydrocoel) is interacting with the vestibule (orange) to form the juvenile. The postmetamorphic juvenile (about 4 days postfertilization) is shown on the right.

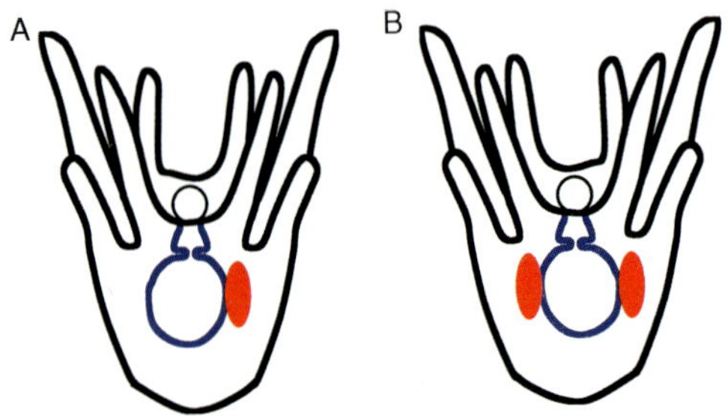

Rudolf A. Raff and Margaret Snoke Smith, Figure 7.5 Schematic of the effect of inhibiting Nodal after the prism stage in indirect developers. This treatment results in a larva with a duplicated rudiment. In control embryos, the rudiment normally forms only on the left side (A), but when Nodal is inhibited, rudiments develop on both the right and left sides (B) (based on data of Duboc *et al.*, 2005).

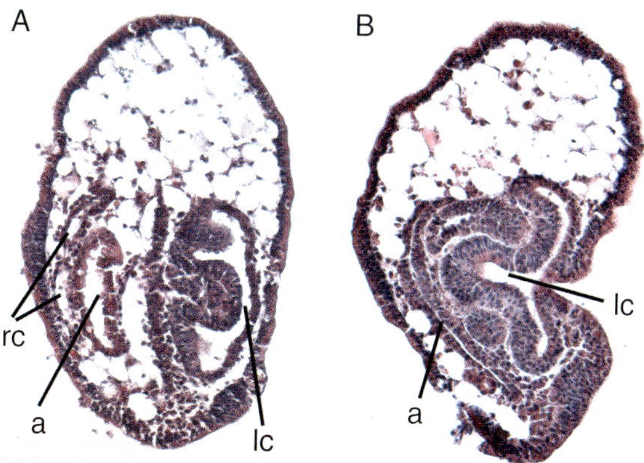

Rudolf A. Raff and Margaret Snoke Smith, Figure 7.6 Inhibiting Nodal later in the development of *H. erythrogramma* (after ~16 h) results in a hypertrophied rudiment and lack of any right coelom. The large left coelom derivative likely results from a conversion of right-sided fates to left side as occurs in indirect developers (from Smith *et al.*, 2008c, used by permission of John Wiley & Sons, Inc.).

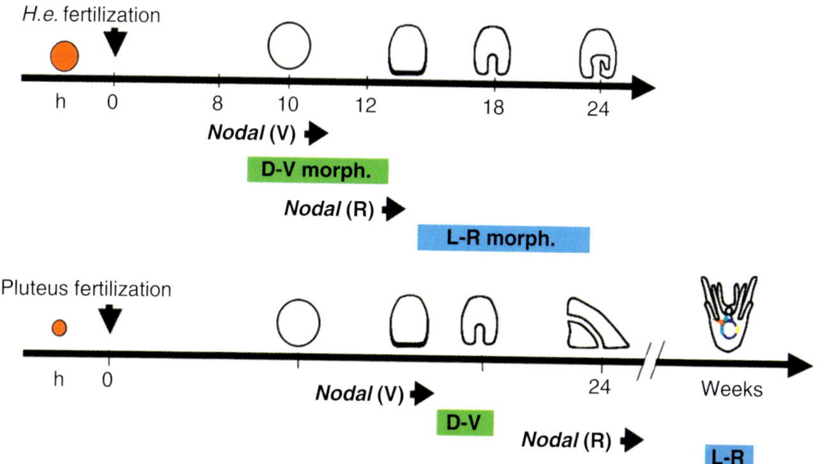

Rudolf A. Raff and Margaret Snoke Smith, Figure 7.7 Comparison of timing of *nodal* expression (arrows) and function (colored blocks) along the D–V ((V) and green block) and L–R ((R) and blue block) for *H. erythrogramma* (top) and indirect developers (bottom). Expression of *nodal* is at roughly similar stages across developmental modes, but execution of the function of *nodal* along the L–R axis has been shift drastically earlier in development (hours vs weeks) (based on the data of Duboc *et al.*, 2004, 2005; Ferkowicz and Raff, 2001; Flowers *et al.*, 2004; Smith *et al.*, 2008c).

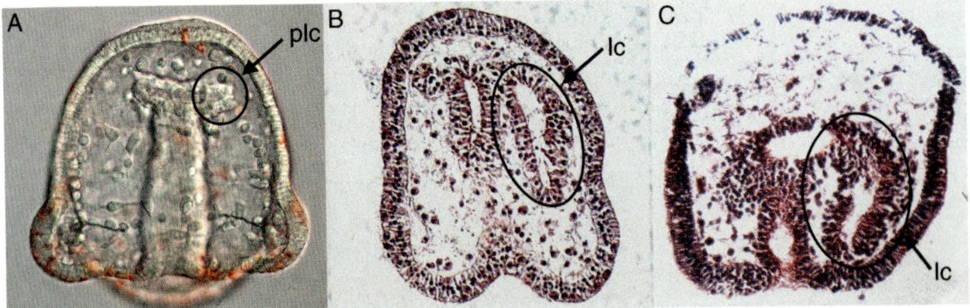

Rudolf A. Raff and Margaret Snoke Smith, Figure 7.8 Early larval stages of *Clypeaster subdepressus* (A), *C. rosaceus* (B), and *H. erythrogramma* (C) show that *C. rosaceus* forms a large left coelom early in development more consistent with coelom development in *H. erythrogramma* (direct developer) than *C. subdepressus* (indirect developer) (from Smith *et al.*, 2007, used by permission of John Wiley & Sons, Inc.).